BLUE GEMINI

BLUE GEMINI

A THRILLER

MIKE JENNE

TALOS

Talos Press books may be purchased in bulk at special discounts for sales promotion, corporate gifts, fund-raising, or educational purposes. Special editions can also be created to specifications. For details, contact the Special Sales Department, Talos Press, 307 West 36th Street, 11th Floor, New York, NY 10018 or info@skyhorsepublishing.com.

Talos Press® is an imprint of Skyhorse Publishing, Inc.®, a Delaware corporation.

Visit our website at www.talospress.com.

10 9 8 7 6 5 4 3 2 1

Library of Congress Cataloging-in-Publication Data is available on file.

Cover design by Haresh R. Makwana for Yucca Publishing
Illustrations © 2017 by E.R. Jenne

Print ISBN: 978-1-94586-305-9
Ebook ISBN: 978-1-63158-057-4

Printed in the United States of America

For Anna Kate Hindman,
I hope that you're able to make your home in the stars.

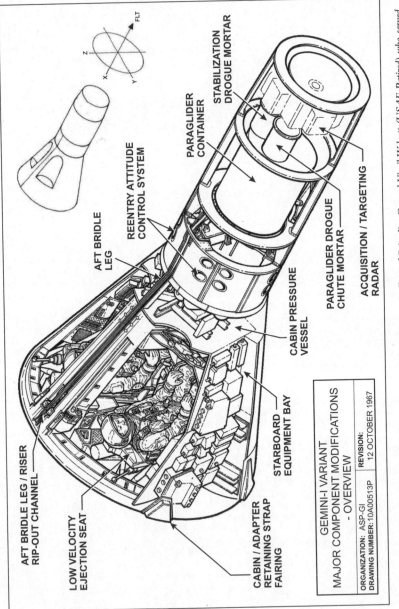

AFT BRIDLE LEG / RISER
RIP-OUT CHANNEL

LOW VELOCITY
EJECTION SEAT

CABIN / ADAPTER
RETAINING STRAP
FAIRING

STARBOARD
EQUIPMENT BAY

CABIN PRESSURE
VESSEL

AFT BRIDLE
LEG

REENTRY ATTITUDE
CONTROL SYSTEM

PARAGLIDER
CONTAINER

STABILIZATION
DROGUE MORTAR

PARAGLIDER DROGUE
CHUTE MORTAR

ACQUISITION / TARGETING
RADAR

FLT
Z
X
Y

GEMINI-I VARIANT
MAJOR COMPONENT MODIFICATIONS
- OVERVIEW

| ORGANIZATION: ASP-GI | REVISION: |
| DRAWING NUMBER:10A00513P | 12 OCTOBER 1967 |

This technical illustration is one of several recently discovered in the personal effects of Brigadier General Virgil Wolcott (USAF Retired) who served as the Deputy Director of the Aerospace Support Project, Wright-Patterson Air Force Base, Ohio, from January 1967 until June 1974. Illustration © 2017 by E.R. Jenne

AUTHOR'S NOTE

This is a work of fiction. Although many aspects of this story are true, the vast majority of this story is purely the result of the author's imagination. Only open source unclassified materials—readily accessible to the public at large—were used as reference sources. Absolutely no classified materials—in any way, shape, or form—were used in the preparation of this story.

Readers should be aware that while this appears to be a tale about a military program that did not actually exist, it's really a story about how lives intertwine and intersect over space, time, and generations. It's about duty, honor, courage, commitment, life, death, and love. And most importantly, it's about the necessity and difficulties of keeping secrets that are larger than ourselves.

Readers should be aware that *Blue Gemini* is the first installment of a trilogy. A website for the book series is located at www.mikejennebooks.com. It contains an extensive compendium of references for pertinent space/aero-space technology of the Cold War era, to include an illustrated glossary and a set of detailed illustrations (like the samples at the end of this edition) of technology unique to the fictional Aerospace Support Project described in the story.

1

PROLOGUE ONE: THE MAN WHO WOULD BE BUZZ ALDRIN

Sixteen miles southwest of Lincoln, Nebraska
8:32 a.m., Monday, April 25, 1966

As ordered, Scott Ourecky planted his hands in his lap, away from the aircraft's controls, and clamped his eyes tightly closed. He felt perspiration beading on his forehead. Through his earphones, he heard the muted drone of the Cessna 150's Continental engine. Wary of what was to come, he recited emergency procedures in his mind.

Ourecky was an Air Force ROTC cadet at the University of Nebraska at Lincoln. Just a month away from graduation, he was undergoing the compulsory "pressure hop"—the flight training aptitude exam—to assess how effectively he might handle the stresses of pilot training. Assuming that he passed his pressure hop, and certainly he would, he would soon matriculate to primary flight training at Randolph Air Force Base in Texas, shortly after he received his Air Force commission and pinned on his gold bars.

Administered by one of the ROTC instructors, a former fighter pilot named Major Dan Bell, the pressure hop was a subjective evaluation. Although it was certainly beneficial, previous flight training was not required; a few days prior to his hop, a prospective pilot was issued a study sheet with several emergency procedures to memorize. On the day of a student's hop, Bell would take him up, familiarize him with controlling the aircraft, and then run him through a few emergency drills. Bell apparently was far less concerned whether a would-be pilot might flub the details of an unfamiliar procedure, but much more focused on determining whether he might choke under pressure.

But the hop was still *highly* subjective. In his capacity as the de facto gatekeeper to the brotherhood of aviators, Major Bell was granted considerable discretion. As rumors had it, Bell was just as likely to grant an "up" ticket to a back-slapping, boisterous frat boy—particularly if the cadet was a member of Alpha Tau Omega, his own fraternity—as he was to pass a serious and studious cadet like Ourecky. So, as in all of his academic pursuits, Ourecky had intently studied his advance sheet and everything else that he could glean concerning the Cessna 150 and piloting procedures in general, until he had the information absolutely down pat. Moreover, over the course of the past two years, he had quizzed other cadets concerning the details of their pressure hops. Although he had only flown three times in his life, he was theoretically as prepared to fly the Cessna as any man could possibly be.

From what other cadets had related, Bell typically took them through no more than two emergency drills before returning to land at Lincoln. After all, everyone was aware that the pressure hop was a cursory check intended to identify the panic-prone before the Air Force expended a lot of money training them to be pilots. But if the other cadets had only been subjected to two emergency drills—three at most—Ourecky was on the verge of executing his *fourth* drill of the morning. *What gives here?*

Ourecky sensed the little aircraft climbing sharply, and then heard the kazoo-like buzzing sound of the stall warning as the plane abruptly lost lift and started to backslide out of the sky. He almost succumbed to an urge to laugh out loud; Bell was obviously setting him up for a simple drill, a recovery from a power-on stall, a common malady of inexperienced pilots who tried to climb too quickly after takeoff.

"Open your eyes, cadet," said Bell curtly. "You have the controls."

"I have the controls," replied Ourecky, quickly grasping the control yoke. "Executing power-on stall recovery." He swiftly executed the drill and recovered from the stall. Tugging back gently on the yoke, he resumed the designated altitude and steered back onto the assigned heading. "Two-Seven-Zero degrees, five thousand feet altitude," he announced.

"Good recovery. Maintain heading."

Grinning, confident that he would ace the hop, Ourecky glanced at Bell. Briskly chewing a wad of gum, with his bloodshot eyes concealed behind the teardrop-shaped lenses of aviator sunglasses, the instructor wore a faded brown leather flight jacket festooned with colorful embroidered patches of fighter squadrons he had flown with. From all accounts, he was a very competent pilot, but had effectively destroyed his military career with excessive drinking, wanton behavior, and a corresponding string of disciplinary infractions. Although he had repeatedly applied for combat duty in Vietnam since arriving at Lincoln, he wasn't going anywhere. Absent a miracle, Bell would remain at the ROTC department, enduring a monotonous routine of shepherding cadets, fated to retire in his current rank.

While it was tragic that Bell's life had stalled, Ourecky was confident that he would ascend on a much longer and higher trajectory. He had grown up in the American heartland, less than a hundred miles from here, in the humble and tiny town of Wilber, Nebraska. Renowned as the "Czech Capital of the USA" because so many Czechs had settled there as a result of the Homestead Act of 1862, Wilber was home to just over a thousand souls.

Like so many others of his generation, Ourecky was infatuated with space exploration. As a teenager, he had haunted the small library at school, incessantly poring over Wernher von Braun's visionary articles on space exploration in *Collier's* magazine, along with books by Willy Ley, Fred Whipple, and other aerospace pioneers. He studied the details of paintings by Chesley Bonestell and Fred Freeman, picturing himself aboard their massive doughnut-shaped space stations and needle-prowed rockets.

He spent countless nights peering through a telescope at the moon, planets, and swirling galaxies. Startlingly brilliant in the Nebraska sky, the stars beckoned; his destiny was there in the heavens, piloting a rocket, not tending corn on the 160 acres of black earth that had once been his great-great-grandparents' homestead. But unlike countless other would-be star voyagers, he had focused his eagerness to concoct a plan that would ultimately place him behind the controls of a vessel bound for space.

It was a scheme of two key interlocking parts. First, since he was convinced that test pilots would be standing at the front of the line to become astronauts, flying lessons were absolutely essential. Second, he vowed to diligently apply himself to his academic studies. It seemed logical to Ourecky that an engineering background would be required for long duration space missions, since engineering skills would be crucial for repairing systems and solving complex problems on the way to the moon or Mars.

As he trudged through high school, it was apparent that his aviation training would have to wait, since no airfields near Wilber offered flight lessons. In his senior year, in a class comprising just eight students, Ourecky won an Air Force ROTC scholarship to the University of Nebraska at Lincoln. He still wanted to fly, but his academic course load—a double major in electrical engineering and mathematics—left little time for extracurricular activities. He made time for elective courses in astronomy, and also gained a working knowledge of orbital mechanics through independent studies with a sympathetic professor. His social life was virtually non-existent; while his classmates urged him to join them in the indulgent joys of collegiate life—keg parties, outrageous pranks, chasing coeds and the like—he committed almost every waking hour to his studies.

And now he was merely a few weeks away from flight school. After earning his wings, as soon as he gained adequate flying experience, he would apply for test pilot school. Ideally, with any luck, he would acquire an advanced engineering degree in the process. But despite his carefully laid plans, there was always the possibility for outcomes that he could not influence. As the United States was becoming progressively more entangled in the war in Southeast Asia, Ourecky was abundantly conscious that his path might entail a detour to Vietnam. If that was the case, he would certainly do his duty but would assume no undue risks.

It would certainly be a long journey, but Ourecky was sure that he was on the right path towards his objective. He had closely followed the astronauts from the very outset of the space program, when the seven test pilots had been chosen for Project Mercury in 1959. He admired all of them, but his personal hero was Edwin "Buzz" Aldrin. Although he joined NASA's astronaut corps relatively late—as a member of the third class of astronauts selected in 1963—and had yet to fly into space, Aldrin stood out sharply from the pack. A genuine American hero, he had shot down two MiGs and won the Distinguished Flying Cross in the course of flying sixty-six combat missions over Korea. If that wasn't enough, Aldrin was also a brilliant theoretician who wrote a groundbreaking thesis on orbital rendezvous to earn his doctorate at the renowned Massachusetts Institute of Technology. From what Ourecky had read, Aldrin was endowed with an irrepressible work ethic, constantly striving to improve the efficiency of spaceflight procedures. While Ourecky had immense respect for great thinkers, he had even more respect for men of action like Aldrin, who would soon fly into space and literally insert himself into his theoretical equations on the biggest blackboard that ever was. Ourecky was absolutely certain that his hero would be the first man to walk on the Moon.

Through the intercom, Bell's voice interrupted his thoughts: "I have the controls. Close your eyes."

A *fifth* drill? Ourecky obediently closed his eyes and tucked his hands over his lap belt.

"Open your eyes," said Bell curtly. "You have the controls. Your engine is on fire."

"I have the controls," replied Ourecky. He glanced sharply to his left and picked out a large wheat field. "Emergency descent. I've selected a landing site."

"Good," noted Bell. "Keep going."

"Fuel selector off." Ourecky reached down and switched off the control. "Primer in and locked. Throttle closed. Mixture to idle cut-off. Cabin heat off. Cabin air off. Overhead vents open." The propeller stopped turning as the engine shut off.

"Good job so far," observed Bell.

Gently pulling back the yoke to maintain as much altitude as he could, Ourecky aimed the plane at his selected landing site as he recited the procedures for an emergency landing. He concluded with: "We're going in for an emergency landing. Make sure your seatbelt is snug. When I tell you, just before we land, unlatch and crack open your door so it doesn't get jammed if we crash. In an actual emergency, I would come up on the radio and declare an emergency…"

"That's good, cadet," said Bell. "That will do. Reset and restart, and then let's head for home."

Ourecky breathed a sigh of relief; he had been genuinely concerned that Bell would have actually allowed him to fly the Cessna all the way down to make an unpowered landing in some farmer's field. He smiled; it looked like he had nailed his pressure hop.

Ourecky switched the fuel selector lever from "Off" to "On" before resetting the other controls to their normal flight positions. He turned the ignition key to restart the engine, and it immediately came back to life. He brought the Cessna around and set a course for the airport at Lincoln.

Only a few minutes passed before Bell spoke: "One more thing. I have the controls. Close your eyes."

Another drill? Certainly this had to be a record. Trying hard to hide his aggravation, Ourecky placed his hands in his lap and shut his eyes.

Almost immediately, Bell said "Open your eyes. You have the controls."

As Ourecky opened his eyes and grasped the yoke, he watched Bell reach out to turn the ignition switch to "Off." The engine quickly stopped.

"Engine on fire?" asked Ourecky.

"Nope," answered Bell. "It just stopped."

As he had before, Ourecky adjusted the controls to maintain as much altitude as he could, and then looked for a place to set down the Cessna if the engine failed to restart. "Engine restart," he said. "Carb heat on. Mixture to full rich. Fuel selector to On. Master switch to On. Throttle set. Primer in and locked. Check magnetos. Ignition switch to Start." Ourecky turned the key and the engine restarted. Checking to see how far he might have deviated off heading, he looked up at the compass. Suddenly, the engine inexplicably sputtered out. The plane gradually nosed over, silently gliding over endless acres of corn and wheat.

"What are you going to do *now*, cadet?" asked Bell.

"Uh, I guess we're going in for an emergency landing," replied Ourecky.

"Good answer, but the underlying problem is that you failed to check the fuel selection lever during your restart procedure," observed Bell, using his pencil to gesture at the control. "You called it, which is admirable, but you didn't look at it or put your hand on it."

Ourecky looked at the fuel selection lever. Sure enough, it was set to "Off," so the engine wasn't drawing fuel from the tanks in the wings. Bell must have switched it off while his eyes were briefly closed. He swiveled it back to On, re-checked the other controls, and restarted the engine.

"But, sir, I just checked the fuel selection lever after I recovered from the engine fire," argued Ourecky. "I reset it to On before I restarted the engine." Even as the words left his lips, his stomach sank. Bell was right; if he followed the drill to the letter, he should have physically checked the control instead of assuming that it was On. But on the other hand, he *had* checked it, just mere moments before, so what was the likelihood that it could have suddenly slipped out of place? Could some nefarious gremlin have boarded the little plane to shut off the gas?

"Doesn't matter," replied Bell. "Procedures are procedures, Ourecky. We have them for a reason, and you're obligated to follow them."

Grimacing, Ourecky kept his tongue in check. He knew better than to debate with Bell. And besides, that was just one drill out of six. Bell had put him through the wringer and he had certainly aced the other five, so it still likely that he passed his hop. In any event, he wouldn't know the verdict for at least another week, when Bell's evaluation report was posted to his official records. He also knew that absolutely nothing could be gained by pestering Bell for his decision, or arguing about the last drill.

For the next several minutes, they flew on in silence. As they drew close to the airport at Lincoln, Bell assumed control of the Cessna and formed an approach for landing. Between radio calls, as he manipulated the controls, he asked: "So, cadet, Buzz Aldrin's your hero, huh?"

"Yes, sir. Very much so, sir," answered Ourecky, still hoping for a favorable outcome on the hop.

"And from what I hear from your fellow cadets, you have aspirations of going to test pilot school as soon as you possibly can, so you can go on to be an astronaut. Is that correct, Ourecky? You want to follow in Aldrin's footprints?"

"Uh, very much so, sir."

"Well, if you had done your homework, Ourecky, you would know that Buzz Aldrin had *never* been a test pilot. He might be an astronaut now, but Buzz was a *fighter* pilot, and a damned good one at that."

2

PROLOGUE TWO:
THE EAVESDROPPERS

Karamürsel Air Station, Turkey
6:30 a.m., Sunday, November 20, 1966

Assigned to the elite United States Air Force Security Service, Technical Sergeant Vic Rybalka was a "Squirrel," a skilled Russian linguist with extensive training in Soviet military technology and operations. Officially classified as "voice intercept operators," he and his fellow Squirrels spent their working hours intently listening to military operations in the Soviet Union.

Although Karamürsel was somewhat isolated, he enjoyed the assignment. It was his second tour here. Like most of his fellow Squirrels, he considered Karamürsel as one of the pinnacle assignments for voice interceptors because the site was uniquely positioned to harvest intelligence from some of the most secretive Soviet activities.

While his work was certainly intriguing, it could also take its toll. It required intense concentration, much more than could be expected of the average man. It was rare that Rybalka left his shift without feeling entirely spent, but it was just as rare that he could lie down for a night's sleep without his head still spinning from the day's activities. He suffered from a chronic backache, a painful memento of sitting for hours in a stiff-backed chair, hunched over a bank of radio receivers. But the aspect that annoyed him most was that his labors were almost entirely passive. He sat, he listened, he made notes, and not much else. Like most men who volunteered for military service, Rybalka was drawn to action, but a Squirrel's life was not an existence for the active. Their bosses praised the interceptors as electronic sleuths, snatching vital secrets from the ether, but most often he felt like an eavesdropper, perpetually leaning against a locked door, vicariously listening to the illicit activities of strangers.

There were some within their tight-knit community who claimed that the entire enterprise would eventually be the domain of machines, and even now, civilian technicians were busy installing massive banks of computers and other equipment at the Karamürsel site. The civilians avowed that that giant machines would systematically vacuum up the radio transmissions and that the time-consuming toils of sifting and sorting would be automated. But Rybalka was confident that his expertise could never be replicated by a computer or even a thousand computers working in unison, because there were some aspects of the job that were beyond the capacity of machines.

As he arrived for his tour of duty, he mused that Sundays were typically very quiet, even on the supposedly atheistic side of the Iron Curtain, with scarce traffic to monitor over the airwaves. He grimaced; a quiet day meant that his shift's Squirrels would be tasked with his least favorite chore: the tedious task of transcribing hours and hours of backlogged tapes. While some genuine intelligence was often gleaned from the glut of recordings, he might also spend hours scribbling down inane drivel. Frowning, he remembered one painful session in which he transcribed an hour-long conversation between a Black Sea Fleet supply officer and his counterpart at a shore depot, in which the afloat officer bitterly complained that his ship's entire stock of flour was infested with weevils.

Although the Squirrels' work could often be monotonous, at least their job wasn't as painful as the "Ditty Boppers" who intercepted Morse code transmissions for the duration of their shifts. Rybalka had known Ditty Boppers who had literally copied code for months on end but had not the slightest clue what any of it meant.

As was his pre-shift ritual, he went to his locker in the break area, stowed the box lunch prepared by the base chow hall, and retrieved his "lucky" coffee mug. He filled the mug with steaming coffee from a huge urn on the counter, stirred in two tablespoons of sugar, and then strolled into the work area. He was immediately taken aback. It was not the somber environment that he expected for a Sunday morning. Far from quiet, the room was abuzz with giddy chatter. It seemed almost festive, like Christmas was arriving a month early.

He walked to his workstation, which was occupied by his counterpart from the previous shift, Senior Airman Paul Smith, a lanky young Squirrel from Kansas. "What's all the hubbub?" he asked.

Rybalka's expectations of an uneventful shift were immediately altered by Smith's one-word reply: "Bluto."

"*Bluto*? No kidding?"

"Yeah," replied Smith. He seemed sullen, as if he had missed the last bus to summer camp. "Bluto's on the pad, ready to launch."

"Hold the fort, pal," said Rybalka. "I'll be right back." He returned to his locker in the break area, retrieved a large Thermos bottle, and filled it from the coffee urn. Carrying the Thermos and a quart-sized mason jar, he returned to the work area as he pondered how the day would proceed. "Bluto"—named for Popeye's hulking nemesis—was the Squirrels' internal nickname for a massive new booster rocket. The giant rocket was apparently still in its testing phases and very prone to failure.

Given the impending launch, all of Karamürsel's considerable listening capabilities would be focused at the Soviet Tyuratam missile launch complex, located in the expansive grasslands of the Kazakh Republic. The Tyuratam facility—often incorrectly described as the "Baikonur Cosmodrome"—was the point of origin for virtually all Soviet flights into space.

As he collected his belongings to go off shift, Smith glanced at Rybalka's Thermos and mason jar. "Ready for the long haul, huh?" he asked, stretching as he stifled a yawn.

"You got that right," replied Rybalka. "I don't want to miss anything." He reached into the bottom drawer of the desk and extracted his personal "dope book." Accumulated over months of monitoring the Soviets, the spiral-bound tablet contained his copious notes concerning frequencies, channel settings, atmospheric conditions, and personality quirks of various Soviet radio operators he listened to on a regular basis. Armed with the information, he could precisely fine-tune his equipment, dialing in minute adjustments to compensate as the Soviet transmitters gradually—but predictably—"wobbled" a few cycles over the course of the day.

"Hey, Vic," muttered Smith. "Would you mind if I…"

"Hang out for a while?"

"Uh, yeah. I've never heard an actual launch."

"Sure," said Rybalka. "There's an earphone jack splitter in my locker. You can stay at the desk with me and listen as long as you want, but just stay out of my way. And don't—*don't*—touch the knobs. They're all mine."

"Thanks, Vic. I really appreciate it."

Rybalka smiled to himself. Smith, like most of the neophyte Squirrels, held him in awe; during his first stint at Karamürsel, five years ago, he had witnessed—over the airwaves—cosmonaut Yuri Gagarin's historic flight into orbit. He had monitored more Soviet launches, by far, than any other voice interceptor assigned to the station. He particularly relished the drama that surrounded the manned flights, since there were obviously lives at stake, and although he was curious about the looming flight, he knew that it would be years—if ever—before Bluto hoisted a crew to orbit.

During the shift change briefing, he learned that the previous crew had been monitoring the launch preparations for several hours, and Bluto's lift-off appeared imminent. After the briefing, he poured himself a cup of coffee, settled into his chair, adjusted his headphones, and prepared for the vigil that was sure to ensue. This wasn't his first rodeo; months of experience had taught him not to be overly anxious. It might be hours or even days before the Soviets launched; quite often, the launch crews proceeded right to the very last tense seconds of the countdown, only to shut down because of some seemingly insignificant technical issue. Sometimes, the problems were swiftly resolved and the countdown resumed within minutes, but often the entire process had to be reset for another attempt.

But all indications were that this was a day when things—at least on the far side of the Black Sea—were going exactly according to plan. Besides Smith, several of the previous shift's younger Squirrels remained on station, clustered around speakers at a back console. Rybalka chuckled; it looked like the fledgling interceptors were in for a good show.

His brow furrowed as he strained to hear a key frequency. He hushed some of the exuberant kibitzers, consulted his dope book, and then carefully adjusted his equipment. Rybalka was an accomplished radio tuner, absolutely adept at twisting and tweaking the dials until he acquired almost absolute fidelity.

In addition to his technical prowess, Rybalka had the uncanny ability to listen to and comprehend several conversations at the same time, so he was able to simultaneously monitor multiple channels. And unlike most of the other Squirrels, he didn't struggle with Russian; he'd been fluent in the language long before joining the Air Force, courtesy of immigrant parents and an upbringing in Brighton Beach.

As he patiently listened, he came to realize that there was something very unusual about this launch. In monitoring distant events as they occurred, he didn't just listen to the substance of the radio conversations, but he also divined content from nuance and emotion. In this instance, the Soviet flight controllers' voices carried a clearly discernible measure of trepidation, as if they were perched on top of Bluto instead of being entrenched in a concrete bunker a safe distance from the launch pad. Rybalka was perplexed. *Was this a manned launch? Were the Soviets actually risking a crew on such a largely unproven rocket? Why else were the controllers so apprehensive?*

Rybalka had been on shift for less than an hour when the countdown went into its final moments, and he held his breath as the controllers succinctly reported that the booster engines had ignited correctly and the rocket was underway.

Within minutes, things abruptly went awry. Even as the controllers were speaking, reporting the rocket's altitude and orientation, he heard a commotion on one of the secondary channels. The transmissions carried a barely masked undertone of panic. It was obvious that the usually unflappable Soviet controllers were concerned for their *own* safety. Their normally deadpan voices were agitated. Something was definitely amiss.

He heard the launch director frantically order that the flight be immediately terminated, and then he listened as the flight controllers initiated their procedures to destroy the platform. But then there was a twist that he hadn't anticipated: A flight controller excitedly declared that the emergency rocket had successfully fired. Rybalka gasped. *Emergency rocket?* An emergency rocket was a launch escape system, which implied that the launch *had* to be a manned flight.

Like the American Gemini, the Soviets' first manned spacecraft—the *Vostok*—was fitted with an ejection seat to save the cosmonaut in the event of a launch accident. Like the Apollo lunar spacecraft, the Soviets' latest spacecraft—the *Soyuz*—was outfitted with a launch escape rocket at the very top, which would quickly yank the spacecraft—and crew—away from a malfunctioning booster.

Although they had monitored a multitude of launch accidents in the past, this was the very first one Rybalka had heard involving a manned flight. And it wasn't resolved quickly; the incident went on for several hours after the ascent was scuttled. Lingering long after his shift was complete, he continued to monitor the frequencies as search and rescue parties were deployed to locate the spacecraft. It was long after midnight when he listened to a relayed transmission that airborne search crews had detected the faint pinging of a locator beacon in the rugged country northwest of Lake Balkhash. Dawn had barely broken over the Black Sea when Rybalka heard the search coordinator gleefully announce that a helicopter crew had discovered the capsule intact. Although the Soviets were the enemy, he breathed a sigh of relief; if the capsule was unharmed, then certainly the crew must also be safe.

3

PROLOGUE THREE: MARCHING ORDERS

The Pentagon
3:25 p.m., Tuesday, January 31, 1967

Air Force Colonels Mark Tew and Virgil Wolcott quickly clambered out of a dark blue official sedan and followed a waiting escort officer into the Pentagon. Last night, both men had been called at their homes in California and directed to report here for a classified briefing. Their instructions were vague, but the matter was clearly urgent.

Tew and Wolcott had worked together, in one capacity or another, for almost three decades. They had both begun their military careers as B-17 bomber pilots in World War II and had gone on to fly fighters during the Korean War. Both were accomplished engineers; after Korea, they had labored on several aerospace research endeavors.

They were currently assigned to AFSC, Air Force Systems Command, in California, where they worked on the Air Force's Manned Orbiting Laboratory. The Manned Orbiting Laboratory—more commonly known as the MOL—was a military space station that would be inhabited by a two-man crew. It was scheduled to be operational in 1972.

Unfortunately, their time with the MOL project was swiftly drawing to a close, since both men had been selected for promotion to brigadier general and consequently would have to be assigned elsewhere.

Physically, the two couldn't be more different. Wolcott was balding, tall and rangy; his skin was permanently dry and cracked from his earlier days in windswept Oklahoma. The corners of his blue-green eyes were crinkled with deep crow's feet, and his teeth were chipped and crooked.

At six feet, Tew was a couple inches shorter than Wolcott, but not nearly as slim; his age and sedentary lifestyle had deposited more than a few adipose pounds around his middle. Golfing was now his sole form of exercise, but his demanding workload left scant time for even that. Both men were fifty-two, but the weather-beaten Wolcott looked younger than Tew.

Tew was developing heavy jowls, and his prominent nose was a garden of red gin blossoms, a testament to his earlier days closing down the bars at officers' clubs around the world. Incessantly pestered by his wife, Tew had reluctantly taken up a vow of sobriety. Silently, she gloated over winning the battle over the bottle, but it was really Tew's job at AFSC that turned the tide against an early death from alcohol; the harsh reality was that there was just too much work and not enough time for drinking.

Neither man had any clue why they had been called to the Pentagon on such short notice.

"I'll tell you, Mark, it's that damned Apollo fire," drawled Wolcott. He referred to the tragic incident at Cape Kennedy just last Friday, in which the Apollo One primary crew—astronauts Gus Grissom, Ed White and Roger Chaffee—had burned to death when their spacecraft's cabin was engulfed by a flash fire during a "plugs out" pre-flight test on the launch pad.

"*Apollo?*" wheezed Tew, struggling to keep pace with their escort officer.

"Yup," asserted Wolcott. "Everyone's clamorin' for an inquiry, and the big hats at NASA are goin' to have to rustle up some impartial investigators to figure out what went wrong. Since we just got promoted and have to ride off the ranch in California, it makes perfect sense that the Air Force would loan us to NASA for the investigation."

Tew nodded. The temporary assignment did make perfect sense. Moreover, he and Wolcott possessed the appropriate credentials, engineering expertise, and aerospace background. They were ideally suited for the task.

Their escort officer paused at the entrance to a secluded conference room deep within the bowels of the Pentagon. "They're waiting for you, gentlemen," he said, opening the door and stepping to the side.

Walking into the room, Tew saw that there were two flag officers present, Lieutenant General Hugh Kittredge of the Air Force and Admiral Leon Tarbox of the Navy. Both were senior representatives to the Military Spaceflight Task Group, a joint services steering committee that recommended spaceflight policies and strategies to key leaders. Ultimately, the primary purpose of the committee was to alleviate bickering among the services by fairly—to the greatest extent possible—parceling out the limited resources available to the Pentagon for manned spaceflight activities.

Short and stocky, Kittredge was fair-haired, with a ruddy complexion that looked as if he had spent a considerable amount of his life outdoors. He walked with a pronounced limp, courtesy of a crash landing in France during the War, followed by two years in a German POW camp. Standing, he greeted Tew and Wolcott and then gestured toward Tarbox, who remained seated at the table.

"Gents, I'm sure that you know Admiral Tarbox," said Kittredge. "He'll be sitting in with us today, representing the Navy's interests."

"Admiral," said Tew icily, extending his hand. He had known Tarbox for well over a decade. Like Tew and Wolcott, Tarbox worked on the MOL project in California, but he was primarily concerned with ensuring that the Navy garnered their fair share of the effort.

"Colonels," replied Tarbox, in an equally distant tone. Broomstick-thin, with white hair and a narrow face permanently dried and puckered by the sun, he looked like a wizened gnome. His shrill voice sounded less like a human's, and more like the terse screech of an owl swooping to pounce on a baby rabbit. Almost lacking a personality, entirely devoid of humor, Tarbox was so insidiously corrosive that battery acid likely oozed through his veins.

He was decades beyond normal military retirement age. MOL workers had dubbed him The Ancient Mariner; one wag joked that the admiral probably still harbored sour memories of the day when he was ordered to finally furl his sails and surrender to steam power.

Tarbox obsessively insisted that the Navy, not the Air Force, should be leading any military manned spaceflight programs, including the MOL. He asserted that for strategic implications, the heavens were a logical extension of the oceans, so military operations in space should be the dominion of the Navy. There was some validity to his argument. Like the trackless oceans, space was immense and not subject to international boundaries. Moreover, on a very practical note, astronauts would navigate by the stars, using essentially the same techniques and tools employed by mariners for centuries.

Ironically, Tarbox and his Navy probably would hold sway over military space operations today, save for a series of unfortunate events. In 1955, the armed services presented competing proposals to develop and orbit the world's first artificial satellite. The Navy's concept—named Vanguard—was selected over an Army plan that would have used Wernher von Braun's Redstone rocket as a booster. The Air Force didn't even have a horse in the race; their plan, which relied on using a modified Atlas ICBM as a launch vehicle, wasn't granted serious consideration because the Atlas was still in development.

Vanguard would have beaten the Soviet's *Sputnik* into orbit, had it not suffered numerous technical problems. With considerable fanfare, the Soviets launched *Sputnik* in October 1957. In the end, the Army's *Explorer 1* was the first American satellite, launched into orbit atop a Jupiter-C—also designed by von Braun and his largely German team—in January 1958. The Navy's Vanguard didn't make it into space for another two months.

Despite the admiral's outlandish opinions and outspoken behavior, Tew was mindful to tread lightly around Tarbox. It was easy to dismiss the acerbic admiral as an annoying kook, but Tew was well aware that he had the ears of some very powerful people, and his sphere of influence extended far beyond the military. Like Hyman Rickover, the brilliant but abrasive dean of the Navy's nuclear program, Tarbox wielded vast political power.

Kittredge sat down and asked, "Mark, Virgil, do you have any idea why you're here?"

"You called us here to work on the Apollo fire investigation, right?" asked Wolcott smugly.

Kittredge frowned and replied, "Not quite, Virgil. We have an entirely different crisis for you two to deal with." He dropped a ponderously heavy binder on the table and added, "This is the executive overview summary of a program that we want you two to head up. We don't have time to peruse the whole book, so I'll just summarize. Just consider this the *Reader's Digest* condensed version."

Kittredge tugged an index card from the interior pocket of his uniform coat. "This is from a speech that Nikita Khrushchev made in 1961, just after the Soviets orbited their second cosmonaut." He slipped on a pair of reading glasses and quoted from the card: "*You do not have 50- or 100-megaton bombs, we have bombs more powerful than 100 megatons. We placed Gagarin and Titov in space, and we can replace them with other loads that can be directed to any place on Earth.*"

Tew clearly remembered the speech, as well as the furor that ensued. Khrushchev's vile threat of an Orbital Bombardment System—OBS—caused great consternation in the American military, and several programs quickly sprung up as possible options to counter the OBS.

In time, even after the US military had expended millions to accelerate anti-satellite research, the OBS threat diminished. As Khrushchev faded from power in 1964, so too did the threat of nuclear weapons raining down from space, with most American military planners assuming that it had been merely empty bluster on the part of the former Soviet premier.

"As both of you are aware, we haven't been too concerned with this threat for quite some time," stated Kittredge, removing his reading glasses. "But unfortunately, there have been some very startling developments as of late, and we are compelled to respond."

Kittredge continued: "Two months ago, our Security Service station near Istanbul monitored the launch of the Soviet's latest booster, the UR-500, from their launch complex in Tyuratam. I assume that you're familiar with the UR-500?"

Tew nodded. He was *very* aware of the UR-500. Developed by the Chelomei aerospace design bureau, it had first been launched in 1965. It was *huge*, on scale with NASA's Saturn I-B booster that would soon launch the early Apollo manned missions into earth orbit. Like the Air Force Titan II ICBM, also used to launch NASA's Gemini spacecraft, the UR-500 relied on hypergolic fuels that did not require an igniter. It had initially been developed as an outsized ICBM, capable of heaving enormous nuclear warheads at the United States. Even though it was being used to launch various payloads, particularly the Soviets' "Proton" research satellites, it was still apparently very prone to failure.

"This was a failed launch," said Kittredge, interrupting Tew's thoughts. "We're not exactly sure what happened, but the Soviets intentionally destroyed the vehicle just a few minutes after liftoff."

"Well, General, *that's* obviously good news," noted Wolcott.

Kittredge frowned and added, "Initial indications were that it was a manned flight…"

"*Oh,*" Wolcott muttered, shaking his head. "Sorry." Despite the vast differences between East and West, one notion was sacrosanct: when it came to the exploration of space, *any* loss of life—American or Soviet—was lamentable.

Kittredge paused to sip from a glass of water, then continued. "As I said, the *initial* indications were that it was a manned flight. One of our voice interceptors recognized that the UR-500's payload had been separated by an escape rocket system just as the booster was being destroyed, and he came to the conclusion—rightfully so—that it was a manned flight. Our people continued to monitor the search and recovery effort, and were able to ascertain that the payload—a very large capsule of sorts—was eventually recovered intact."

"And if it wasn't manned, then what?" asked Tew.

Kittredge's reply was ominous. "We have substantial evidence to indicate that the Soviets were placing a nuclear warhead—a *very* large nuclear warhead—into orbit. Moreover, it's very apparent that this launch was intended to be the first of many more, and that the Soviets are still aggressively pursuing an operational Orbital Bombardment System."

Kittredge slid the binder across the table and flipped open the cover to reveal the first page. Tew and Wolcott read it together.

Looking up, Wolcott groaned. "Tarnations, Hugh, this danged boondoggle just looks a whole lot like Blue Gemini," he lamented. "And we all know how *that* turned out."

Tew nodded in agreement. Blue Gemini was a proposed program that would have run parallel with NASA's follow-on to Project Mercury, launching military astronauts in the two-man Gemini spacecraft. Under

Blue Gemini, the Air Force would have launched seven missions to research military applications of manned spaceflight. Because of its questionable objectives and because he viewed it as a duplication of NASA's efforts, Secretary of Defense Robert McNamara axed the project in January of 1963.

Tew and Wolcott had worked on Blue Gemini for nearly three years; after it was cancelled, Tew walked away without a second thought, but Wolcott still seemed infatuated with the thought of sending military astronauts into space.

"Virgil, this *is* Blue Gemini," answered Kittredge. "Or at least it's the resurrection of Blue Gemini, with much more specifically designed objectives. There's absolutely no pretense of conducting any quasi-scientific experiments. The mission is to surreptitiously intercept, inspect and—if warranted—*destroy* hostile satellites. NASA has shown us that the Gemini is more than capable for this task, and while they're busy going to the Moon, we plan to exploit the Gemini for *this* mission. We've had this under wraps for a while, but with this new OBS development, we're moving forward."

Kittredge continued: "We're going to preserve all the positive virtues of NASA's Gemini, while adding new features specific to the mission. We'll rely on a specially configured interceptor variant that we're calling the Gemini-I. The hardware is already being built, as we speak."

"The plan calls to amass a contingency stockpile of the Gemini-I spacecraft and launch vehicles, so that we have a strategic standby capability to intercept and interdict the Soviet OBS, when and if we determine that it's a viable threat."

"So there really ain't plans to actually launch any of these critters, right?" asked Wolcott. "You're just plannin' to salt them away, just in case."

Kittredge nodded. "That's correct, Virgil. Except for a couple of unmanned flights to certify the hardware, we might be able to fly *one* manned mission to validate the concepts and technology, but that would be subject to approval at the highest levels of government, and I doubt that they would buy off on a practice shot with such a high degree of risk."

"So you're askin' us to sink our hearts and souls into this, and there's scarcely any chance that it will even leave the ground, right?"

Kittredge nodded again. "Mark, Virgil, I know that you two have put your hearts into the MOL," he confided. "I know it has to be a huge disappointment to leave that program just as it becomes operational, but this is a matter of the highest strategic importance. You two should know that Bennie Schriever personally handpicked you to lead this project, over a year ago, before he retired."

Tew swallowed; their former boss at Air Force Systems Command, General Bernard "Bennie" Schriever, was one of the most powerful men in the United States Air Force, if not the entire country. A brilliant four-star visionary, Schriever was personally instrumental in the development of the nation's entire arsenal of strategic missiles. That Schriever would have so much confidence in him and Wolcott spoke volumes.

"We're in," uttered Tew. "*All* in."

"Good. At this point, everything is falling into place, but there's one loose end," Kittredge said. "We haven't identified a launch site."

"Loose end?" drawled Wolcott. "A clandestine space program without a launch site is just a tad more serious than a danglin' thread."

"Vandenberg and Canaveral wouldn't be suitable?" asked Tew.

Shaking his head, Kittredge sipped from his water glass. "Besides the obvious security concerns surrounding this effort, we anticipate firing into some very high inclination orbits. Actually, if at all possible, we would like to be able to shoot directly into any inclination."

Tew saw the logic. Ideally, to reach any potential orbital inclination, the launch site would need to be as close to the equator as possible. Additionally, it should be at a remote location far removed from human habitation. It was far too risky to fire a rocket over a populated area, which was why Cape Canaveral was constrained to eastbound launches out over the Atlantic, and Vandenberg could only shoot to the south or slightly to the southeast.

As if waiting for an opportunity to pounce, Tarbox cleared his throat and interjected, "My staff has already examined this launch site issue at length, and we have an option. We're confident that the Gemini-I could be launched from a surface ship. As a matter of fact, we're already making plans to initially test-fire our new Poseidon ICBM from a surface ship."

"Good point, Leon," Kittredge said. "And we've reviewed the feasibility study that you forwarded to us, but we think it would be more advantageous to launch from a fixed site, provided we can find a suitable location."

Tew grimaced; he had witnessed Tarbox at work before, and knew that once that admiral had even the slightest grip on the project, he would dig in his claws and it would only be a matter of time before he wrested the entire effort from the Air Force.

Tew looked toward Wolcott, hoping that his counterpart might have a quick answer to the launch site dilemma. Grinning broadly, Wolcott nudged him with his elbow, scrawled two words on a scrap of paper and slid it in front of Tew.

Tew glanced at the note, raised his eyebrows, nodded, and said, "General Kittredge, we have a launch location that should support all of your requirements. We'll just need to verify that it can be made available to us."

"Excellent," said Kittredge.

"And coincidentally, it just happens to be real estate that the Air Force already holds claim to," boasted Wolcott, grinning at Tarbox. "Of course, we *might* need a little assistance from the Navy to move our hardware there."

"I'm sure that the Navy will be happy to oblige," said Kittredge.

"I'm still not entirely clear how we fit into this plan," interjected Tew, "if this project has essentially been underway for the past two years and the hardware is already being fabricated."

"You two will assume the management role over the entire program," answered Kittredge. "Including fiscal oversight and coordination of all the various moving pieces. The project will be based at Wright-Patterson. As I've said, virtually all of the pieces are already falling into place, except for one significant aspect. You will oversee the training and preparation of the flight crews and mission control team. To this end, to ensure that they're operational in a timely manner, we've established a set of milestones for training." Kittredge flipped through the binder's contents until he came to a red-bordered page.

Tew and Wolcott scanned the page. "Whew," exclaimed Wolcott. "This is some mighty demandin' stuff." He pointed at a line and added, "Here you're expectin' these boys to fly an intercept profile for forty-eight continuous hours. Is that even humanly possible?"

"We have no choice," replied Kittredge. "Our preliminary studies show that it will likely take at least forty-eight hours, from launch to intercept, with an absolute minimum of assistance from the ground, to reach these targets. So, in order to declare the program as operational, we have to validate the concept."

"And we have just a year to do this?" asked Tew.

Kittredge nodded.

"And what happens if we don't make the cut?" asked Wolcott.

"Then I suspect that the program will be handed off to another entity," stated Kittredge, looking toward Tarbox. "One that *can* make the cut."

"When do we get our personnel?" asked Wolcott. "Are we goin' to be compelled to go out and shake the bushes lookin' for someone to do this thing?"

Kittredge answered, "We've identified twelve pilots—six crews—but we anticipate you trimming the roster down to four crews by the time this project is operational. All of these men are unmarried, to minimize personal problems that might interfere with the training, and have undergone extensive vetting for security and reliability issues. They're top notch men; of the twelve, nine had previously applied to NASA for the astronaut program."

"And they were rejected?" asked Wolcott. "That doesn't exactly fill me with confidence, General. It's a lot like getting the pick of the litter, but only after NASA already has picked over them."

"They weren't selected for very minor reasons," explained Kittredge. "Suffice it is to say, most weren't chosen because NASA perceived them as *too* military, and they were obviously concerned that they wouldn't effectively blend into NASA's culture or be suitable for the public relations work that comes with the astronaut job. Obviously, since we'll be working primarily in the shadows, we're not too concerned with these pilots' ability to make polite speeches to the local Rotary Club."

Adjusting his tie as he looked at the clock on the wall, Kittredge said, "Mark, Virgil, I'm a little pressed for time. Why don't you read that material in greater detail, and we'll chat tomorrow morning?"

Tew and Wolcott nodded in unison, and Tew began dividing up the pages in the briefing binder.

"If you have just a minute, General, I think we have an answer for the electrical power issues on the ocean surveillance MOL variant," said Tarbox, removing a report from his leather attaché case. The ocean surveillance MOL was the admiral's pet project; if realized, it would be a Navy-specific MOL mission crewed by Navy astronauts and controlled entirely by the Navy.

Kittredge smiled as he shook his head. "Nice try, Leon, but we're here to wrestle just one bear today," he observed, standing up and adjusting his coat. "We're here to issue Mark and Virgil their marching orders and get them on their way. We'll just have to table your project until we convene our next meeting."

4

THE LETTER

Base Exchange Snack Bar, Eglin Air Force Base, Florida
12:07 p.m., Thursday, April 11, 1968

Clutching a manila envelope tightly in his left hand, First Lieutenant Scott Ourecky leaned forward and kept a quick pace. He was only authorized an hour break for lunch, and the BX snack bar was a fifteen-minute brisk stroll from the Armament Lab facilities where he worked.

The letter had landed on his desk this morning, but he had yet to open it. After being turned down for flight training *four* times, he had requested a fifth waiver a month ago. He returned the salute of a passing airman and then examined the unopened envelope; it bore the return address and office symbol of the Air Force Military Personnel Center at Randolph Air Force Base, Texas. A rubber-stamped notice on the front stated "OFFICIAL BUSINESS" in red block letters. Surely, it contained the long-awaited reply to his appeal.

Arriving at the snack bar, he folded his blue garrison cap and neatly tucked it under the left epaulet of his "1505" khaki shirt. Next to the entrance was a full-length mirror under a sign that declared: "Look Sharp—Look Proud—Fly High—Air Force!" Ourecky paused to check his appearance. He adjusted his open collar before checking the "gig line" alignment of his shirt, belt and trousers.

There was nothing particularly distinctive about his appearance: He was a very ordinary-looking man from the American Midwest. His short black hair was precisely parted and combed neatly in place. His facial features—green eyes, moderately high cheekbones, short nose and sharp chin—reflected his Czech heritage.

Physically, he really hadn't changed much since he had graduated from high school six years ago. Although he ate virtually anything that didn't get out of his path, he seemed incapable of gaining weight. He wasn't exactly frail, but his reflection looked like he was still awaiting the growth spurt that had eluded him in adolescence.

Satisfied that he was suitably presentable, Ourecky took a plastic tray from a stack and fell in line behind a queue of hungry airmen. He had hoped to beat the lunch rush but hadn't escaped his desk in time. His heart thumped in his chest, partly from the hurried walk to get here, but mostly in anticipation of opening the letter.

As he came alongside the grill, he shouted his order to the harried cook. In truth, he wasn't that fond of the snack bar's fare. The burgers were consistently overcooked and saturated in grease, the French fries were typically undercooked and cold, and the pre-packaged sandwiches were so tasteless that they could have been stamped out of damp cardboard. The snack bar wasn't exactly convenient for him, either. It would have been much simpler and faster for him to make a dash to the burger joint just off base.

Pushing his tray forward along stainless steel rails, he gazed toward the cash register at the end of the serving line and thought of his ulterior motive for enduring so many crappy meals. The cashier was a pretty brunette named Sara. Ourecky had been making small talk with her for the past three weeks but had not yet mustered sufficient courage to ask her out.

In their abbreviated conversations, he had gathered Sara's biography in terse snippets.

Two years out of high school, she had aspirations of attending a cosmetology school as soon as she saved up enough money. She lived with her parents in nearby Niceville. Her father worked as a lumberjack for St. Joe Paper, clear-cutting pines to feed the company's insatiable pulp mills. Her mother spent her days spooling blue denim into bolts at the Vanity Fair Mills textile plant in Milton. A Capricorn, Sara liked puppies, Corvairs, the Beatles, and Elvis. She disliked split ends, rude people, and country music.

A brusque shout interrupted his romantic musings. "Hamburger, extra onions?" queried the short order cook, thrusting a paper plate toward Ourecky.

"Uh, yeah, that's mine," replied Ourecky. He reached out for the plate, but another hand darted in to swiftly snatch it away. Noting that the man wore a major's oak leaves on the shoulders of his flight suit, he politely said, "Uh, sir, uh, I think that's my…"

"You don't *mind*, do you, Lieutenant?" asked the major, quickly wielding a plastic fork to scrape away the extra onions. "I have to be back out to the flight line in a few minutes. I'm a test pilot. No time to waste. Know what I mean?"

The broad-shouldered interloper was roguishly handsome, with dense brown hair and matching mustache. His eyes were a startling shade of crystalline blue, almost unnaturally so, as if glistening particles had been snatched from the sky and embedded beneath his brows. He looked like someone destined to fly, who could not be anything else but an aviator. It was as if Steve Canyon had somehow clawed his way out of the two-dimensional confines of the Sunday comics, arriving here to lay claim to Ourecky's hamburger.

Like Ourecky, the pilot wasn't particularly tall, but then Scott remembered that NASA's astronauts—*his* heroes—were also short of stature. The pilot squirted catsup on the hamburger before cutting all the way to the front of the line, directly to the cashier's station. After paying the check, he leaned toward Sara and said something quietly. Giggling, she nodded and then smiled broadly as she jotted something on his ticket. The square-jawed pilot casually tucked the slip of paper into his pocket, picked up his tray, and then strode away to enjoy his purloined lunch.

Ourecky sighed as he selected a cellophane-wrapped egg salad sandwich from a nearby cooler. A faded label indicated that it had been prepared last week and that it should be eaten or discarded by today's date. There were about twenty men in line in front of him, so it took almost five minutes for him to make it to the cash register.

"Hey, Scott," said Sara, smiling as she made change from his dollar. "How's it going today?"

"Just busy," he replied, looking at the floor as he stuffed the coins into his front pocket. "*Really* busy. I guess I'll see you tomorrow."

"Yeah," she replied. "Guess so. See ya then." Sara kept glancing toward the handsome pilot, who was seated at a table about thirty feet distant. Finishing Ourecky's hamburger, he wiped his lips with a napkin and grinned at her. She winked and smiled back.

Ourecky found an unoccupied table in a far corner. After taking his seat, he placed the letter on the table, flat and unopened, and studied it as he slowly consumed his bland sandwich. He contemplated the envelope's significance; in a sense, it contained all of his boyhood dreams and all that he had worked for.

After his graduation and commissioning nearly two years ago, he was posted to Eglin Air Force Base, where he worked in armament design. Located in the Florida Panhandle, Eglin was a sprawling expanse of pine thickets, swamps, and bombing ranges. Pilots trained here on their way to Vietnam, perfecting their bombing and strafing skills. The huge complex also provided a perfect environment for testing the cutting edge ordnance concocted by Ourecky and his peers.

Although his path to the stars had taken a detour, he fervently poured himself into his labors at Eglin. There could be no man more perfect for the tasks assigned to him. When it came to understanding trajectories, the paths that objects follow through the air, he was a computational machine incarnate. He could literally picture complex calculations in his mind to study their nuances. Capitalizing on these abilities, he became instrumental in designing "stand-off" weapons that would enable a pilot to safely attack a target from a considerable distance away.

He finished the sandwich, washed it down with tepid milk, crumpled the wrapper and gently nudged the plastic tray aside. Even though he diligently applied himself to his work at the Armament Lab, Ourecky still stubbornly clung to his dream of becoming a pilot and eventually an astronaut. But would the Air Force grant him a waiver to attend flight school? The answer obviously resided in the envelope.

He could barely contain his excitement. He had recently read a best-selling self-help book, *The Power of Positive Thinking*, so he had convinced himself that he knew *exactly* what the letter would state. It would declare that the almighty United States Air Force had come to its collective senses and now understood the necessity of his continued service as a flying officer. Now, it was just a simple matter of opening the envelope to confirm that its contents matched the reality within his thoughts.

But opening the envelope wasn't a simple matter. He just couldn't will himself to do it. As strange as it might seem to others, mathematics had a calming effect on him, so he worked a few equations in his head to settle down. Slightly more focused, he decided that it was time to know the truth.

His fingers trembled as he slowly tore open the flap. He held his breath as he unfolded the crisp white paper, then forced himself to read the dispatch. The very first sentence curtly summed up the Air Force's decision: *"The applicant's waiver request for flight training is hereby DENIED."*

As if to heap burning coals on his scorching disappointment, the letter succinctly added: *"This office considers this matter CLOSED. In the interest of effective stewardship of the Air Force's limited resources, this office admonishes the applicant to make NO subsequent attempts for similar waivers. Any future waiver attempt will be immediately referred to the applicant's chain of command with a recommendation for potential disciplinary action."*

Tears welled in his eyes as he barely resisted the urge to curse out loud. As his face burned with anger and embarrassment, the paper slipped through his fingers. Like Icarus falling into the sea after flying too close to the sun, the single page—bearing Ourecky's dreams of flight—slowly fluttered to the scuffed linoleum floor. He glanced up to see the handsome pilot, now standing at the register, chatting with Sara as he bought a slice of apple pie. Swaggering back to his table, the aviator winked at him.

He closed his eyes and gritted his teeth. He couldn't remember when he had suffered so much humiliation in a single day. A few minutes passed before he felt someone tapping on his shoulder. He turned to see a heavyset man with features that were slightly Oriental, maybe Japanese or Korean or partly so. The man looked vaguely familiar; Ourecky was sure that he had recently seen him around the base.

"Lieutenant, I think you dropped this," stated the stranger quietly, handing Ourecky the letter. "I thought it might be important."

"Uh...thanks," mumbled Ourecky, regaining his composure. He took the paper and turned away. He stuck the letter back in the envelope. Now that his fate was effectively sealed, his choices were much plainer. Since he could not be a pilot, it made little sense for him to remain in the Air Force. He would serve out the remainder of his term, complete the projects he was working on, and then apply for an advanced degree program, perhaps at MIT. Afterwards, he would submit an application to work at NASA; if he could not ascend into space himself, then perhaps he could help others to get there.

5

NIGHT DROP AT EGLIN

Armament Evaluation Range Three-Alpha
Eglin Air Force Base, Florida
11:45 a.m., Monday, April 15, 1968

Crouched in a heavily fortified bunker, Scott Ourecky listened to the F-4 fighter make its run-in to the bombing range. Muffled by his earplugs and several feet of tightly packed dirt, the jet's roar was barely audible. Looking out through an observation port, he gazed at a rusted tank hull—the target of today's test—through a set of massive binoculars.

A radio speaker squawked: The pilot announced that he was thumbing the "pickle" switch to release the prototype robotic bomb. If all went well, the bomb would sail through the air on a pre-set trajectory, automatically adjusting its course with slight flicks and twitches of its tail fins.

Seconds later, the derelict tank disappeared in a sudden wash of flame and smoke. Whirling like a child's toy, its decapitated turret sailed several yards into the air. Ourecky winced as the bunker was rocked by the shock wave of the bomb's detonation. A cloud of fine dust quickly saturated the air; he covered his mouth and nose with a white handkerchief. Clods of dirt and chunks of scrap metal spattered the bunker's reinforced earthen roof like hail. It took a few minutes for the smoke and dust to clear, and for the world to settle back to normalcy.

"Excellent!" declared the man standing next to Ourecky. "Dead on!"

"Huh?" asked Ourecky, cupping his ear as he swiveled around. "Sir, I can't hear you." Then he remembered the waxy cotton plugs jammed in his ears and tugged them out.

"Dead on," repeated the man. He was Ourecky's boss, Colonel Ron Paster, the director of Advanced Armament Design at Eglin Air Force Base. He was a tall man and had to stoop to fit inside the confined bunker. "I don't think we could have nailed it any closer."

Scribbling notes on the test, Ourecky nodded in agreement. They had refined the steering system just about as much as it could be refined; soon, the new bomb would make its lethal debut in the skies over Southeast Asia.

Smiling to himself, he reflected on how suddenly things had changed since receiving his rejection letter last Thursday. An opportunity had arisen that caused him to question his decision to leave the Air Force. Unbeknownst to him, at least until late Friday afternoon, he had been the subject of some intensive negotiations between his boss and an influential retired general at Wright-Patterson Air Force Base in Ohio. The former general wanted to shanghai him for a highly classified research project but was meeting significant resistance, since Ourecky's current work was deemed vital to the ongoing war in Vietnam.

Eventually, Paster and the general hammered out a time-sharing agreement in which Ourecky would spend two weeks here, then the following two weeks at Wright-Patt, alternating between the two bases until at least one of the crucial projects was completed. Of course, all of this was contingent on whether he made it through his initial interviews and tests in Ohio.

Ourecky didn't savor the notion of shuttling between projects and bases, but he was a miniscule cog in a large and complex machine. Besides, they promised to pin captain's bars on him—over a year ahead of his peers—so there was something to be said for the deal. If nothing else, perhaps there was a remote chance he could etch out a career in the Air Force without ever wearing the silver wings of a pilot.

As Paster, Ourecky, and three other officers emerged from the bunker, momentarily dazed birds and shock-addled squirrels looked on in confusion. The acrid scent of cordite wafted heavily in the air, and the budding leaves of scrub oaks were coated with a patina of powdery brown dust.

The range phone jangled in a plywood shed. A sergeant answered it and called for Colonel Paster. After a short conversation, Paster spoke to Ourecky. "Scott, you're supposed to meet Virgil Wolcott out at Auxiliary Field Ten this evening. He's visiting the base for some sort of test. He's expecting you out there at eighteen hundred. You need to wear civilian clothes suitable for being outside."

Eglin's expanse was dotted with outlying sub-bases called auxiliary fields, and Ourecky really wasn't sure where Aux Field Ten was or how to drive there. In any event, he thought, he had a few hours to find out. "I'll be there, sir," he answered.

"Colonel, is that *General* Wolcott?" asked a major standing next to Paster. The major's face was a mixture of surprise and confusion. The other officers appeared perplexed as well. "Ourecky is meeting General *Wolcott? The* General Wolcott?"

Paster nodded. "One and the same," he answered, patting dust from his khaki uniform. "Except ol' Virgil is no longer a general. He retired last year, and now he's working on some hush-hush project up at Wright-Patterson. And gentlemen, let's exercise a little discretion and keep this matter among ourselves."

Eglin Air Force Base, Florida; 2:35 p.m., Monday, April 15, 1968

The hop from Ohio to Florida was exactly the kind of whirlwind hell-for-leather jaunt that Virgil Wolcott relished. Although he had retired over a year ago, the Air Force had granted him special authorization to continue flying, so long as he maintained proficiency and received his annual "up" slip from a flight surgeon.

Remaining on flight status was only one of several concessions he had been given to stay on as the deputy director—as a government civilian executive—of Blue Gemini. When he filed his retirement paperwork last year, he intended to spend the rest of his days riding horses and mending fences back on the family spread in Oklahoma, but a chance to jump behind the controls of a T-38 Talon whenever he desired was just too tempting to pass up. He and his back-seater, Mark Tew, had flown the T-38 down from Ohio to inspect some facilities at Eglin.

Following the ground marshaller's signals, he deftly taxied the supersonic trainer to the spot where a transient alert crew awaited their arrival. Their Plexiglas canopies whirred open as the ground crew positioned ladders to facilitate their descent. Wolcott peeled off his damp flight gloves and dug a foil-lined packet of Red Man chewing tobacco out of a breast pocket of his flight suit. As he tucked a damp lump of tobacco in his mouth, he heard a rustling noise from the back seat. "Are you *still* doin' paperwork back there, pardner?" he asked. "You couldn't just relax and enjoy the scenery for a change?"

"No rest for the weary, Virgil," replied Tew, hurriedly jamming a stack of documents into a leather attaché.

A tech sergeant climbed the ladder and threw Wolcott a salute almost as stiff as his starched utilities. Ticking off the last items on his post-flight checklist, Wolcott glanced up and casually returned the salute.

The tech sergeant related further instructions. "General, you can change clothes in the locker room in the Operations shed. My people will secure your flight gear. There's a sedan and driver waiting at Operations that will take you on to your destination."

Wolcott allowed the sergeant to unfasten the last few connections that secured his parachute harness to the ejection seat and then pushed himself up out of the narrow seat pan. His body ached like he had spent the past few hours tossing hay bales. He had always been a bit too tall for jet cockpits, and the T-38's was no exception, but now he was starting to face the grim truth that he was getting a little too old for demanding cross-country flights.

Trying his best to appear fresh and nonchalant, he clattered down the ladder. While Wolcott flew every chance he had to keep current, Tew wasn't that familiar with the T-38 and took slightly longer to disembark. As Tew gingerly descended, Wolcott opened an under-wing luggage pod and wrestled out the canvas kit bags that contained their clothes and sundries.

Half an hour later, they had doffed their flight gear and were in civvies. With the exception of flight gear and when he was summoned to the

Pentagon, Tew rarely wore any semblance of a uniform. The same was true of the other military personnel assigned to Blue Gemini.

Wolcott stepped out from behind his locker door. He was dressed in starched blue jeans, a white shirt with pearl snap buttons, and well-worn cowboy boots. His neck was adorned by a bolo string tie bearing a gaudy slide fashioned of silver and turquoise. His slender waist was encircled by a brown leather belt fastened with a massive hand-tooled silver buckle.

In contrast to his compatriot, Tew had packed more conventional garb in his kit bag. He wore tailored khaki trousers, a light blue knit sport shirt, and brown penny loafers. "I see by your outfit that you are a cowboy," chided Tew, paraphrasing Marty Robbins. "Virgil, do you suppose that there will ever come a day when you don't dress like an extra in a spaghetti Western?"

"Mi amigo, now that I'm retired, I don't much give a hoot what anyone thinks," replied Wolcott, donning a genuine white Stetson hat to complete his ensemble.

3:15 p.m.

In front of the Operations building, Master Sergeant Jimmy Hara patiently waited beside an official blue Ford sedan. Short, bulky, with a crew cut and sparse moustache, Hara was a Japanese-American counter-intelligence specialist with eighteen years experience in the Air Force. He slid in behind the wheel and donned sunglasses as Tew and Wolcott quickly took their places in the car's back seat.

"So, Jimmy, you've been keepin' a watchful eye on this kid Ourecky?" asked Wolcott, rolling down his window before lighting a cigarette.

"That I have, Virgil," replied Hara, yawning as he pulled away from the curb. "Care for a report?"

"Sure, pard," answered Wolcott. "What sort of skeletons did you find rattlin' in his closet?"

"None. Not a single bone. If you must know, Virgil, this has been the most boring two weeks of my life. I've known Shinto monks who lead more exciting lives than this egghead."

"How so, Jimmy?" asked Tew, opening a spiral-bound report in his lap.

"Well, sir, he usually works at least twelve hours a day, seven days a week. If he's not at work, he's at his room at the BOQ. He doesn't hang out at the Club and rarely goes off base. Quiet as a church mouse. I'm not even sure that he drinks."

"Girlfriend?" asked Wolcott.

Hara shook his head. "Not that I can tell. This guy is definitely not a social butterfly. Very much the loner. It looked like he had his eye on a little local number who works at the base snack bar, but I think she's a lot more interested in pilots than engineers."

"And what red-blooded American woman wouldn't be more partial to pilots, pard?" asked Wolcott, cupping his hand around his cigarette to shield it from the breeze. "Anything else?"

"Not really, Virg. I've had his phone tapped for the past two weeks, but he's only placed two calls. Both were to his parents in Nebraska."

"So he's clean?" asked Tew. "No problems? No outside commitments? Not married?"

Hara looked in the rear view mirror and smirked. "No distractions. And I don't see him getting hitched anytime soon, General."

"Excellent. Good job, Jimmy. As always."

As he drove northwest along an isolated base road lined by pine forests, Hara looked back momentarily and said, "Virgil, even though it's none of my business, Ourecky sure looks like just an ordinary lieutenant. He sure doesn't seem to be anyone special, not like some of the pilots you've had me pull surveillance on."

Chuckling, Wolcott said, "Quite the contrary, Jimmy. For starters, besides bein' a crackerjack engineer, he's a bona fide math genius with a practical background in ballistics and trajectories. Personally, I don't claim to be too danged smart, but I can pick horses, and if my instincts are right, Ourecky is the best bet to solve our biggest problem."

"He's your fix for the Block Two computer?" asked Hara, adjusting his sunglasses.

"You've got your nose in everything, don't you, Jimmy? Yup, I think he's the man to fix it."

"Let's hope you're right, Virgil," Tew interjected, glancing up from his report. "This Block Two situation is starting to wear on me."

"Yup, with any luck, this kid Ourecky could be our great salvation," Wolcott said. "Hey, Mark, after we meet with Isaac Fels, how about we grab an early dinner? I know a great little rib shack over in Milton. Best danged fall-off-the-bone barbecue on the Panhandle. It's just a few minutes away."

"Sure, but can you make it back in time for the drops tonight?" asked Tew. "At least one of us should be there."

"So ain't you comin' out to watch tonight?" asked Wolcott, removing his left boot and scratching his toes. "It should be a good show."

"No. Paperwork, Virgil, always more paperwork. I have an all-nighter ahead of me."

"Suit yourself. I'm hankerin' to see the drops, of course, but I ain't particularly relishin' the thought of seein' Agnew again."

"Problem?" queried Tew, penciling notes in the margin of a budget summary.

"Yup. Agnew's usual guff. Him and Carson didn't exactly geehaw at the desert survival course out at Stead, and now Agnew's spoutin' off about a transfer out of the Project again. His bellyachin' is really starting to dig under my hide, pardner."

"You'll handle it, Virg. You always do."

Auxiliary Field Ten, Eglin Air Force Base, Florida; 5:45 p.m.

The waning sun hovered low over the horizon as Ourecky arrived at Auxiliary Field Ten. Towering pines swayed in the evening breezes blowing in from the Gulf of Mexico. He was surprised at the extent of activity at the field even this late in the day. A typical auxiliary field was just a patch of bare ground, with a rudimentary airstrip and little else, but this one was almost like a miniature airbase unto itself. Secluded on the distant fringes of Eglin, it contained an interlocking pair of paved runways, as well as a small complex of buildings and hangars.

As Ourecky got out of his car, a rangy man strolled out of the hangar to meet him. Wearing blue jeans, a faded denim jacket and a cowboy hat, the tall stranger looked like he was on his way to a rodeo or perhaps a cattle drive. "Lieutenant Ourecky?" he asked.

"Uh, that's me. I'm Lieutenant Ourecky, sir."

The man chuckled. "I'm Virgil Wolcott. Any problems drivin' out here?"

Ourecky was taken aback; the rawboned, mufti-clad man standing before him was not at all what he had expected. "None at all, sir. It's an honor to meet you, General."

"Just call me Virgil," drawled Wolcott, extending a hand. "The stars fell off when I shed the blue suit. Care for some coffee, Lieutenant? There's a fresh percolator brewin' inside."

"Sir, I'll pass if you don't mind. It's a little late in the day for me to be drinking coffee."

"Suit yourself, pardner. We're conducting some drop tests this evening, so it's going to stretch into a mighty long night for me. Let's head inside."

Within the hangar, several men conferred beside the open rear ramp of a C-130 cargo plane. Most were attired in standard gray-green flight gear, but two wore pumpkin-orange flight suits and parachute harnesses.

Ourecky glanced inside the transport's capacious cargo bay, and was surprised to see a full-scale mockup of a Gemini two-man spacecraft. Strapped down to a plywood skid resting on two parallel tracks of conveyor rollers, the capsule mock-up appeared ready for ejection from the C-130, in the same manner as vehicles or heavy equipment bundles dropped by cargo parachute. Since NASA's Gemini program had ended two years ago, he was curious what system was being tested this evening.

The meeting broke up, and the C-130 crew went to their stations to prepare the plane for takeoff. "Let me introduce you to a couple of our top hands," said Wolcott, leading Ourecky toward the men in orange. Both were deeply tanned, as if they had just returned from a prolonged vacation at the beach, and appeared to be in excellent physical condition. "This here is Major Drew Carson and Major Tim Agnew. Gents, this is Lieutenant Scott Ourecky. If the stars line up, he'll be doin' some special engineering work for us."

Carson looked extremely familiar to Ourecky, and finally he realized where their paths had crossed. "Uh, Major Carson and I have already met."

Momentarily puzzled, the pilot studied the engineer's face. He grinned and noted, "Yeah, Virgil, the lieutenant and I shared lunch last week, down at the main base snack bar."

"So you two are already acquainted?" asked Wolcott. "Ain't it just an itty-bitty little world?"

Ourecky shook hands with both pilots. In contrast to the poster-perfect Carson, Agnew was plain-featured, about six feet tall, with thinning blonde hair. Strapping a navigation kneeboard to his right thigh, he appeared apprehensive about tonight's planned activities.

Ourecky noticed that Agnew made a point to slip off his flameproof Nomex gloves to shake hands. On the other hand, the aloof Carson didn't take off his gloves. He treated Ourecky with blatant disdain, like the subtle contempt that aristocracy reserves for the masses, as if he were of some servile subclass not worthy to rub shoulders with the likes of a vaunted test pilot.

"All up, gents?" asked Wolcott. "Ready to drop?"

Agnew frowned and answered furtively, "I would be lying if I told you I was thrilled to ride this sled again, Virgil. I'm just not entirely confident that all the kinks have been worked out. It's only a matter of time before…"

Carson smiled and interrupted with, "We're ready, Virg. Tim's just experiencing a few butterflies. They'll pass. He'll be just fine."

"Well then, time's a' wastin'. Let's saddle up and drive this herd," declared Wolcott. He glanced up at the cockpit, grinned, and flashed the pilot a thumbs up.

Carson tugged on his helmet and made some minor adjustments to his parachute harness. He turned to board the C-130 through the crew door, just aft of the cockpit.

"A moment of your time, sir?" asked Agnew sheepishly, donning his helmet.

"What's on your mind, buck?" replied Wolcott.

"I was wondering if you've had an opportunity to consider my reassignment paperwork, Virgil. If you recall, I gave it to you right after we came back from Stead, about a week ago."

"Yeah, I recollect you givin' me that transfer request, pard." Obviously perturbed, Wolcott spat tobacco juice into a nearby waste can. "But you need to come to grips with the fact that it ain't happenin'. First, if you had trepidations about this business, you should have aired them before you signed on the dotted line. Second, you know danged well that we're down to three flight crews, so you shouldn't feel too offended if I ain't overly disposed to hasten your departure. Now, mount up and ride, brother. That's what you volunteered for, so snap to it."

"If I can't transfer, sir," said Agnew, snugging the chest strap of his parachute harness, "would you at least consider shifting me to another crew? Please?"

Wolcott's face gradually took on a shade of red. "And break up another crew? Like I said, buster, we only have *three* crews. I can't shift you without first wranglin' someone to fly with Carson, and you know full well the likelihood of that happening. Look, Carson's a tad obnoxious and a mite volatile, but he's a danged good pilot. You couldn't be in any better hands. Can't you find a way to make peace with him?"

Agnew rubbed his left shoulder and replied, "I'll try, Virgil, but you're asking a lot. Maybe more than I have to give." He pivoted away and climbed aboard the transport plane.

Turning to Ourecky, Wolcott said, "Sorry you had to witness that, son. Agnew's a good hand, but he needs to learn that he can't always choose who he flies with. In any event, I would appreciate it if you kept that little transaction to yourself."

"Will do, sir."

Wolcott walked over to a table and filled a Thermos from a large metal urn. "Let's mosey on out to the strip, pard," he said, screwing the lid on the insulated bottle. "We'll watch the show from out there. We can chat on the way."

The two men strolled out of the hangar as the C-130's engines roared to life. They walked along a grassy strip adjacent to a taxiway. By now, the sun had all but retired from the sky; the buildings and trees were bathed in the red-hued light of near dusk.

With turboprop engines groaning, the C-130 emerged from the hangar and rolled slowly along the taxiway. Doffing his Stetson, Wolcott waited for the plane to pass and for the noise to subside. He pulled a small envelope of chewing tobacco from a breast pocket of his denim jacket, spat out a depleted wad, and replaced it with another. "Chaw?" he asked, offering the pouch to Ourecky.

"Uh, pass, sir."

"Son, I want you to know that Colonel Paster speaks very highly of you. As much as he doesn't want to share you, it sounds like you're just the man we're lookin' for."

"Thank you, sir."

"I s'pose you want to know what you're getting into. The *official* name for our outfit is the Aerospace Support Project. We run it out of Wright-Patterson, but we have facilities all over, like this field here. We have a big charter, and we're always on the lookout for new talent." Without elucidating, Wolcott walked on.

After a few minutes, Ourecky broke the silence. "Umm, that sounds very intriguing, sir."

Wolcott chuckled and spat. "If I was walkin' in your boots, Lieutenant, I would say it sounded a trifle vague. You ain't inclined to nibble on the hook and ask what we do?"

Ourecky shook his head. "Sir, uh, I assumed you would tell me when you saw fit."

Wolcott smiled. "Not askin' a lot of questions is an admirable quality, son. Especially with the sort of chores we do. You might fit in just fine."

At the end of the runway, the C-130 taxied into takeoff position and paused for several seconds as its engines roared up to full power. Gathering speed, the plane rolled down the runway, lifted off, and disappeared into the growing darkness.

Half an hour later, the two men reached the test site, a dirt "assault strip" that ran parallel to the main runway. Assault strips were commonly used by C-130 crews to practice the tactical landing and takeoff skills that would be so critical to them in Southeast Asia. Ourecky discerned the outlines of a pickup truck and a jeep, and heard the voices of several men.

"Excuse me, pard," said Wolcott. "I need to go chat with these folks. I'll be right back."

Ourecky watched the angular general saunter away in the darkness and then lifted his eyes toward the heavens. The sky was startlingly clear, and he picked out familiar stars to orient himself. He quickly found his old friend Cassiopeia and used its lazy "W" shape as a guide to look down and to the right to find Ursa Major—the "Big Dipper"—and Polaris, the North Star. Looking toward the east, he found the two stars—Gomeisa and Procyon—that comprised the constellation Canis Minor, and then he glanced to the west to locate the four stars—Markab, Algenib, Scheat and Sirrah—that marked the corners of the big box in Pegasus. With his bearings fixed, he jammed his hands deep in his trouser pockets and waited for Wolcott.

A few minutes later, Wolcott returned. Obviously comfortable in the tranquil darkness, he guided Ourecky to the rear of the pickup truck, opened the tailgate, and sat down. "Hunker here with me, buckaroo," he drawled. "Let's jaw while we're waitin' for the festivities to begin."

Ourecky sat next to Wolcott on the tailgate. Shuddering, he regretted that he hadn't brought a jacket. The temperature had dropped at least five degrees since the sun had disappeared.

"I saw you stargazing over here," noted Wolcott. "Your college transcripts said you took several astronomy courses. Ain't that a might peculiar for someone majorin' in engineering?"

Ourecky was taken aback. *College transcripts? They've looked at my college transcripts?* "Just something that I picked up back at home, sir," he explained. "You can see a lot of stars in the sky out there on the farm, and I just wanted to learn some more." He paused and added, "And I had wanted to eventually go to work for NASA, so I thought it might come in handy."

"Well, hoss, believe it or not, we might be able to put that knowledge to use." Wolcott's cigarette glowed bright red as he drew in deeply; his crinkled face was illuminated in the dim glow. "Hey, Ourecky, if you know your stars, maybe you could help me out. As far back as I can remember, even when I first started ridin' the range as a young buck, I used to spy an itty-bitty group of stars and always thought it was the Little Dipper." Wolcott pointed out into the night sky. "Right over there, just above the trees. What's that constellation?"

"Uh, that's the Pleiades, sir. It's not really a constellation; it's a star cluster. It's my favorite."

Wolcott smiled. "Your favorite? Really? It's pretty durned bright, ain't it?" He looked at the luminous dial on his watch, spat a long string of tobacco juice into the darkness, and then took a sip of coffee. "Now, Ourecky, I can't really tell you much about what you'll be doin' up at Wright–Patt, but I did want to chew the fat before you came up for your formal interview. Next week, you'll be meetin' with me and my compadre, Mark Tew. I want to caution you that Mark's still on active duty, and he's a lot more formal than I am, so don't let my behavior be your guide."

"I appreciate the heads up, sir."

Wolcott topped off his cup from the Thermos. "Paster tells me that you're interested in going back to school after your Air Force stint is up. Is that correct?"

"Yes, sir," replied Ourecky. "I did some independent study on orbital mechanics while I was earning my undergraduate degree, and I would eventually like to go back for a PhD. Hopefully, I'll go to work with NASA after that."

"Well, I sure ain't going to fault you for wanting to hitch a leg up, pardner. Do you have any particular school in mind?"

"Uh, I'm looking at MIT, sir. That's where Colonel Aldrin did his thesis on rendezvous. Are you familiar with Aldrin? He's an astronaut now."

In the darkness, Wolcott grinned. "Yeah, pard, I'm familiar with Buzz. Quite a thinker, that one. Have you pondered about possibly seekin' your PhD on the Air Force's nickel?"

"Sir?"

"Look, son, if you do a good job for us, then we can certainly exert some influence on your behalf. The Air Force has plenty of money to pay for advanced degrees, and Mark Tew has the connections to loosen those purse strings. Of course, that would mean remainin' in the Air Force for a few more years before you have an opportunity to work for NASA, but the Air Force also has space projects you might be able to work on. You ever heard of the MOL?"

"MOL? Uh, the Manned Orbiting Laboratory, sir?"

"One and the same," replied Wolcott. He finished his coffee and screwed the plastic cup back onto the top of the Thermos. "I have some friends workin' on it. If you're hankerin' to do space work, I'm sure that we could wrangle you a slot there later."

A man walked up and spoke. "Virgil, we're just about to open the curtains. Taco Two Six should be closing in on the communications checkpoint at any moment."

Wolcott buttoned up his denim jacket and said, "Well, it's time for me to earn my paycheck."

Not too far from the rear of the truck, a technician shined a flashlight on an instrument and fidgeted with some dials. The device was about the size of three footlockers stacked atop one another, crowned by a round object that looked like a large hatbox. "That's a portable TACAN beacon," explained Wolcott. "We've been workin' on it for quite a while. With that critter, our boys can set up an instrument approach on just about any airfield in the world."

Ourecky heard the faint drone of the C-130 in the distance. The radio crackled. "Taco Two Six is at CCP."

"Taco Two Six, this is Exercise Control. I copy you are at CCP," answered a sergeant monitoring the radios. "Sir, they're at the communications checkpoint. Five minutes to drop."

"Thanks, Jack," Wolcott said. He turned to Ourecky. "Just so you know, we're runnin' a night flight test on a Rogallo paraglider. You ever heard of a Rogallo wing, son?"

"I have, sir. It's kind of a cross between a parachute and a glider, isn't it? I think NASA considered them for the Gemini program. But didn't they abandon the whole concept?"

"NASA dropped it, but the Air Force didn't," replied Wolcott. "We're still lookin' at the paraglider for the MOL, among other things. We're really trying to avoid having our Gemini capsules splash down at sea. For one thing, the Air Force doesn't own fleets of ships that we can scatter all around the world, and we would just as soon limit our reliance on the Navy."

"Good point, sir," noted Ourecky. Now the Gemini mock-up made perfect sense to him; the test was obviously associated with the Air Force's ongoing MOL program.

"That ain't all. A spacecraft is built of metals that are highly vulnerable to corrosion, plus it's loaded to the gills with high-dollar electronics and wiring. Soaking one in salt water is a surefire way to ruin it. It you could return them to dry land instead of dunkin' them in the ocean, you can reuse them. Granted, the paraglider ain't perfect and we still have plenty of kinks to work out, but we're making good headway on refining the concept."

Their conversation was interrupted by the radio squawking, "Control, this is Taco Two Six. Angels Two Zero. Two minutes from release."

The sergeant replied, "Taco Two Six, I copy you are at Angels Two Zero and two minutes from release."

"Lights," ordered Wolcott quietly.

The sergeant spoke a few words into the radio, and almost immediately the entire area was dark. All the runway markers were doused, as well as all the lights on the compound next to the runways. Several men lit red railroad flares to mark the edges of the dirt assault strip.

"Taco Two Six is one minute out."

With the lights extinguished, the sergeant's face was barely visible in the glow of the radio dials. "Taco Two Six, this is Exercise Control. I copy that you are one minute from release point. Break, break, break…All stations, all stations, all aircraft in the vicinity of Eglin Air Force Base, be advised that paradrop operations are currently in progress in a five-mile radius of Aux Field Ten, twenty thousand feet AGL to surface. All aircraft are directed to remain clear of this airspace."

"Control, this is Taco Two Six. Thirty seconds. Request clearance for release."

Gazing up into the darkness, Ourecky glimpsed the faint red and green anti-collision lights of the C-130. It was about four miles away, flying west to east above the Florida Panhandle.

"Taco Two Six, Control, I copy you are thirty seconds away. Airspace is cleared. You are cleared for release."

"Control, this is Taco Two Six. Copy cleared for release. Thank you."

A new voice came over the radio: "Control, this is Chase One. In position."

"Chase One, I copy that you are in position." The sergeant turned and translated the chatter for Ourecky. "Taco Two Six is the C-130 making the drop. Angels Two Zero means that he's at twenty thousand feet. Chase One is an OV-10 Bronco trailing Taco Two Six. He'll observe the drop. Call sign for the paraglider test vehicle is Ultra Zero One. Here at Eglin, the Ultra call sign designates an unpowered aircraft."

The C-130 pilot spoke. "Control, this is Taco Two Six. Count down to release. Five, four, three, two, one, Mark. One package away. Stand by, stand by, stand by."

"Control is standing by for confirmation of release," stated the sergeant, keying the radio handset.

"Control, this is Taco Two Six. My loadmaster reports that the package is clear."

"Control, this is Chase One. Package in view. Drogue is deploying… drogue is deployed. Paraglider is deploying. I see longitudinal struts inflating and cross strut inflating." After a few tense moments elapsed, Chase One reported, "Paraglider appears fully deployed."

Except for a slight buzz of static, the radio was silent for a few minutes. A terse voice finally broke the silence; Ourecky recognized the speaker as Agnew. "Control, this is Ultra Zero One. Receiving TACAN on Channel Six. Command pilot reports that paraglider deployed correctly and is fully controllable with nominal steering response."

"Ultra Zero One, Control, copy TACAN on Channel Six and paraglider deployed normally."

Several minutes passed as the men gazed out into the night sky. Ourecky saw the running lights of the twin-engine Bronco spotter plane, but had yet to spot the paraglider. Straining to find the unpowered craft, he cupped his hands around his eyes to shield out the red glare from the railroad flares.

"Don't look directly at the chase plane right now, pardner," advised Wolcott. "Scan more to the right. They're flyin' a right-handed pattern to our north. Carson will fly parallel to us west to east, then swing around to land east to west." Wolcott unzipped a fabric case and pulled out a PVS-2 "starlight scope." The starlight scope—which amplified ambient light from the stars and moon—was brand new technology and was only now being delivered to combat troops in Southeast Asia. He flipped a switch on the device; it hummed audibly as he held it up to his eye.

Ourecky looked to the north and finally distinguished a dark shape as it passed over and blotted out bright stars. Against the stark black sky, the paraglider and Gemini mock-up weren't discernible shapes, but just a vague blob sailing through the night.

"Control, this is Ultra Zero One, turning right to base." Ourecky recognized Carson's voice.

"Ultra Zero One, I copy that you are turning right to base leg," replied the sergeant.

A few moments later, Carson spoke again. "Control, this is Ultra Zero One, turning right to final. Field in view. Gear down and locked."

"Control, this is Chase One. I confirm that Ultra Zero One's gear are down and locked."

The sergeant responded: "Chase One, good copy. Ultra Zero One, Control, copy gear down and locked. Winds from Two Two Zero at three knots. Surface is packed earth. Ultra Zero One, you are cleared to land on Assault Two Seven."

"Control, this is Ultra. Copy winds from Two-Twenty at three, cleared to land on Assault Two Seven."

Finally, Ourecky glimpsed a shadowy object gently gliding toward the assault strip. He heard a gentle popping sound of the fabric wing. He could barely make out the outline of the Gemini mock-up suspended under the paraglider. The dark apparition reminded him of an immense raptor, returning from a nocturnal hunt with a terrified rabbit clutched in its talons.

"Contact light. Flaring paraglider," reported Carson.

As it drew nearer, the rear of the paraglider drooped sharply down-wards as the pilot intentionally stalled it to increase its braking effect. Suddenly, a loud scraping noise confirmed that the object had made contact with the runway.

"Releasing paraglider," stated Carson. As the jettisoned paraglider fluttered to the ground, the Gemini mock-up continued to slide down the dirt strip, trailing a shower of sparks.

Startled, Ourecky nudged the sergeant and asked, "Did something happen to his wheels?"

"Skids," replied the sergeant. "He doesn't have wheels. Just skids."

Once the mock-up had come to rest, Carson's voice came over the radio. Laughing, he said, "Ultra Zero One on Assault Two Seven, waiting for taxi instructions." A few seconds passed, and he added, "Be advised my right-seater got kayoed on the landing."

"Jack, ask him if Major Agnew is injured," said Wolcott, switching off the starlight scope and slipping it back into its rubberized case.

"Ultra, is your right-seater injured?" asked the sergeant.

"Control, this is Ultra. Yeah. It looks like he didn't brace properly and smacked his face on the controls. He's conscious now but sustained a pretty nasty gash to his forehead. I think he'll need a few stitches."

Wolcott whistled lightly and commented, "We'll forgo the remain-der of this evening's drops, buckaroos. Jack, I'll ride with Agnew to the dispensary. After you dismantle everything, would you be so kind as to shepherd the lieutenant back to our hangar?" He slapped Ourecky lightly on the back. "Ourecky, I'll see you up in Ohio next week."

"I'll see you there, sir." Shivering slightly in the cool night air, Ourecky looked toward Wolcott; the older man's silhouette was outlined by the red glow from the sputtering marker flares. In the strange light, he looked almost surreal, like a demonic apparition. Ourecky felt an awkward urge to salute the retired general but let it pass. The two shook hands and parted company.

6

INSOMNIACS

Aerospace Support Project
Wright-Patterson Air Force Base, Ohio
11:53 p.m., Thursday, April 18, 1968

Air Force Brigadier General Mark Tew sputtered awake from a sound slumber. He rubbed his bleary eyes before consulting his Timex watch: the faint luminous hands reflected that it was seven minutes before midnight. He had been asleep for slightly less than an hour. As had been his routine for the past five years, he would lie down for another nap in five hours and then repeat the cycle throughout the day.

Yawning, he pushed aside a thin blanket, swiveled upright on the canvas Army cot, pulled on his blue cotton bathrobe, slipped his feet into soft suede moccasins, and then padded into the adjoining office that he shared with his deputy, Virgil Wolcott.

Tew opened a Mosler safe behind his desk, removed a highly classified intelligence summary of some recent ominous developments in the Soviet Union, and resumed his nocturnal chores. He sat down, switched on a desk lamp, stretched, closed his eyes, and thought about how he came to be at Wright-Patterson and why he felt so compelled to work at all hours of the day and night.

He attributed his irregular sleeping habits and intense work ethic to his former boss at Air Force Systems Command, General Schriever. Besides his legendary intellect and movie star looks, Schriever was renowned for his uncanny ability to function for days on end without regular sleep. He maintained his inexorable pace by periodically sneaking catnaps. He conditioned his closest subordinates to do likewise, counseling them to always grab a few winks before making any crucial decisions or attending any important meetings. During their stint at AFSC, Tew and Wolcott had relentlessly pushed themselves, striving to keep pace with their exalted boss, snatching rest when they could, like combat infantrymen who learned to doze on the march.

In truth, to imply that they merely worked for General Schriever was not nearly an accurate description of the relationship; they were sworn acolytes of the man. AFSC was the development hub for the nation's strategic arsenal of ballistic missiles, but Schriever had directed Tew and Wolcott to focus their energies on the military's role in the rapidly expanding universe of space exploration and exploitation.

In their eyes, if there could be a god of military space programs, like a focused god of Greek mythology, then Schriever was the one who set the stars in the firmament—and then adjusted their brightness and clarity so that they could be seen by lesser men. But all comparisons to ancient deities aside, a more tangible indicator of Schriever's tremendous reach and power was the fact that he personally controlled almost forty percent of the Air Force's budget before he retired two years ago.

Though their recent work was largely theoretical, Tew and Wolcott were intimately familiar with the sacrifices of war and the fragility of peace. As B-17 bomber pilots in World War II, they had witnessed close friends blown to pieces right before them. They had seen massive planes instantly transformed into shredded aluminum, flame and smoke. And while they didn't see it with their own eyes, they were very conscious of the death and devastation wrought by the ordnance they rained on cities below.

After the war, Tew shifted from bombers to fighters. He much preferred the personal one-on-one nature of fighter combat to the cold impersonality of strategic bombing. In 1950, assigned to the Fifth Air Force in Japan, he flew an F-80 Shooting Star against North Korean MiGs after Communist forces crossed the Yalu River. He later transitioned to the F-86 Sabrejet, and by tugging a few strings, he succeeded in bringing his friend Wolcott over as well, and the pair piloted F-86's for the remainder of the war.

After Korea, in the post-war heyday of aerospace research, Tew's career followed in Wolcott's wake. Despite the homespun demeanor reminiscent of his Oklahoma upbringing, Wolcott established a reputation as a practical thinker who could envision new ways to exploit the new technologies arriving at the forefront. What he lacked was the capacity to match fiscal realities and bureaucratic constraints to his ideas. In counterpoint, Tew had mastered the arcane arts of managing budgets, designing organizations, and doing those things necessary to bring Wolcott's lofty concepts from the drawing table to the skies. They existed as a symbiotic pair, closer to one another than they were to their own wives.

But much to their chagrin, the two had been burdened with a string of troublesome projects. In the late fifties, they had toiled on an effort to build a nuclear-powered bomber, capable of flying from the United States to the Soviet Union and back without refueling. But extensive technical problems—especially the lack of a nuclear reactor light enough to be safely carried aboard an airplane—eventually led to the program's cancellation in 1961.

After the demise of the original incarnation of Blue Gemini, the pair was transferred to Ohio's Wright-Patterson Air Force Base to work on the X-20A Dyna-Soar. Dyna-Soar, short for "Dynamic Soaring," was envisioned as a winged, reusable space plane that would be launched on a rocket booster, execute missions in space, and then return to earth as an unpowered glider. Had it come to fruition, Dyna-Soar would have performed several potential roles: Hypersonic flight research, inspection of hostile satellites, strategic bombing, space logistics and reconnaissance.

McNamara closed down Dyna-Soar in December of 1963. Afterwards, Tew and Wolcott were assigned to the Manned Orbiting Laboratory project at AFSC. The MOL was one of the Air Force's premier projects; it was considered of such strategic importance that General Schriever himself was tapped to be director of the project, in addition to his regular duties as the commander of AFSC. Finally, they had hitched their wagon to a rising star, and returned to California from their exile in Ohio.

Even after Schriever retired, the MOL was still gathering momentum and was well on its way to becoming reality. Unlike the other pie-in-the-sky schemes that Tew and Wolcott had labored on, MOL hardware had actually left the ground; launched from Cape Canaveral, a Titan IIIC had carried a mock-up MOL into space in 1966. Thirteen military astronauts—ten Air Force pilots, two Navy aviators and one Marine aviator—were diligently training for MOL missions.

As much as Tew and Wolcott hated leaving the MOL project, Blue Gemini was a task tailor-made for the duo. The very notion of a Soviet Orbital Bombardment System had terrifying implications. Once it was even marginally operational, the Soviets could rain down bombs on the United States without warning. The whole idea scared the crap out of US strategic planners.

As luck would have it, even as Khrushchev uttered his tacit threat in 1961, the Air Force had already been developing a robotic satellite interceptor—SAINT—to counter Soviet reconnaissance and other potential threats from space. Unfortunately, the intercept mission was proving to be far too complex and unwieldy for an unmanned system. A technological nightmare, the incredibly complicated SAINT was like a Rube Goldberg contraption gone awry. By the time it was cancelled in late 1962, SAINT was well on its way to becoming a bigger flop than the Ford Edsel.

Even though SAINT fizzled out, everyone still recognized the need to shoot down hostile satellites. The Army devised an anti-satellite system on Kwajalein Atoll in the Pacific, called Project 505, based on their Nike-Zeus surface-to-air missile. In a competing effort, known as Project 437, the Air Force positioned Thor IRBMs on Johnston Island near Hawaii. Both systems would launch nuclear warheads on sub-orbital trajectories to destroy hostile satellites.

The Army's Project 505 was eventually shelved in favor of Project 437. Project 437 was actually tested with live nuclear warheads launched in 1962 under a program called Starfish Prime. Despite a launch accident that contaminated a large portion of the remote island with radioactive debris, the lessons of Starfish Prime were myriad, but not entirely positive. When one of the tests inadvertently caused a massive electrical blackout in nearby Hawaii, it lent scientists their first glimpse into the crippling effects of the electromagnetic pulse—EMP—that accompanied nuclear explosions.

Another lesson was that while it was entirely feasible to use a nuclear warhead to destroy an enemy satellite, it was by no means a precision option. A nuclear detonation in space carried the risk of destroying everything in its vicinity, including friendly satellites. This was very significant, because outer space was increasingly becoming an extremely crowded place. Besides potentially hostile satellites, there were also legitimate research satellites, weather observation platforms, and communications relays. Additionally, there was an abundance of sheer junk—discarded boosters, miscellaneous debris, and dead satellites—whizzing around overhead. Looking up from earth, even with the most advanced optics and technology, it was virtually impossible to discern what was a valid threat and what was not. But despite its limitations, Project 437 was designated as an operational system, with a fixed launching pad and support facilities permanently stationed at Johnston Island.

In the years after Khrushchev's ominous speech, a general consensus evolved that it was just another saber-rattling bluff, mostly because no one believed that the Soviets possessed the technology to follow through on an OBS. After all, just last year, they had signed the Outer Space Treaty, which unequivocally banned nuclear weapons in space. It made perfect sense for them to sign the Treaty if they didn't have the technological horsepower to orbit weapons in the first place. Of course, thought Tew, the Soviets' gesture was rather meaningless, since they hadn't demonstrated any great propensity to abide by any other treaties that they'd signed.

In recent years, the game had shifted to peaceful competition in space. But although the Soviets fervently asserted that they had no desire to militarize space by stationing nuclear weapons in orbit, intelligence was slowly filtering out that indicated they still intended to do just that.

As Hugh Kittredge had indicated, recent evidence clearly showed that they had been pursuing an OBS the whole time, and up until very recently, they had successfully hidden their efforts from the intelligence services of the Free World. Why were they so intent on orbiting nuclear weapons? It appeared that the Soviet leadership was deathly afraid of the growing strategic arsenal possessed by the United States. They were particularly concerned about our sophisticated ICBM capabilities, especially the submarine-based missiles that they could not effectively monitor or counter. Desperate, the Soviets realized that a space-based system was the only way that they could ever hope to achieve strategic parity, especially if the United States abided by the treaties and they did not. And so, since desperate measures call for equally desperate counter-measures, Blue Gemini was born.

Blue Gemini was by no means a casual endeavor. To elude notice by even the most diligent of outside auditors, its massive budget was carefully dispersed across hundreds of "black" funding accounts. Tew's expertise in managing exceedingly complex projects was indispensible to the effort, but although he was handpicked by Schriever, Tew was an odd candidate to lead such an expansive but clandestine space program. Unlike virtually all of his Air Force contemporaries, Tew would never willingly kneel at the altar of nuclear weapons.

Within AFSC, Tew was renowned for his vocal opposition to McNamara's apocalyptic vision of Mutually Assured Destruction. As the leader of the nation's ballistic missile program, Schriever was effectively the premier architect of Armageddon, so his choice of the misfit Tew was a mystery to the few AFSC insiders who even knew about Blue Gemini. But despite Schriever's own affinity for nukes, he was still a practical man who recognized talent.

As a professional warrior, the very notion of wholesale atomic anni-hilation sickened Tew. To him, war was an endeavor only for armed combatants. Despite this, he was also a realist; he knew that the modern world had grown so complex that it was necessary to strike deep within an opponent's territories to destroy their means and will to make war. His years of strategic bombing, of pulverizing factories and rail yards from the air, had hammered that lesson home.

He also understood that there was no practical means to escape collateral damage, but to him it was entirely unacceptable that cities—teeming with thousands, if not millions, of defenseless civilians—could be targeted for destruction. He could not imagine any legitimate justification for such inhuman actions as the Dresden fire bombings, where thousands of civilians had been burned alive. While he hadn't flown in the Dresden raids, since he had been a POW at the time, he had participated in the Hamburg bombings in 1943. The firestorm sparked by the Hamburg raids rivaled Dresden, virtually leveling the city and killing an estimated 50,000 civilians in the process. To this day, he could not speak of it.

To a limited extent, Tew understood Truman's decision to subject Japan to the Bomb. The campaign to capture the Japanese mainland would have cost over a million lives on both sides, so it was absolutely essential to break the national will of the Japanese people. But at what cost? Could it have been done in some other manner? Couldn't a simple demonstration of the Bomb, detonated on an isolated and uninhabited island, have sufficed? And was it absolutely necessary to drop Fat Man on Nagasaki after Little Boy had been unleashed on Hiroshima?

Oddly, Tew had been rather ambivalent about the Bomb until he was stationed in Japan just before the onset of the Korean War. After all, it was extremely likely that he would have been dispatched to the Far East at the end of the War, to participate in the final push on Japan, so there was a strong possibility that the Bomb may have saved his life, just as it had theoretically saved millions of other lives.

What changed Tew's mind was a chance encounter with a Japanese-American orphan when he visited the ruins of Hiroshima in 1949. The half-white teenager—Jimmy Hara—was the son of a Marine embassy guard and a Japanese housekeeper, born ten years before Pearl Harbor. After the War started in the Pacific, he endured countless beatings as a brutal consequence of his mixed heritage. Shunned by his embarrassed grandparents, cast out into the streets, he and his mother wandered the fringes of Tokyo, barely eking out an existence.

After his mother had succumbed to typhoid, a kindly aunt and uncle took him to live in the town of Kumano-cho, roughly eight miles southeast of Hiroshima. He was there when Little Boy decimated the city on the morning of August 6, 1945. Before dawn on the day of the Bomb, Jimmy's aunt had walked into the city to buy food at a market. Hours after their world had been permanently altered, even as buildings still burned and the earth smoldered, Jimmy and his anguished uncle ventured into the rubble to find her. Oblivious to latent radiation and other hazards, they searched for two days, futilely probing the depths of the man-made Hell, but found no trace of her. Like so many others, she had just ceased to exist.

As the years passed, because he was marginally fluent in English, Jimmy became sort of a tour guide to the scores of American servicemen who made a pilgrimage to Hiroshima to witness and study the aftermath of the Bomb. Four years later, retracing his steps with Tew and several other visitors, he calmly described a surreal landscape, the scorched and broken remains of a once-thriving city, populated by the dead and not-yet-dead.

Tew's stalwart aversion to the Bomb solidified as he listened to Jimmy describe that fateful day and the terrible days that followed. Before coming to Hiroshima, Tew had read the official transcripts and documents that detailed the effects of atomic weapons, but the boy's vivid accounts could not be neatly packaged into the sterile terminology of a field manual.

Although Tew could not condone the wanton sacrifice of civilians, he was most appalled by the incomprehensible suffering of those left living after the Bomb. To him, the radii around Ground Zero were not unlike Dante's Circles of Hell, with each successive concentric ring yielding its own unique brand of suffering. Obviously, right in the vicinity of Ground Zero, most victims were instantly and completely vaporized. Moving slightly away from the detonation, others were mercifully rendered into desiccated lumps of charcoal. At the outermost Circles, thousands of victims were spared from immediate death but were exposed to doses of radiation that would doom them to die years later of leukemia, cancer or other insidious diseases.

For everything that Jimmy told him, one gruesome vignette remained indelibly lodged in Tew's mind. As Jimmy and his uncle approached within a mile of the city's center, they happened upon several children aimlessly wandering in shock. Miraculously, they were alive, but their faces—along with every square centimeter of uncovered skin on their bodies—were *gone*. Scraps of shredded skin dangled from their raw flesh, like a flimsy curtain ripped to tatters by a hurricane's winds. Their mouths were agape, as if they were shrieking in agony, but they made no sound.

Judging by their relative size, Jimmy guessed that they were roughly his age or slightly older; on any other day, they probably would have spat on him and pummeled him with sticks or their fists. For a brief moment, he actually laughed at their suffering. The high and mighty Japanese children who had tormented him for being an illegitimate half-breed were now without their precious yellow skins.

Later, Tew discovered the cause of the children's horrific dilemma. They were unfortunate enough to have been trapped in one of those Circles of Hell where they initially survived, at least for a day or so, but where their very identities were stripped away in the blink of an eye. How? A nuclear detonation sends out an intense pulse of thermal energy which literally causes the subcutaneous fat layer of exposed skin to spontaneously boil and explode; the bomb's shock wave, following just fractions of a second later, instantly peels the loosened skin from the underlying muscle tissue.

Tew shuddered as he recalled the horrific image of the children deprived of their faces, but perhaps even more terrifying was Jimmy's emotionless recollection, told in matter-of-fact monotone. As he spoke, the boy's eyes were blank and dry, as if his soul had been excised and he had no more tears to cry.

Thankfully, Jimmy Hara's story didn't end there in the ruins of Hiroshima. After learning the details of his lineage, Tew had arranged for him to obtain American citizenship and join the Air Force, where he was currently a counter-intelligence agent assigned to Blue Gemini. Although he had regained at least a portion of his humanity, he was still a spooky character. A lethal master of jiujutsu, learned from his uncle, he was ultra-patriotic; as someone raised as a stranger in a strange land, he appreciated the United States as few naturally born Americans could.

Blinking, Tew opened his eyes and returned from the shattered ruins of Hiroshima. He cringed at the dire thought that these nuclear monstrosities could soon be overhead, if they weren't already there. He thought of Schriever's decision that brought him here; instead of overlooking Tew's disdain for nuclear weapons, Schriever may have chosen him *because* of it. After all, who could possibly be more perfect to lead such a concerted effort to swat Soviet nukes from orbit?

Tew yawned broadly. Although his peculiar sleeping pattern was a long-engrained habit, he still hovered on the verge of exhaustion most of the time. Right now, he wanted no more than to retreat to his cot to sleep for a few hours, perhaps to grant himself the indescribable luxury of dozing until the dawn broke over Ohio. But despite that temptation, he clearly understood the strategic implications of Blue Gemini and the potential consequences if he failed. And so he stayed at his task, conscious that his vigilance might be the vital bulwark to save America from nuclear annihilation.

Tyuratam Cosmodrome, Kazakh Soviet Socialist Republic, USSR
3:12 a.m., Friday, April 19, 1968

Just as he had countless times before, Rustam Abdirov emerged from the concrete blockhouse and strolled toward the launch pad where the experimental rocket waited. He had inadvertently left his favorite slide rule—a precision-crafted Dietzgen, confiscated from a German rocket engineer at the end of the War—in the bunker and had momentarily left the pad to retrieve it.

The test wasn't going well, and technicians raced to replace key components so the rocket would be ready for liftoff. Ignoring even the most fundamental of safety precautions, the rocket was still fully fueled even as it was hastily repaired. Worse yet, scores of men milled around at the base of the rocket as the work was underway.

Abdirov had not taken more than a few steps when the rocket inexplicably exploded, igniting an immense inferno that engulfed the pad and most of the men gathered there.

He instinctively threw up his hands to guard his face. The blast's shock wave slammed him like a massive fist, literally picking him up off the ground and pitching him backwards over a hundred meters, into a retaining pond brimming with filthy, stagnant water.

Initially knocked unconscious, he almost drowned, but came to in time to claw his way out of the pool. Flailing with his arms, he struggled to the surface, only to discover that the air was choked with dense toxic smoke. Preferring a death by drowning rather than asphyxiation from the noxious fumes, he dunked his head back under. For whatever reason, even as he was on the verge of death, he groped for his treasured slide rule, but could not find it. He held his breath until his lungs burned and then came up again. The flames had subsided somewhat, but the air was still vile. Sputtering, wiping gunk and mud from his eyes, he crawled out of the pond and over the earthen revetment that encircled it.

Wracked with pain from a fractured pelvis and a broken arm, he heard agonized screaming. He pushed himself to his feet, ran into the flames—twice—and retrieved two badly burned men who had somehow survived the initial conflagration. As he emerged from the flames, he fell to his knees. The exposed skin of his head crackled with burning fuel; as he swatted at his face to abate the flames, he felt a foreign object protruding from his right eye socket—and realized that it was his missing slide rule.

Drenched in sweat and trembling, Abdirov was jolted awake from the horrific nightmare. When his heart stopped pounding in his chest, he pushed the bed covers aside, switched on the table lamp, and sat up. A Lieutenant General in the RVSN—Soviet Strategic Rocket Forces— he nightly worked until he was absolutely exhausted, simply because he dreaded the thought of falling asleep. Every night, almost without fail, not long after he drifted into slumber, his subconscious would deliver him to a place that he was loath to visit.

He slowly climbed out of bed and went to the mirror to remind himself that his recurring dream was not a vivid figment of his imagination, but a regurgitation of painful memories of an event that had changed him forever. He grimaced as he gazed at his ghastly reflection. Simply stated, he looked like a creature from a Western horror movie. Most of the right side of his head was encased in crinkled pink scar tissue. His right eye and right ear were gone, and he was missing two fingers on his right hand and three on his left. And there was more: his pajamas, tailored from white silk, concealed scar tissue that covered over fifty percent of his body. He didn't wear silk out of extravagance, but rather because it was the only fabric that was even somewhat tolerable when worn against his badly damaged flesh.

Abdirov's appalling visage was a reminder of the constant hazards inherent in working around rockets. He had sustained his scars in 1960, when he barely survived the infamous Nedelin catastrophe at Tyuratam, a disaster of such horrific magnitude and tragic consequences that it was still a tightly held state secret. In a freak accident, the second stage of an experimental R-16 ICBM prematurely ignited on the launch pad, spawning a massive fire that incinerated Abdirov's boss, Chief Marshal Mitrofan Nedelin, Commander in Chief of the RVSN, and nearly a hundred others. The R-16 was one of the first Soviet rockets fueled with a potent hypergolic concoction—nicknamed "Devil's Venom" for its dangerously corrosive properties—much like the self-igniting mixture that Americans successfully used in their Titan II ICBMs.

Had he not gone back to the blockhouse for his slide rule, Abdirov would have also been spontaneously immolated like Nedelin and the other unfortunates. In the aftermath, Soviet Premier Nikita Khrushchev dispatched none other than Leonid Ilyich Brezhnev to investigate the incident.

After the incident, Abdirov languished for nearly three years in a Moscow burn ward. In the first days, he screamed as nurses used stiff-bristled brushes to scour away charred tissue. As the weeks and months passed, he suffered through endless rounds of agonizing surgery to excise dead flesh and graft on new. He stubbornly remained alive, but if he had it all to do over, knowing the torture that he would eventually endure, he would have walked back into the flames to die with Nedelin and the others.

After emerging from the hospital and returning to the RVSN, his courageous actions were quietly hailed by those aware of the disaster. Although a closely held document read only by a few, the official report of Brezhnev's investigation acknowledged Abdirov as a significant hero of the incident, and noted that he would not have suffered his terrible burns had he not chosen to run into the flames—not once, but *twice*—to rescue others. Brezhnev personally commented that Abdirov's selfless actions reflected that he was the epitome of a true Soviet officer. In a secret ceremony in the Kremlin, Abdirov was formally honored as a Hero of the Soviet Union, receiving his gold star from none other than Nikita Sergeyevich himself, only a few months before Khrushchev was ousted by his rivals.

Abdirov was offered a substantial pension, by Soviet standards, and a comfortable place to retire. Although enticed by the proposition, he fought to remain in military service. Personally backed by Brezhnev, who had since succeeded Khrushchev as the First Secretary of the Communist Party, he was successful in his bid. He made up for lost time by swiftly shooting up the ranks as he excelled in one difficult assignment after another. His contemporaries would have been grateful if he just went away—permanently—to live out his remaining years, but he refused to be shunted into seclusion.

Anxious to make excuses for their own shortcomings, Abdirov's detractors quietly claimed that his meteoric progression could be at least partially attributed to the fact that he was not distracted by a wife and family. To some extent, they were probably right; in his younger years, Abdirov had put his career ahead of all else, and it was highly unlikely that he would find a bride in his current state. Even he conceded that only a blind woman could share his bed. To make matters worse, he joked, he couldn't marry a Muslim—even though he hailed from the Kazakh Republic, where submission to Allah was at least tacitly accepted by the Soviet regime—because he had received so many grafts of pigskin that he was now at least part swine.

In a sense, Abdirov also effectively became an orphan on that terrible day. His close relationship with Nedelin had dated back over two decades. Abdirov had served in the Soviet Army since the mid-thirties, starting his military career as a cavalry officer, since as a Kazakh he had practically grown up on a horse. As comfortable as he was atop a steed, Abdirov also possessed solid mathematical skills, and found himself—much to his chagrin—drafted into the horse artillery. Rising up through the ranks of the artillery, he became involved in the design and testing of missiles at the outset of World War II, and eventually became a favored protégé of Nedelin, a renowned artillery officer who ultimately rose to become one of the prominent leaders in rocket development.

But his loyal allegiance to his mentor also proved to be highly detrimental. As Nedelin's acolyte, Abdirov had no allies within the powerful aerospace design bureaus. This was very significant, because within the Soviet system, the design and development of rockets and related hardware was confined to a very insular world of aerospace design bureaus. As the Commander in Chief of the RVSN, Nedelin personally directed the actions of the bureaus, so Abdirov found himself bearing orders to the bureaus on behalf of his boss.

While the development of military technology was certainly critical, all the design bureaus desperately vied for the most prestigious prize of all, control of the manned spaceflight program. The premier Korolev design bureau—OKB-1—headed by the brilliant and all-powerful "Chief Designer," Sergei Pavlovich Korolev, held sway over manned space flight. Abdirov fervently believed that the design bureaus should be working toward a common goal instead of their own interests, and counseled Nedelin to compel them toward cooperation.

Now, Abdirov wanted to join the crusade toward manned spaceflight, but his path was inextricably blocked. With his mentor gone, Korolev and his bureau wanted nothing to do with him, and Korolev's main rival, Vladimir Chelomei, ostracized him as well. Certainly, when Nedelin was still around, they were compelled to curry favor with his emissary, but now that the Chief Marshal was dead, Abdirov was all but a nonperson.

But even if he couldn't grasp the stars, Abdirov still wasn't the least bit interested in retirement. After all, retirement was for those who desired to rest, and for him, there was too much work yet to be done. He vowed that he would remain in service to the Motherland, and to the ideals of Socialism, until the moment he drew his last breath.

7

AUX ONE-OH: UNTIL YOU CAN'T

Auxiliary Field Ten, Eglin Air Force Base, Florida
10:35 a.m., Friday, April 19, 1968

The convoy of buses pulled into the small encampment and creaked to a stop. Seated at the rear of the second bus, Airman First Class Matthew Henson watched in anticipation. He had been in the Air Force five months, just long enough to finish his initial training, and so he was accustomed to being rushed. Listening to the brakes hiss, he waited for the inevitable.

Henson had recently graduated from the Air Force's security police training course at Lackland Air Force Base, Texas. While he and a handful of others were new to the Air Force, the majority of the men on the bus had already served at least one hitch. During the hour-long ride from Eglin's main base, Henson feigned sleep as he listened to the others compare notes about their backgrounds and military service. At least three came here from the pararescue training pipeline, and several others were security police who arrived at Eglin after guarding ICBM missile silos in remote locales.

Henson wasn't the average Air Force enlistee. Formerly a student at Louisiana State University, he had been a semester away from a degree in Business Administration when his savings dried up. His mother tried to help with his tuition; she owned a small cafe in New Orleans, but her cash flow wasn't much better than Henson's. Suddenly out of school, with the draft and the dismal prospects of a Vietnam tour looming before him, he volunteered to join the Air Force.

But dodging Vietnam wasn't merely as simple as enlisting; despite his educational background, all the Air Force could offer was a three-year stint in the security police, and it was still likely that he might go overseas. Despite those odds, he felt he had at least slightly more control over his existence than the average Army infantryman slogging through the jungles of Southeast Asia. So, twenty-four years old, with two years laboring on an oil rig in the Gulf of Mexico, followed by nearly four years of college, he found himself at Lackland, training alongside pimply-faced kids only weeks removed from their high school homerooms.

Henson applied himself to the training, and in so doing discovered that he was a natural-born leader. After graduating at the top of his class, he was approached by an officer who offered him an opportunity for advanced training and a special assignment. He volunteered, and after enduring over a month of background checks and aptitude tests, he was finally here.

An instructor clambered aboard the bus. Coiled to spring, Henson warily clutched his belongings close to his chest, anticipating a screaming rant that would incite the arriving candidates to pile off the bus in shortest order. All of his worldly possessions—at least for the moment—neatly fit into a duffle and gym bag. He travelled light, and was poised for any sort of wacky ass-chasing scramble, but what ensued was certainly not what he expected.

The instructor wore a distinctive uniform bearing a tiger-striped camouflage pattern of jagged black slashes on a green background, along with a matching bush hat. He did not scream or yell, but spoke clearly and with just sufficient volume to be heard throughout the bus. "Gentlemen, welcome to Auxiliary Field Ten, also known as Aux One-Oh. Dismount the bus in an orderly fashion and dump your stuff out there on the gravel. Captain Lewis will meet you outside to give you the nickel tour."

The men climbed off the bus and milled about as they waited for more definitive instructions. The instructor called them to attention as another man approached the group. The wiry newcomer wore the same tiger-striped uniform as the first instructor, but he also wore a black and gold Ranger tab—signifying a graduate of the Army's highly demanding Ranger course—sewn on his left shoulder, as well as Army jump wings above his left breast pocket.

"Gather in. Relax, gentlemen. I'm Captain Lewis, commander of the Training Flight. I oversee the assessment training phase. In front of you is your humble home for the next four months or as long as you're attending the training cycle," he announced, gesturing at the single-story corrugated metal building to their front. Broad-shouldered, Lewis stood roughly five feet eight in his spit-shined jungle boots. Moving with the casual grace of a natural athlete, he was obviously in remarkable physical condition. With his crew-cut blonde hair and dark tan, he could have easily been mistaken for a California surfer. "Follow me," he said, walking inside the building.

With no air conditioning, it felt like a brutally hot sauna. The open bay was filled with four long rows of gray metal bunk beds and matching wall lockers.

Henson glanced around and swiftly realized that he was the solitary black man reporting in for the course. He was also conscious that several men looked at him like he got off the wrong bus. He was long accustomed to being an outsider, but couldn't decide whether he was currently unwelcome because of his race or because he was a new enlistee. He surmised that it was probably a little bit of both.

"Gents, this isn't basic training," announced Lewis. "My instructors won't inspect your living quarters, but we do ask that you keep them clean and tidy. If this place starts to resemble a hippie commune or a hobo jungle, we'll evict you outdoors for the remainder of your training cycle. Here at Eglin it rains a good bit, so unless you want to spend the next four months wallowing in the mud under a leaky poncho, you'll heed my advice."

Lewis guided the men into the latrine at the opposite end of the bay. In a sweeping gesture, like a salesman displaying his wares at a trade show, he waved his hand toward the array of plumbing fixtures lining the walls.

"You folks will be operating on a very tight schedule, so you need to get organized quickly so that you use these facilities effectively."

"If our statistics hold out, at least four of you will come to blows because someone lingers in the shower or the toilet too long, or takes too much time scraping whiskers off his chin. And so there's absolutely no confusion, fighting is grounds for immediate and uncontestable relief from the course. Get caught fighting, and your fate is permanently sealed. You'll be whisked away to a life much more mundane. Any questions?"

Lewis directed the men to a room adjacent to the latrine. "That's the bulletin board," he noted. "My first sergeant will post the training schedule, as well as any specific study assignments or special duties. It's constantly updated, but it's up to you to keep up with the changes. Gentlemen, we own the time on the training schedule, but the other hours in the day—and there won't be too many—belong to you. Use them wisely."

He pointed at a series of metal racks bolted to the wall under the bulletin board. "These are your weapons racks. You'll draw weapons tomorrow morning when you're issued your field gear, and you'll keep those weapons until you leave here. They will either be on your person at all times, or they will be secured in this rack under guard. You will provide the guard."

Slowly rocking back and forth on his heels, Lewis added, "One other thing: down the road, there's a slop chute where you can drink a beer at night if there's time and you're not absolutely dog tired. Also, if you so desire, you can buy cookies, candy bars, and other lickie-chewies. Feel free to stuff your face down there, but *do not* smuggle food into your billets. When you squirrel away chow in your wall locker, it draws mice. Mice attract snakes."

Lewis pointed out toward a window. "There's an abundant supply of snakes happily slithering out there in the woods. Three to five of you will be bitten by poisonous snakes in the next few weeks. I assure you that we can readily meet this quota with just our outdoors snakes, so we don't need to bring any additional reptiles into the living spaces. Enough said on that topic, gentlemen?"

"Excellent. Let's go outside then." Donning his hat, Lewis guided them out through a side door and motioned toward several long wooden tables.

"Those are your weapons cleaning tables. We expect you to clean your weapon at the end of every training day, but my instructors aren't going to stand over you to make sure that you do it, and they won't line you up to inspect your weapons every morning. But if you're ever out on a range and your piece jams because you haven't given it the tender loving care that it deserves, then it's grounds for immediate and uncontestable relief. The same goes for all your gear. You either maintain it in good working order, or you'll be destined for a less demanding assignment elsewhere. Fair enough?"

There was no comment from the candidates.

"Fabulous," noted Lewis, swatting at a mosquito on his forearm. "Follow me." He walked them around to the front of the building and stopped in front of four pull-up bars constructed of telephone poles and steel pipes. "Last stop. Gather in close, men. It's a Training Flight tradition that you don't enter the billets until you do pull-ups. You had your one and only freebie when we went in for the tour, but from here on, you will mount the bars and crank 'em out before you cross the threshold." He paused, waiting patiently for the inevitable question.

And then it came. A red-haired man, new to the Air Force, raised his hand. "How many pull-ups do we need to knock out, Captain?" he asked. "How many is enough?"

Lewis smiled, bent down, and pointed at a small granite marker set in the ground directly in front of the pull-up bars. "Come here, airman, and read this."

The man stepped forward, squatted down, and read aloud the simple inscription literally etched in stone: "Until You Can't."

Lewis nodded. "*Until you can't.* That's what's expected on these bars and in all aspects of training. Give it your all, and give it your best. Now, gentlemen, you have thirty minutes to move your gear inside, choose a bunk, and make yourselves at home. The Wing commander will talk to you just after lunch chow, at thirteen hundred, so you need to be seated in the bleachers at least two minutes prior. Be late, and you'll be *gone*. Until then, gents, the clock belongs to you."

12:59 p.m.

The intense Florida sun beat down mercilessly on Matt Henson and the fifty-three other men perched in the unshaded bleachers. He diligently fought not to nod off; not only had he been on the move since the wee hours, but his effort to remain conscious wasn't being helped by the generous serving of greasy chili mac weighing heavily in his stomach.

There was a slight but persistent breeze; it smelled subtly of pine pollen and spring blooms, with an underlying essence of aviation fuel. The bleacher seats were old and warped at the ends; years ago, the planks had been painted dark green, but were now mostly sunbaked gray wood with only chips and flakes of faded enamel remaining.

Henson glanced at his watch as the Wing commander strode before the bleachers just mere seconds before the appointed time. The candidates spontaneously sprang to their feet; the derelict bleachers swayed and groaned, threatening to collapse at the sudden movement.

"Take your seats, gentlemen. I'm Colonel Isaac Fels. I command the 116th Aerospace Operations Support Wing. Welcome to Aux One-Oh." Captain Lewis stood quietly to the side as Fels addressed the incoming candidates. The two men were a study in contrast; unlike the short blonde Captain with wide shoulders, Fels was tall, dark-headed and built similar to a broomstick.

"All of you are here because you desire to be part of this organization, and we want you to eventually join us, provided you meet our standards," said Fels, pausing momentarily to take off his hat and wipe sweat from his bald crown. "Most of you only have the vaguest notion of what we do. Today, I'll give a slightly better idea of what awaits you during your assessment cycle, and also what you can expect once you're assigned to the 116th on operational status."

"All of you have been subjected to an extensive battery of background checks," he said. "Most facets of our mission are highly classified, and we will not discuss any of those today, but everything else that we talk about must remain here at Aux One-Oh." Far to the south, there were a series of explosions and staccato bursts of gunfire.

"To my knowledge, none of you are here under duress. You're here because you *want* to be here, but less than half of you will successfully emerge from the other side of this wringer. This is arduous and stressful training, both physically and mentally. It will take its toll. Some of you will quit, some of you will fail to achieve our standards, and some of you will fail to abide by the rules that we set."

Passing low over the pines, a pair of massive CH-53 helicopters roared by. Fels waited for their noise to fade before continuing. "Captain Lewis and his cadre will treat you as adults, and we expect you to behave like adults. If you fail to comply with our rules or cannot meet our standards, you are subject to immediate and uncontestable relief. So it's abundantly clear what that entails, it means that you will be expelled from the assessment cycle without any recourse or appeal, and that you will be reassigned to a distant and less-than-pleasant location."

Fels continued. "And regardless of whether you successfully pass this cycle or whether you wash out, you are *never* to speak of this place or what we do here. If you ever feel a pressing urge to blab about it, either a year from now or twenty years into the future, then you should always imagine that there is someone listening, because there will be. Understood?"

"*Everything* we do here revolves around the mission of this Wing." Fels cleared his throat as he referred to an index card. "Simply stated, this Wing's mission is to locate specified objects and personnel of strategic value in denied or non-permissive areas, and to rescue and/or recover the same, by force if necessary, with a low operational signature and a minimum of external support from other U.S. military forces or agencies."

"That's a mouthful, so let me break it apart for you." He slipped the card into a breast pocket. "First, our mission is to *locate* objects and personnel. That probably sounds simple, but it's not. Far from it, in fact. In a more conventional environment, the Aerospace Rescue and Recovery Service takes the lead for most searches. They normally have the luxury of putting plenty of aircraft and other search resources in the area."

"But we operate under different circumstances," noted Fels. "You'll be hunting for objects and people in denied or non-permissive areas, where there is no U.S. presence or where a U.S. presence is not formally permitted. Keeping a low operational signature means that you folks will be working on the ground, mostly by yourselves, with a minimum of air or other support."

Despite his sleep-inducing gutful of chili mac, Henson was wide awake, now having some second thoughts about what he had signed on for. For a draft-motivated volunteer striving to avoid potential danger, it looked like he had inadvertently inserted himself at its innermost core.

Fels continued. "In the coming weeks, you'll be exposed to the ground search techniques that we employ to quickly find objects and personnel. Then, you'll spend most of the next four months perfecting those techniques before you're assigned to an operational squadron within the Wing. Additionally, our mission may require the use of force, if necessary, to effect a rescue or recovery. Ideally, of course, we like to quietly slip in and slip out without drawing any undue attention to ourselves, but that may not always be a feasible option."

"As for describing '*specified objects of strategic value*,' I'm afraid I can't be very explicit. At any given moment, there are a lot of objects that might fall out of the sky, and a sizeable percentage of these objects may be of strategic importance, so it would be in our national interest to find them and recover them as swiftly as possible."

Seemingly for dramatic effect, the lanky colonel paused for several seconds. "Men, it's important for you to understand that just because an object may be of *strategic importance* to the United States, it doesn't necessarily mean that it's the *property* of the United States. At least not initially." He smiled slightly and winked.

"But let's focus on those objects that *are* US property. Without delving into details, we are in constant coordination with those agencies that deploy such strategic assets, so we have a reasonably good idea—well in advance— of the potential locations in the world where a search effort may be required. Consequently, in many instances, we are able to pre-position resources in preparation for a deployment or mission involving a strategic asset."

"For you gentlemen, *if* you make it through the training cycle and join an operational squadron, that means that you'll spend a lot of time on the road, staging at different locations. Journeying from one country to the next may sound exciting, but it's not. Typically, when you're pre-positioned, you'll be stashed in a hangar somewhere, out of sight. It's not a sightseeing spree. There's no shopping for souvenirs, meeting the locals, practicing the local language, etc.

"Now you have a slightly less murky picture of what we do. The next four months will be challenging for you, and if you graduate to an operational squadron, life isn't going to be that much more pleasant. So what are the benefits to you?

"For one, you're here to be part of something truly special. Even though they're designed for a very specialized purpose, the operational squadrons are some of the most highly trained fighting formations in the world. During your cycle, you'll undergo weeks of combat training, and you'll become extremely proficient with weapons and explosives. Our mission mandates that we be able to use force as necessary, and we don't take that notion lightly.

"Because of this potential combat mission, you'll have the opportunity for some unique training that's not available to the average airman. All of you—except for you pararescuemen who've already been there—will attend the Army's jump school at Fort Benning.

"Many of you will attend the Army's Ranger school as well. Most of you will rotate to Vietnam for combat tours with the Combat Security Police Squadrons. There are other opportunities for advanced training as well. Beyond training courses conducted by the military, we send airmen to the Forest Service's smokejumper school in Montana as well as training conducted by other government agencies. If you prove your mettle in cycle and spend some time with an operational squadron, you'll leave here with a fairly impressive resume.

"Finally, gentlemen, I am authorized to offer you three other unique incentives. First, once you're assigned to the 116th Wing, you may remain in the Wing as long as you wish, provided you continue to meet our standards.

"Second, when and if you elect to leave the Wing, you will have your choice of next assignment, provided that your performance has been up to snuff. Third, unless you've been promoted in the past year, you will be automatically promoted one grade in rank before transferring to your next assignment. But I should also make you aware that despite these incentives, not one man has ever *voluntarily* elected to leave the Wing." Fels paused and then asked, "Gentlemen, what are your questions?"

At that point, Henson had no question. Only moments ago he wasn't sure if he had come to the right place, but now there was no doubt in his mind.

8

THE SKINNY MATH WHIZ FROM WILBER, NEBRASKA

Atlanta Municipal Airport, Atlanta, Georgia
9:15 a.m., Monday, April 22, 1968

Ourecky dashed up the steps and boarded the plane with just minutes to spare, only to learn that the Delta flight—bound for Dayton, Ohio—would be delayed until two extra carts of luggage could be loaded. Wearing his short-sleeved Air Force khakis, he negotiated the narrow aisle, and took a place next to a large man studying sales brochures for industrial refrigeration.

The man, smelling heavily of Old Spice cologne and last night's liquor, stuck out an oversized hand. "Bill Jeffers. Dayton Industrial Supplies," he announced, loud enough for the entire plane—and possibly the airport— to hear. He shook out a cigarette from a pack of Camels, and deftly lit it with an ancient Zippo lighter.

"Uh, Lieutenant Scott Ourecky." He forgot the salesman's name almost immediately.

A passing stewardess stopped and leaned over Ourecky. He read her name, "Bea," engraved deeply in the brass plate pinned to the light blue polyester of her snug-fitting jacket. "You'll have to put that out," she said, pointing at the fat man's cigarette. "No smoking 'til we're airborne." She leaned back, smiled at Ourecky, and asked, "Can I bring you anything? Pillow? Magazine? I think we have the latest *Life* and *Look*. Maybe *Esquire*?"

"Uh, no thank you. I'm fine," answered Ourecky, fastening his lap belt. "Maybe later I might want to read something, but not just now."

"Sure, honey. So, are you in the Army? Back from overseas?"

"No, ma'am. I'm in the Air Force, stationed at Eglin Air Force Base. That's in Florida, right on the Panhandle."

"Really? I've just been dying to go down there to the beaches. Is that near Panama City?"

"Very close," answered Ourecky.

"That's a very interesting name you've got there," she commented, gazing at the name tag on his tan uniform. "How do you pronounce that?"

"It's *Oh Wreck E*, ma'am. Ourecky. It's Czechoslovakian. My great-great-grandparents immigrated to America in 1864 and settled in Nebraska."

"Fascinating," she replied, displaying the most dazzling smile that Ourecky had ever witnessed. It was so bright, he felt like he was looking into the sun.

To Ourecky's left, the salesman chimed into the conversation. "Darling, you sure look mighty familiar. Haven't I seen you somewhere? Maybe on television?"

"I don't think so," she answered, shaking her head and wrinkling her nose at his overwhelming aftershave.

"Wait, wait, I knew it," the salesman exclaimed, slapping the armrest. "You're a dead ringer for that girl who plays the genie…"

She held up a palm and shushed him. "I get that all the time. So much so, I think I might invest in a frilly little harem outfit and wear a jewel in my navel." Smiling again, she turned her attention back to Ourecky. "I should get back to work. I'll swing by later to check on you."

As she strolled up the narrow aisle, the salesman elbowed Ourecky and muttered, "Coffee, tea, or me?" He chuckled as he took a furtive draw from his cigarette before stubbing it out in the armrest ashtray.

"What?"

"Oh, man, she's sweet on you. Can't you see that?"

"No, she's just doing her job," answered Ourecky, reviewing the safety card for the airliner. He looked around the cozy cabin, noting the nearest exits. "She's supposed to be friendly."

The corpulent salesman shook his head and laughed. "If you say so, buddy. So, you're in the Air Force, huh? I was Army, back in the War. Paratrooper. *Screaming Eagles*, by God," he declared, holding up his tarnished Zippo, which was emblazoned with the famous shoulder patch of the 101st Airborne Division.

The salesman slipped the lighter into his pocket, and glumly lamented, "Guess you couldn't know that now, looking at this damned big gut of mine. But I jumped into France on the night before D-Day. I was eighteen years old and scared completely out of my wits. The damned Air Corps scattered us all over Normandy, everywhere but where we were supposed to be."

"I'm sure that must have been terrifying," Ourecky said, opening one of the three thick spiral-bound notebooks perched in his lap. Trying to be polite, he focused on a series of equations that had been vexing him. The problem, which he had copied from a college text on orbital mechanics, involved two satellites placed in dissimilar orbits. Specifically, it required him to calculate the means to gradually shift the planes and phases of their orbits to allow them to eventually meet. Glancing up, he noticed the stewardess solemnly presenting a set of junior stewardess wings to a young girl.

"So you're a pilot?" asked the salesman, derailing Ourecky's train of thought.

"Uh, no," replied Ourecky, gazing down at his penciled diagrams in the notebook.

"Really? Since you were an officer, I just figured you for a pilot." The big man reached in his shirt pocket, pulled out his pack of Camels, and tapped out a cigarette. He started to light it with the Zippo, thought better of it, and stuck the cigarette behind his ear.

"No. I'm an engineer," explained Ourecky, slipping a slide rule out of a worn leather case. He manipulated the slide rule and then used a pencil stub to jot numbers in his notebook. "I'm involved in ordnance design down at Eglin. It's interesting work, very challenging."

Seeming satisfied with Ourecky's answer, the man pulled an in-flight magazine from the seat pocket in front of him. Toying with his Zippo, he studied a map of Delta's domestic routes.

As Ourecky verified some of his previous calculations with the slide rule, he had the unsettling feeling that he was being watched. He looked up and saw the stewardess standing motionless in the aisle, staring at him as if she were mesmerized or in some sort of trance. Her mouth was fixed in an odd smile, like she suddenly recognized him from somewhere in the past when they might have shared a brief moment. Painfully self-conscious, he shyly smiled back. The salesman was right: tall, slender, stunningly attractive, with a waterfall of blonde hair that cascaded over her shoulders, she certainly looked like she belonged on television.

Even though she was light-years beyond his grasp, he felt compelled to scribble her name—*Bea*—in his notebook's margin, next to a diagram depicting converging orbits of two satellites. His face suddenly felt warm, and he realized that he was blushing. He glanced away, as if to avoid the blinding glare of oncoming headlights. He looked up to see her slowly pivot away and gracefully saunter toward the first class compartment. *Probably a case of mistaken identity*, he thought, *surely she confused me with someone else.* Ourecky resolved himself to put the glamorous stewardess out of his thoughts, to focus on more pressing matters, like being prepared for his meeting at Wright-Patterson Air Force Base.

The salesman spoke, interrupting the awkward moment. "But I guess you'll be going to flight school soon enough, right? I know you want to earn those wings, don't you?"

"No. I really like what I'm doing. Now, if you don't mind, I have to finish this set of problems. They need to be ready before my meeting today." Ourecky didn't particularly enjoy lying, especially since he only vaguely knew why he was going to see Virgil Wolcott in Ohio, but the notion of a three-hour gab session with the salesman was not something that he relished.

Ourecky closed his eyes, pretending to concentrate on a particularly tough problem. He had to admit to himself that the salesman's infuriating question about pilot training had struck a nerve. His boyhood aspirations of travelling into space had never faded completely from his thoughts. But some men were meant to fly, and some weren't; the harsh reality was that the arcane equations in these notebooks were as close as he would ever come to orbit.

Aerospace Support Project
Wright-Patterson Air Force Base, Ohio
3:30 p.m., Monday, April 22, 1968

Even though he had visited Wright-Patterson several times in the past, it took Ourecky almost an hour of diligent searching to locate the offices of the Aerospace Support Project. He was finally directed to a three-story brick building of World War II vintage, not very far from the flight line, adjacent to a row of equally old hangars.

He expended another twenty minutes filling out paperwork before being issued a temporary badge that granted limited access to the building. Accompanied by an airman assigned as his escort, he strolled down several corridors before entering an outer office that served as a holding area for the offices and workspaces for Tew and Wolcott. Punching a button on a desk intercom, a sergeant announced his entrance and asked him to take a seat while he waited.

Ourecky adjusted his uniform before taking a seat in a stiff-backed metal frame chair. His stomach growled audibly; he wished he hadn't skipped lunch, but he'd just been too nervous to eat. Quietly tapping his left foot, he was anxious to discover why he had been summoned by the generals. What on earth could they possibly want him to do?

He had looked for clues on the way in, but there were none. It definitely wasn't what he had imagined. There was no evidence of any unusual activities. While the people in the hallways weren't exactly chatty, no one seemed the least bit furtive or guarded. The Project's headquarters could have been any government office building on any Monday afternoon.

After a few minutes, the intercom buzzed. "Lieutenant, you can go right in," stated the sergeant, looking up from his typewriter. "The generals will be right with you."

Ourecky rapped lightly on the door and then stepped inside to find a spacious but relatively bare room. There were two small windows and two doors, aside from the one to the waiting room, apparently opening to separate offices for each general.

The austere space was devoid of plants, pictures or frivolous decoration. From the green linoleum floor to the bland white ceiling tiles, the appointments were strictly government issue. At one side of the room were two battleship gray metal desks. One was adorned with a plain wooden nameplate that read "Virgil Wolcott" and the other held a nameplate that stated "MG Marcus Tew, USAF." A large conference table, surrounded by matching chairs, took up most the floor space in front of the desks. Several large gray metal cabinets occupied the wall behind the desks; the cabinets were completely jammed with black leatherette binders. All other available wall space was covered with large blackboards. The room smelled of pine oil cleaner, cigarettes, and chalk dust.

Ourecky had imagined that Wolcott would dress more formally for the office, but he was mistaken; with the exception of a bolo string tie, Wolcott's attire was scarcely different than what he had worn when they first met at Eglin. The two generals were engrossed with their work, and seemingly didn't notice his entrance. He quietly cleared his throat to draw their attention.

Briefly looking up, Tew motioned for Ourecky to take a seat at a conference table. As the two men leafed through their stacks of paperwork, he studied two meticulously detailed spacecraft models at the center of the table. He recognized one as the two-man Gemini spacecraft and the other was a Titan II rocket topped by a Gemini. The Gemini model appeared slightly different than the NASA version; the nose was slightly longer and wider. Additionally, "USAF" was emblazoned in white letters above an American flag. He assumed that it was a special variant that would be flown as a return vehicle on the upcoming MOL missions.

Tew finally paused long enough to introduce himself. "I'm General Mark Tew," he said, taking a seat at the table across from Ourecky. As the general sat down, Ourecky clumsily jumped to attention. His chair, equipped with rollers, skittered off in the opposite direction. "I understand that you've already met my deputy, Virgil Wolcott."

"Good to see you again, Ourecky," said Wolcott, joining them at the table. "Let's get right down to brass tacks. We've been on the hunt for someone with some unique skills, and you might be just the man we're lookin' for. Now, you need to bear in mind that what I'm goin' to describe is classified well beyond your current security clearance."

Ourecky grinned. "General, I hold a Top Secret clearance," he said smugly.

Frowning, Tew's temperament seemed to change in an instant. "Lieutenant, we know *precisely* what type of clearance you have. Virgil was being polite, but I'm not so predisposed. You are not to discuss this Project, ever, with anyone outside of this office, regardless of whether you work on it or not. If you do, then you will serve out the remainder of your Air Force tour in a *very* unpleasant place. Do you understand?"

Swallowing, Ourecky nodded meekly.

Wolcott laughed, trying to dispel the awkwardness. "Now that we've exchanged pleasantries, pard, let's talk business. So we don't squander a lot of valuable time, I'm going to cut right to the chase. For starters, while the sign outside says 'Aerospace Support Project,' the true name of our operation is 'Blue Gemini,' but we only use *that* name under *this* roof."

Wolcott leaned forward. "Why are we here? Way up at the stratospheric levels of our government there are folks gravely concerned about what kind of hardware the Soviets are shootin' into orbit. While we're sincerely hopin' that they're abiding by the treaties they've signed not to militarize space, we have to assume the worst. With all that said, Ourecky, Blue Gemini's mission is to develop contingency plans to intercept and destroy hostile satellites, specifically those that might carry nukes."

"Of course, this is all purely theoretical at this point," Tew interjected. "As it stands, this country already maintains an unmanned capability to destroy enemy satellites, but it's a scattergun approach that leaves a lot to be desired. We've been tasked to develop a more surgical approach using a manned platform."

Wolcott brandished the Gemini spacecraft model, turning it slowly in his wizened hands, and said, "We've been handed a tough mission, but we have this critter to make it happen. The Gemini is an operational space vehicle. It's been proven in orbit, and it's as capable as any machine we're flyin' closer to the ground."

Wolcott continued. "Our theoretical missions would require a Gemini spacecraft and crew capable of executing the intercept scenarios autonomously, without relying on extensive assistance from the ground. That's a mighty tall order, but we have MIT and some contractors developin' an onboard computer to pull it off."

"Begging your pardon, sir, but doesn't the Gemini spacecraft already have an onboard computer?" asked Ourecky.

"Correct," replied Wolcott. "But that computer is based on NASA's specifications. You might recollect that all of their rendezvous missions involved a cooperative target in a pre-fixed stable orbit. To support those missions, NASA cobbled together a worldwide data and tracking network. So even though they had an onboard computer, virtually all of the NASA's rendezvous missions were assisted with big tracking radars on the ground. But in our theoretical scenarios, it ain't highly likely that NASA's tracking system will be available for our use, and it's almost a sure bet that the radars ain't going to be positioned where we would need them."

"So we're working on an improved onboard computer for the Gemini," explained Tew. "We call it the Block Two. It's going to be a much more robust computer than what the Gemini has now. Ultimately, we want a guidance computer like the one MIT is building for the Apollo lunar flights, one that constantly tracks its position in three-dimensional space, as well as the intercept target's position. Our crews will be able to extract maneuvering guidance from the computer, as well as instructions for star sightings so that they can verify the computer's fix at any time."

"There's another reason that the Block Two computer is necessary," noted Wolcott. "When and if we do launch, our boys won't be flyin' a standard Gemini. They'll fly a hot rod version that we call the Gemini-Interceptor, or Gemini-I. We're reducing the weight in some areas, like removing the fuel cells and other hardware specific to NASA's missions, so we can add other capabilities. The stock Gemini spacecraft was equipped with two rear-facing hundred-pound maneuvering thrusters; the Gemini-I will have *four*, so it can make more radical orbital corrections."

"See how the nose of this Gemini is larger than NASA's version?" asked Tew, gesturing at the model.

"I noticed that, sir," replied Ourecky. He also noticed that Tew's demeanor had softened considerably in the past few minutes.

Tew explained. "It's bigger because our Gemini-I will also have a much more powerful acquisition and rendezvous radar to detect and track targets at longer ranges. It will also have an instrumentation package that will allow the pilots to photograph and collect readings on target vehicles.

Consequently, we need the Block Two computer so we can take full advantage of these new capabilities."

Tew said, "And although this is largely a theoretical exercise, we have been allocated sufficient funding to build critical systems that would otherwise require too much lead time to construct. Obviously, we're already pursuing the Block Two computer we've discussed. We're also fabricating a very small number of Gemini-Interceptor spacecraft. All of this hardware will be maintained in secure storage, so if a significant threat presents itself, we can respond in a matter of weeks instead of waiting months for equipment to be produced or modified."

"That ain't all, bub," added Wolcott. "Besides saltin' away the hardware, we're also training stand-by crews. We operate a full-blown mission simulator, the same as NASA's, but modified to meet our requirements. Our crews are constantly working in there to evaluate and refine the intercept procedures. As a matter of fact, the two boys you met at Aux One-Oh—Carson and Agnew—are in the Box today."

Ourecky resisted an urge to gasp. It was all almost overwhelming. Not only were the two generals describing a program to secretly launch military astronauts into space, but there were actually pilots currently training for the flights. Granted, it didn't sound like it was likely that they would ever fly, but the two men he had met at Eglin were genuine *astronauts!*

Wolcott added: "Last but not least, pard, we need to explain where you'll be riding in this great big rodeo. We want you to wrangle two functions. The first is to refine the intercept profiles and distill the guidance procedures down to the point where they can be written into a computer program for the Block Two."

Ourecky quietly whistled. "Uh, sir, I'll do my best, but I don't have much experience with computers. There's a computer at the Ordnance Lab at Eglin. It's an IBM 360 Model 40, top of the line. It's available to us to run our calculations, but I've found that by the time the keypunch girls punch and stack all my cards, I can work all of those equations myself."

"Duly noted," Tew said. "Now, let's discuss your second task. We're assuming that we'll receive our new Block Two computer in time, but there's always an ugly possibility that we may be ordered to intercept a hostile satellite even before the Block Two is available."

Wolcott interjected, "And to address that contingency, pard, we're going to ask you to formulate protocols to enable a Gemini-I crew to go up on their own, with just the old computer and with minimal assistance from the ground. Of course, we'll send a sextant up with them to make star shots for their celestial nav, but they'll be relyin' almost entirely on what you provide them. Do you feel up to the task, young lieutenant?"

"I think so, sir," answered Ourecky. Now, he wasn't entirely confident that he wasn't in over his head. Perhaps *way* over his head.

"Good." Wolcott leaned forward in his chair, pointed at the leather briefcase in the chair next to Ourecky and said, "Ronnie Paster tells me that you always tote around some notebooks, and that whenever you land a spare moment, you're workin' equations. Rendezvous and what not. Is that true, pardner?"

"Uh, yes, sir. I'm trying to stay on top of it for the future, just in case I land a shot at MIT."

"Well, is there any chance those are what you're totin' in your satchel there?"

"They are," replied Ourecky proudly. "I brought them because Colonel Paster said you might want to look at them." He dug into his leather attaché to retrieve the three notebooks filled with several years' worth of his work on orbital mechanics. He placed them on the table before the generals. They perused them for several minutes as he watched the clock on the wall.

Scratching his head and squinting his eyes, Wolcott examined a diagram that Ourecky had been sketching this morning on the flight from Atlanta. His face bore a perplexed expression.

"Something wrong, sir?" asked Ourecky.

"Yeah. B…E…A. What in tarnations is a BEA?" asked Wolcott. "I ain't seen that one before. Is it some sort of formula or engineering shorthand?"

BEA? thought Ourecky. *What was Wolcott looking at?* Finally he figured it out. Grinning, he replied, "Nothing, sir. Just a name I was trying to remember."

"Very impressive," said Tew, scanning over the notebooks. "And this is your own work?"

"Sir, yes it is. All of it."

"Fascinating. Virgil, do you have anything else for Lieutenant Ourecky?"

Wolcott was right in the middle of jamming a wad of Red Man chewing tobacco in his mouth. He sputtered, and waved his hand. "Nope, no, nothing here," he said.

Ourecky nodded. He stood up and then reached over the table to pick up his notebooks, but Tew scooped them up first.

"We'll maintain these here," said Tew sternly. "For safekeeping. Just so we're perfectly clear, intercept missions against hostile targets, rendezvous, orbital mechanics and space vehicles are not to be discussed outside of this facility, at any time, under any circumstances. These things are no longer your hobbies, they are your assignment, and that assignment is not to be spoken of outside of this building. Not out there on the sidewalk, not at the Officer's Club, and definitely not at Eglin. Do you understand?"

Ourecky straightened up, and assumed the position of attention, much like he had done as an ROTC cadet back in Lincoln. "Sir, I understand, sir. By your leave, sir?" he asked, throwing his hand up in a salute, formally requesting permission to depart the premises.

Wolcott grinned; miniscule bits of damp tobacco stuck to the left side of his mouth. "Relax, pard," he said. "We ride this herd pretty loose, but we keep everything on the ranch. So long as you cotton to that, we'll be just fine. Now, you need to head downstairs to talk with Ted Seibert, he's going to have some security paperwork for you. I caution you to take your time and read every bit of the fine print. Doin' so may save you an unpleasant holiday at Leavenworth."

"Understood, sir."

"One other thing, Ourecky," added Wolcott, talking around the lump of tobacco. "Unless we tell you otherwise, the next time you come up, leave your uniform back at Eglin, at least for the time being. Got that, pard?"

"Yes, sir."

"Now git," said Wolcott. "Time's a' wasting. Fetch your paperwork upstairs to the Personnel shop and then see what you can do for us in the basement. Stop back by here on Friday. If there's anything you need while you're here, don't hesitate to let us know."

"Welcome to Blue Gemini, son," added Tew. "But always remember: what you see here stays here."

Ourecky saluted again, sharply pivoted around in a snappy about face, and headed out the door and into his bright new future.

9

THE IDEA THAT REFUSED TO DIE

Headquarters of the General Staff of the Soviet High Command
Arbat Military District, Moscow, USSR
9:29 a.m., Friday, April 26, 1968

With instructions that he was to receive a new mission, Rustam Abdirov had been called to appear before the General Staff of the Soviet High Command. The General Staff oversaw the activities of the Ground Forces, the Air Defense Forces, the Air Force, the Navy, the Strategic Rocket Forces, and numerous other entities of the Soviet armed forces.

One of the senior generals audibly gasped as Abdirov entered the ornately decorated meeting chamber. It never failed; his grotesque visage never failed to evoke at least one horrified response, even from hardened men who had known him for years. He marched stiffly to the chair provided for him, stood at attention, saluted briskly, and waited for instructions from the Chief of the General Staff.

"Have a seat, Abdirov," said the Chief cordially. "I hope that you are well."

"As well as can be expected," replied Abdirov, slowly lowering himself into the seat, assisted by an aide. "Much better than I deserve, sir."

"We have a new task for you," declared the Chief. "You are to immediately assume command over all aspects of the *Skorpion* space-based nuclear bombardment system. You will be granted personnel and technical resources to establish a bureau specific to this mission, and you will be allocated a special operating facility at Kapustin Yar for research and testing."

The aide proffered a leather folder and fountain pen to Abdirov. The folder contained the formal orders appointing him to his new duties. Abdirov quickly reviewed them, verified his name and service number, signed the bottom of the document, and handed the folder and pen back to the aide. He waited as the aide gradually made his way around the Staff's table, pausing for each member to sign the document.

As Abdirov waited for the final step—a wax seal embossed by the Chief—to render the orders official, he pondered the assignment he had been given. He resisted the urge to laugh. He had heard of the *Skorpion* project before but mostly in rumors. He was aware that there had been an aborted launch only a few months before, in which a nuclear payload was nearly lost, which could have readily resulted in a significant catastrophe.

Not only was this notion dangerous, it was also highly impractical, but it seemed as if it could not be stopped. It was an idea that refused to die, relentlessly slogging forward, despite concerted attempts to kill it. *How could this be?* While Abdirov was a steadfast believer in the Soviet system, he knew that centralized planning had its limitations. One immense drawback was something that he labeled bureaucratic inertia, in which an idea or plan persists long after it has outlived its usefulness or has been overcome by events, simply because it has gathered such inertia that it cannot readily be halted. Certainly, this concept was a perfect example; Khrushchev had declared to the world that we could put nuclear weapons in orbit, so the River Styx be dammed if we didn't. Never mind that Khrushchev had since been ejected from his position of power and influence, and disregard the fact that all of Uncle Nikita's engineering expertise would likely fit neatly into a thimble with considerable room to spare.

If not because of bureaucratic inertia, there was likely another reason that some high level leaders remained committed to the development of a space-based nuclear bombardment system. While the General Staff itself was composed of very practical men, they were compelled to answer to their superiors within the Defense Council and the Collegium of the Ministry of Defense, who in turn took direction from the civilian leadership of the Politburo of the Communist Party of the Soviet Union. Somewhere within this hierarchy, someone of significant power—more likely a group rather than a single individual—obviously remained obsessively true to Khrushchev's dream of placing nuclear weapons in orbit.

Again, how could this be? The most probable answer lay in the complicated calculus of nuclear annihilation. While there was considerable evidence that the Soviet Union had sufficient land-based missiles and bombers to achieve parity with the United States, if not outright superiority, the United States possessed a powerful card that could not readily be trumped. The Americans had far surpassed the Soviet Union in the development and fielding of undersea technology, and their peerless nuclear-powered submarines—each bearing an arsenal of thermonuclear ICBMs—silently traversed the oceans of the world with absolute impunity.

But if some powerful cabal—perhaps led by Brezhnev himself—persisted in advancing this idea, surely the same buffoons could be made to understand that there were simpler avenues to achieve nuclear parity with the West. If nothing else, the millions of rubles invested in this project could have funded the development of ballistic missile submarines on par with or even superior to the Americans' vaunted Polaris boats.

Abdirov's thoughts were interrupted by the Chief's voice. "You will receive several detailed technical briefings concerning *Skorpion* before you travel to Kapustin Yar to survey your new facilities, but in the meantime, do you have any questions of the General Staff?"

"*Da*," replied Abdirov. "Respectfully, Comrade General, I am greatly honored to be of service, but I must assume that there is a multitude of technical obstacles to overcome. Assuming that I might require resources outside of my own bureau, what authority am I granted to seek resources and assistance from other bureaus?"

"You are granted absolute autonomy over all aspects of this project," stated the Chief. "You may seek resources and assistance as you wish, within reason."

"Within reason, Comrade General?"

"*Da,*" replied the Chief. "But be aware that any significant changes to the existing design must be approved by the General Staff. Any other questions?"

"*Nyet,* Comrade General. By your leave?"

"Go. You have much work to do, as do we. Planning for the invasion of Western Europe demands our present attention."

Assisted by the aide, Abdirov quietly grunted in pain as he rose to his feet. He saluted stiffly, gradually turned, and left the chambers.

10:26 p.m.

After sitting through several excruciating hours of technical briefings, Abdirov retreated to his small suite in a military hotel, slowly stripped out of his uniform, and poured himself a tumbler of *Stolichnaya*. Sitting naked on the edge of his bed, he gulped down two codeine tablets to alleviate the incessant pain that wracked his body. He chased the pills with the vodka, and reflected on the day's events.

It hadn't taken very long for him to comprehend that the *Skorpion* plan was grossly flawed. First, the designers seemed compelled to adhere to an almost exact interpretation of Khrushchev's words, even though his grandiose speech was essentially just propaganda directed at the West. The idiots were literally trying to orbit the biggest warhead they could, which was why they were willing to risk the recent launch with the largely unproven UR-500 booster. The whole notion smacked of inter-bureau rivalry; mightily opposed by Chief Designer Korolev, who thought hyper-golic propellants were too dangerous to handle, Chelomei's bureau built the massive UR-500. If it was so damned imperative to put atomic bombs in orbit as a deterrent to the West, mused Abdirov, wouldn't it be far more practical to send up *several* smaller warheads instead of a single giant one? After all, Korolev's ubiquitous R-7 *Semyorka* rocket had proved its reliability time and again and would be the perfect launch vehicle to loft an entire constellation of nuclear-armed satellites. Moreover, the existing *Zenit* reconnaissance satellite, which also served, with few modifications, as the *Vostok* manned spacecraft, could be readily adapted to the task.

There were other nagging issues with *Skorpion*, but perhaps the most troubling was the control system. Like most Soviet satellites, to include manned spacecraft, the nuclear-armed satellite was controlled from the ground. In the present design, reentry instructions were to be relayed up by radio, which in turn would be automatically executed by *Skorpion's* onboard control systems. At present, they were restricted to launching satellites into relatively high inclination orbits; as such, if the United States was the intended recipient of the warhead, it was possible that at any given time, several hours—up to an entire day, in some instances—might elapse before a satellite actually passed over North America. In the meantime, if the Americans had executed a first strike without warning, or had reacted effectively to a Soviet first strike, the *Skorpion* control stations could very well be obliterated. And even if the control stations weren't physically demolished, there was now plenty of empirical data to indicate that their electronic guts might be destroyed by powerful energy pulses emitted by nuclear detonations.

He had to laugh; in assigning him the *Skorpion* mission, the General Staff solved two problems at once. First, they knew as well as he did that the existing design was faulty, so they were foisting it off on the perfect scapegoat. Without a doubt, they wanted to hang this reeking albatross around his neck and then wash their hands of it. They could be assured that he would fail and that the repercussions—for them, at least—would be minimal. Second, and perhaps more to the point, their action provided them with a means of subtle reprisal against Abdirov, who at first had been Nedelin's pet, and now was Brezhnev's fair-haired child. Although he had been authorized his own special facility at the Kapustin Yar cosmodrome, they might have well banished him to the far side of the moon. He was sentenced to toil in distant seclusion, like a hideous hunchback sentenced to a bell tower.

At this point, given the hand dealt him, he could readily concede that the task was just too damned difficult. Certainly, he could return to the General Staff in the morning, beg for relief due to his poor health and then just slink away into retirement. The *Skorpion* task would be levied on some other unfortunate recipient, who was sure to fail, and then perhaps the seemingly indomitable idea would finally expire.

But what if he stayed the course and made it work? He jotted a few notes as he finished his vodka. While the existing *Skorpion* concept was asinine, there were some aspects that could be salvaged. He felt that the key was in the control system. The ideal weapons platform would be autonomous, to a large degree, so that it was not dependent on ground control stations. While this degree of autonomy was certainly the goal, Abdirov was aware that it would require furnishing the satellite with considerable computing power, which would require a powerful computer much lighter and smaller than what was currently available within the Soviet Union. Where could he find such a machine?

He switched off the lamp by his bed. Clearly, the General Staff had positioned him for his downfall, but Abdirov saw the potential for redemption instead. If this thing was done right, it would not only work, but he might actually give the Americans something to truly fear.

10

UNWELCOME

Auxiliary Field Ten, Eglin Air Force Base, Florida
6:15 p.m., Thursday, May 9, 1968

Matt Henson and the other candidates had just returned from the multi-purpose weapons ranges, where they had spent most of the day firing pistols and shotguns. Piling off the trucks, they formed four neat files, patiently waiting their turn at the pull-up bars. Their training cycle had started with fifty-four candidates, but had since dwindled to thirty.

Waiting for the last man to drop off the bars, Henson glanced down to read the inscription on the small stone marker: *Until You Can't*. When he first arrived, he could nimbly jump up to the pull-up bar, knock out a string of fifteen with scarcely any effort, and then grunt out a few more. Even the whip-tough pararescue "PJs" found it difficult to match his pull-up prowess.

Now, though, he didn't find it nearly as effortless as before. The uprights for the pull-up bars were constructed of telephone poles, with a small wooden step nailed on the inside. Henson slung his weapon around his neck on a strand of parachute cord and swung up to the bar. He strained out five pull-ups and then painfully dangled, gripping the rusty bar with sweaty palms. The impromptu weapons sling cut deeply into his neck, and sixty odd pounds of equipment—carried in his rucksack and web gear—weighed heavily on his shoulders. He exerted himself enough to raise his chin over the bar three more times and then dropped to the ground.

A few hundred yards away, on the asphalt runway adjacent to the row of buildings, a C-130 revved its turboprop engines, drowning out all conversations with a pervasive growl. As most of the candidates turned to watch, the transport plane jolted forward a short distance. Bolted to its fuselage, eight JATO rockets suddenly spewed a torrent of orange flame, like a volcano erupting sideways. With a deafening roar, the squat transport leapt into the air and climbed at a furious rate, disappearing into the sky at a forty-five degree angle.

Before entering the billets, Henson safety-checked his GAU-5 submachine gun—a stubby weapon similar to the Army's "CAR-15" carbine—one more time. Glancing into the GAU-5's chamber, he saw no errant round and let the bolt snick home. If the cadre discovered a loaded weapon in the billets, it was grounds for immediate and uncontestable relief from the course.

Unlike most Air Force training courses, "immediate and uncontestable relief" had dark connotations; at Aux One-Oh, it meant swift reassignment to some distant and dismal place. Henson was aware that some former candidates were now stuck in the barren hinterlands of Montana or South Dakota, guarding missile silos for the remainder of their enlistments. Others had been banished to remote locations in Alaska and Greenland. A candidate could be summarily sent away for a rules infraction—safety violations on the ranges were the most common—or an academic failure, or he could voluntarily request to be dropped from training.

Sore after a long day on the ranges, Henson shed his web gear—a collection of suspenders, belt, straps and pouches that carried water, ammunition, grenades and other gear—and hung it on the end of his bunk. His right shoulder ached from several hours of absorbing recoil.

After storing his gear, he would strip down the GAU-5, clean it until it was pristine, grab a quick shower, check over his field gear to ensure it was ready for tomorrow, and then look over any reading assignments that might have been posted.

If time permitted and if he wasn't too exhausted, he would indulge in a beer at the slop chute. That was pretty much the extent of his existence for the past month, and there were at least three months yet to go. Beyond that, if he made the grade, there was Army jump school and possibly Ranger school, both at Fort Benning near Columbus, Georgia. Then he would be assigned to an operational squadron, either here at Aux One-Oh or elsewhere.

Henson had no idea where the other operational squadrons were based, or even how many there were. Furthermore, there was apparently little to be gained by attempting to wangle the best geographic location or most plum assignment; judging by the likes of the squadron based here, most of his life would be spent on the road. He rarely saw those guys for any more than a week at a time, and then they would be loading up on a C-130, bound for somewhere. And they didn't say much about where they went or what they did.

Last week, he had struck up a conversation with one of the operational guys—a brother, no less—down at the slop chute. Recovering from a broken ankle sustained on a parachute jump, he was one of the few black men assigned to Aux One-Oh. Their shared skin color was a good enough excuse for Henson to buy him a beer and chat him up to find out what life was like after the training was over.

The brother was polite, profusely thanking Henson for the cold beer, but he was as vague and elusive as any man could possibly be. Fifty cents bought Henson little that he didn't already know. He gleaned nothing about what life held in store after training, or even if all the training was really worth it. Until the future arrived, he kept at the training grind, up before dawn for physical training followed by hard days and sometimes harder nights.

His OG-107 jungle fatigues were soggy with sweat from the day's labors. Henson stripped off his shirt and draped it over his web gear to dry. The jungle fatigues were relatively comfortable in the otherwise unbearable heat and had a multitude of pockets to stash gear, but Henson looked forward to the day when he would go to supply and draw the tiger-striped camouflage uniforms worn by the men assigned to the operational squadrons. The tigers could be worn here on the compound and on the ranges; otherwise, they were strictly forbidden to be worn off Aux One-Oh, particularly on the main base of Eglin proper. It was another rule not to be broken or trifled with; even an assignment to a vaunted operational squadron didn't necessarily mean a man wasn't subject to abrupt exile for testing the boundaries.

Slightly more comfortable, Henson tucked his weapon into the crook of his arm and opened his locker to grab his cleaning kit, shaving brush, and can of Three-in-One oil. As he turned around, he saw a small strip of masking tape on his gray metal bed frame, next to a similar strip that bore his name, rank and roster number. But this new piece of tape had a pointed message scrawled in black grease pencil: "NIGGER NOT WELCOME." He cursed quietly under his breath, knelt down, and used his pocketknife to slowly strip the tape from his bunk. Preserving it as evidence, he stuck it in his pocket.

He glanced around, looking at the other men in the billets space. Most were still present, having just got off the trucks from the ranges. The place reeked with the sweaty funk of thirty bodies jammed together in close quarters. Johnny Cash sang from one AM radio, lamenting that he had fallen into a burning ring of fire, while Jim Morrison and the Doors wailed from a second tinny box, encouraging all within earshot to break on through to the other side.

Concealing his anger as best he could, Henson read the taunting script once more: "NIGGER NOT WELCOME." It wasn't the first racial threat he had received since arriving here; among other things, there had been degrading notes jammed into the air vents of his locker, a noose, fashioned of parachute cord, dangling over his bunk, and two holes had been cut into his pillowcase, apparently to represent the intimidating mask of the Ku Klux Klan.

Despite the pressure and isolation, he maintained his composure, never once letting on that it was getting under his skin. And although no one had overtly vocalized any racial slurs at him, the most likely culprits were a close-knit group of five cracker white boys who always hung together in training or at the slop chute. Now, the five mingled around the red-painted butt can in the common space, sharing a cigarette as they discussed tentative plans for the weekend. He scanned each of their faces, expecting to glimpse the smirk or stupid grin of the perpetrator or at least someone in cahoots, but saw nothing. Even for dumb white crackers, they were either the coolest customers alive or they really had nothing to do with putting the tape on his bunk.

But Henson had endured enough. Incensed, he donned his damp shirt and quietly clicked his locker shut. Carrying his weapon and cleaning gear, he strolled out the back door. At the cleaning tables, men joked and sipped Cokes as they used long cleaning rods to plunge oiled patches through the barrels of their weapons. Two men worked together to oil the steel guts of an M-60 machine gun. Seething with anger, gritting his teeth and tightly balling his fists, Henson continued walking, striking a beeline for the Training Flight's headquarters.

Arriving at a corrugated metal building almost identical to their living quarters, Henson knocked three times on the screen door, and waited patiently. Several minutes passed as he waited, allowing him to meditate on the smothering Florida heat, and also to tally—one by one by one—the number of mosquitoes that landed on his face, arms, and every other square centimeter of exposed black skin. But the wait didn't allow him to calm down; residual anger roiled inside him just as powerfully now as it did when he first saw the racist message on his bunk.

"Enter!" cried a shrill voice. Reacting immediately, Henson swung open the screen door and stepped inside. It wasn't much cooler inside the building than outside; the only noticeable difference was that there were two electric floor fans that apparently served little purpose but to move the warm air from one side of the room to the other. The Flight's first sergeant sat behind his desk, and beckoned Henson forward. Captain Lewis stood by silently. Despite his small stature, he was unbelievably fit; the class dreaded the days that he led them in calisthenics.

"Here to *quit*, Henson?" implored the first sergeant. "And you were doing so well. I really suspected you were going to hang on for at least a few more weeks."

"There's no shame in quitting, son," added Lewis. "This job's not for everyone. We're much happier if you recognize it now before you realize it in the field with people's lives at stake." As Lewis spoke, the first sergeant yanked voluntary drop paperwork out of his top desk drawer.

"Sir, I'm not here to quit," boomed Henson.

The first sergeant looked puzzled. "You're not? Then why are you here?"

"Sir, the candidate feels obligated to report a situation not conducive to good order and discipline," blurted Henson. He pulled the strip of masking tape from his pocket, and stretched it out on the first sergeant's desk. "I just found this on my bunk, sir."

Lewis shook his head. "It seems like this happens about every other class. I just don't know what possesses these people. The military is integrated now. Why can't they just accept that?"

"Sir, that's not all," Henson added. He briefly described the other artifacts left on his bunk, as well as the messages stuffed into his locker.

The first sergeant scribbled notes as Henson spoke, and promised that he would personally take action. "I'm surprised that you haven't brought this to our attention already," he commented. "And I'm impressed that you've kept your cool. That's admirable, given the circumstances. Now, we'll see to it that this business comes to a dead stop. In the meantime, Henson, you need to head back over there and drive on with the program. Just three months to go." The first sergeant stuffed the voluntary drop paperwork back into his desk.

Henson snapped back to attention and threw up a salute to Lewis. "Permission to be dismissed, sir?" he asked.

"Dismissed," replied Lewis, returning the salute. "Keep your calm, Henson. You're doing very well."

Henson dropped the salute and hustled smartly for the door, almost knocking over one of the screeching fans in the process.

The first sergeant walked over to the screen door and watched as Henson made his way back to the weapons cleaning tables. Lewis joined him. "I am really surprised that he didn't quit," observed the first sergeant. "He's obviously a very tough customer."

"Agreed," answered Lewis, slapping a mosquito on his neck and then examining the smudge of blood on his fingers.

"So is there any need for me to continue leaving messages for him, sir?"

"No," replied Lewis. "As a candidate so new to the Air Force, I strongly suspect that he'll be an academic failure in due time. If not that, we'll find some other suitable reason to cull him."

11

THE BOX

James M. Cox Municipal Airport, Dayton, Ohio
1:35 p.m., Monday, May 20, 1968

The stewardess placed her hand on Ourecky's shoulder, shook him gently, and softly whispered in his ear. "Rise and shine. It's time to wake up, honey."

Recognizing Bea's voice, he awoke with a start, not sure of where he was. He unfastened his seat belt, patted his lap, and then checked the seat before anxiously looking on the floor underneath.

"Looking for this?" she asked, grinning, holding out a thin brown leather holster. "So what do you measure with this kind of ruler?"

"Uh, Bea, that's not a ruler, that's a slide rule. I use it to solve equations."

Adjusting her wings, she smiled. "I was just kidding, silly. I know what a slide rule is. My dad had one. You dropped it when you fell asleep, so I was holding onto it for you. One of the kids was intent on playing with it. I think he planned to keep it, also."

"Well, thanks. I appreciate you looking out for me." Ourecky glanced around. Except for a few straggling passengers and a cleaning crew just coming aboard, the cabin was empty.

"You were out like a light. Out late last night? Big date?"

Yawning, Ourecky stretched his arms and slowly shook his head. "No. I'm working on two big projects, and I had to finish the one in Florida over the weekend before I flew up here."

"Well now, you're just melting the candle at both ends, aren't you? You really need to slow down and smell the roses. You fly up to Dayton every other week, right? So where do you stay when you come up here? Hotel? Stay with a friend?"

"Uh, I stay at the VOQ—Visiting Officers Quarters—on Wright-Patt." Ourecky gathered up his papers, pencils, and the slide rule, and stuffed them into his attaché.

Bea collected magazines from the seats near Ourecky. "Do you ever go out? I've not seen you in any of the clubs near the base, where the pilots and Air Force guys usually hang out."

"No," he answered, slowly standing up and stretching. "I don't get out. I only come up here to work. Hey, did you get your hair cut? It looks shorter than last time."

Smiling, obviously pleased that he noticed, she pivoted her head side to side. "I did. This style saves me at least thirty minutes in the morning, since it's a lot easier to wash and dry."

"Well, it looks nice. Uh, I would really like to stay and chat, but there's supposed to be a car waiting for me in front of the terminal, and I can't miss it. I had better get moving. Uh, bye Bea." With that, Ourecky walked down the narrow aisle, attaché clutched to his chest.

"Bye, Lieutenant Ourecky," she called after him. "See you around."

Aerospace Support Project
Wright-Patterson Air Force Base, Ohio
9:35 a.m., Wednesday, May 22, 1968

Ourecky had been granted his own little office in the basement. The dreary space was damp and musty, with peeling green paint and a cracked linoleum floor. The only furnishings were an old drafting table and a straight-backed wooden chair. He had tilted down the drafting table to fashion it into a makeshift table; its scarred and stained surface was presently covered with binders, books, star charts, pencils, pens, and various drafting instruments.

Although his modest workspace left much to be desired, working for Wolcott and Tew obviously had other significant benefits. Ourecky looked at a shiny new set of captain's bars, which the generals had bestowed on him Monday afternoon, as they had promised, not long after he had arrived from the airport. The pair obviously held considerable influence with the Air Force's otherwise rigid personnel system; little else could rationally explain how Ourecky—a lieutenant without pilot's wings, combat time or other significant military experience—had been promoted well over a year ahead of his peers.

Chewing on the frayed end of a toothpick, he leaned back in the chair and studied a massive tome—labeled *NASA Project Gemini Familiarization Manual*—that was essentially an encyclopedia devoted to the two-man spacecraft's systems and procedures. Although it was not entirely within the purview of the tasks he had been given by the generals, he had taken it upon himself to become intimately familiar with the workings of the Gemini, so that he could better understand the challenges faced by the men who might ultimately fly the intercept missions.

Right now, his focus was devising a series of "cheat sheets" for the different calculations required at various phases of a rendezvous in orbit. It was not a simple task. Orbital rendezvous was an elaborate three-dimensional exercise in which a space vehicle gradually made the necessary adjustments to match its orbital plane to a rendezvous target, and from there executed more shifts to eventually overtake the target. NASA had refined the practice—essential to the upcoming Apollo moon landings—during their Gemini missions. NASA's crews had rendezvoused with cooperative targets—*Agena-D* satellites equipped with radar transponder beacons and marking lights, no less—but conducting a rendezvous with a non-cooperative target—a *hostile* target—was a complex symphony yet to be played.

In rendezvous, the underlying problem was that every movement, even the slightest correction, required an expenditure of energy, and that energy, in the form of fuel for maneuvering thrusters, was in finite supply. To further complicate matters, particularly at the final stages of the chase, it was extremely easy to overcompensate, which could inadvertently result in the chase vehicle being placed in a higher orbit than the target.

Consequently, even more energy had to be expended to correct the error, if it could be corrected at all.

But as Ourecky struggled to distill the equations into a series of worksheets, he was also very aware of the extreme burdens that would be levied on the crews. Some of the proposed intercept missions could be arduous marathons in which the Gemini-I crew might have to maneuver for an entire day or longer to catch up to their target. Thumbing through his index cards, he wondered how effective the men could be after going that long without adequate rest. Staring at a pale green wall, he was deep in thought when Virgil Wolcott walked in.

"Captain O. How's it goin' there, pardner?" Wolcott sipped from an oversized coffee mug. "Makin' yourself at home, are you?"

Startled, Ourecky quickly extracted the toothpick from his mouth and stood up; he wasn't yet used to being addressed by his new rank. "Good morning, sir." He hadn't showered or shaved this morning, and his clothes were more than a bit rumpled.

"Young Captain, just how long have you been down here working?" asked Wolcott.

Suddenly self-conscious, Ourecky rubbed his stubbly chin. "Uh, since yesterday morning, sir. I have a lot of catching up to do."

Wolcott nodded. "Well, you need to draw some fresh air into your lungs, son. I think it's high time that you and I moseyed out to the back forty, so you can see another special part of our little ranch. Grab your hat, pard, and let's go."

Ourecky followed Wolcott out of the building. Crossing a parking lot, they headed toward a decrepit-looking brick hangar of pre-war vintage. Gesturing at it with his white Stetson clutched in an outstretched hand, Wolcott drawled, "Yonder is our Simulator Facility. Currently, we have one Gemini procedures trainer in there, and we're trying to wrangle a second one from NASA. Right now, we have three crews in training. Of course, we can only run one crew through the simulator at a time, so we cycle the others through technical and contingency training."

"Contingency training, sir?"

"Yup. That's a fancy high-falootin' term for survival training. If you're a pilot in this man's Air Force, bub, you definitely get your fill of survival training. Of course, not so much as we endured back in the fifties, when Iron Pants Lemay was holdin' court and SAC was center stage."

Wolcott waved at a colonel walking by; the man casually waved back. "Myself, I didn't exactly cotton to all that survival training, mostly since the instructors seemed hell-bent on makin' everyone miserable. Most folks don't need any practice to be miserable; they usually get it right the first time.

"But since our pilots could presumably land up anywhere on earth, we owe it to them to be ready to survive in any situation. So we send the boys all over the world: down to Panama, to the Philippines, to Alaska, and out to Stead in California. Anyway, if you're ever stranded on a desert island, you better hope one of my guys is with you. Unless you're lucky enough to be marooned with Ginger or Mary Anne, but even that ain't going to be very entertainin' when you're both starvin' after the first week."

Ourecky considered the desert island scenario briefly, trying to decide whether he preferred the sultry redhead or the more wholesome brunette but couldn't really settle on either.

The two men arrived at the hangar and entered an anteroom where a sergeant kept guard. Wearing fatigues at least a size too large, he was armed with an M3 "grease gun" and a .45 caliber pistol, and appeared anxious to use both. A bandolier of extra magazines was strapped around his skinny waist, perhaps to mercifully finish off anyone who survived the initial volley. Wolcott flashed his credentials, signed an access roster, and stated, "The captain is with me."

A red light bulb, mounted in a fixture by the interior door, blinked on and off continuously. "There's a session in progress right now, Virgil," noted the humorless guard. "I came on duty at six this morning, and they had already been going at it all night."

Ourecky noticed a handmade wooden sign above the door, bearing an inscription in a foreign language. "*Lasciate ogne speranza, voi ch'in-trante,*" he read, stumbling through the pronunciation. "What does that mean?"

"Dunno," replied the guard, shrugging his shoulders, "That Major Agnew tacked it up there. He's kind of a jokester, so I figured it was some-thing funny."

"It's Latin, pard," explained Wolcott, "It's a mite long-winded, but mostly it just says *Welcome to All*."

They passed into the hangar's inner sanctum, a massive space with a curved metal ceiling and a gray-painted concrete floor. At the far end was a life-sized replica of something that looked vaguely like a futuristic aircraft. Ourecky recognized it as a mock-up of the now defunct X-20 Dyna-Soar space plane. It seemed almost forlorn, like it had been hastily shoved to the side and abandoned.

Ourecky marveled at the training facility. Its centerpiece was a full-scale mock-up of the two-man Gemini spacecraft, mounted atop a short platform. The Gemini was upright, so that its blunt nose pointed toward the ceiling. The platform was surrounded by three concentric rings of workstations, metal cabinets, and boxy instruments. The consoles in the innermost ring were manned by technicians who apparently orchestrated the training simulation. The second ring was comprised of ordinary desks and chairs, and the third ring was a bank of metal cabinets that enclosed large computers and other electronics.

"There's the Box," stated Wolcott, sweeping his Stetson in a flamboyant gesture toward the Gemini mock-up. "That's the *main* attraction."

As the guard had previously indicated, a simulated mission was currently underway. There were about a dozen civilian technicians seated at the consoles. In the outer ring, two men were changing a large tape reel on a freezer-sized computer. It looked as if there was a dress code in effect; the civilian technicians uniformly wore short-sleeved white shirts, dark ties, dark slacks, and black shoes. Off to the side, a table held a metal coffee urn and several open boxes of Krispy Kreme glazed doughnuts. Judging by the hefty appearance of some of the men, Ourecky wouldn't be surprised if they subsisted solely on coffee and stale doughnuts.

"Rise and shine, buckaroos," announced Wolcott, striding up to the control area. The technicians appeared exceptionally tired, but they looked up at Wolcott and smiled.

One man—seated at a console marked "SIMSUP"—turned to the technician next to him, and said, "Take it, Chris. I'm going to take a smoke break." With his white shirt stretched to the point of bursting, the man was well beyond portly. His black hair was greasy, slicked across his broad head. He wore black-framed glasses; the thick lenses were smudged with fingerprints. He pulled a pack of Winston cigarettes from his shirt pocket, lit one, and inhaled deeply. He walked over to greet Wolcott. "Good morning, Virgil."

"*Und guten Tag* to you, Gunter. Ourecky, this is Gunter Heydrich, our token German rocket scientist. He's our Simulation Supervisor. He's the honcho who rides herd on this rodeo," declared Wolcott. "Gunter, the good captain is TDY from Eglin, helping us to streamline procedures for the intercept missions."

"Nice to meet you, Gunter," said Ourecky.

"Likewise."

Wolcott interjected, "Captain O, begging your pardon, but Herr Gunter and I are long overdue for a powwow. I need you to hang back here for a few minutes. I'll collect you shortly."

"You're welcome to watch the simulation," noted Heydrich. "But don't bother my controllers. They have a lot to keep track of. They don't need any distractions."

As Wolcott and Heydrich strolled toward the front row of consoles, Heydrich asked, "Well, Virgil, what can I do for you today? I assume that this is not a social call."

"I come bearing good news. I just got a call from the contractor building our paraglider landing simulator. They're in the final stages of their testing. Once they're done with the testing, they'll disassemble it, crate it, and ship it here. We can expect delivery around the middle of next month. Ain't that just splendid, Gunter?"

Heydrich looked at the ceiling and groaned.

"That ain't *quite* the response I expected," Wolcott said. "What's the problem?"

Heydrich gestured at the men working at the consoles. They all looked frazzled, like they hadn't slept in days. "Virgil, I know how anxious you are to get this new machine up and running, but I'm not sure that I have the manpower to handle it, especially if you expect me to run both simulators simultaneously."

Heydrich took a long draw from his cigarette and exhaled. "As it is, I'm running two twelve-hour shifts, seven days a week. A lot of these men haven't seen their families for over a month, and some of them don't even bother to go home anymore." He pointed at a row of Army cots at the far end of the hangar, under a drooping wing of the disused Dyna-Soar mock-up. Swaddled in blue wool blankets on loan from the base hospital, two men slept there.

"What if we wrangled you some extra hands?" asked Wolcott.

Heydrich shook his head and replied, "Mark has promised me that he would hire extra folks, but I haven't received any reinforcements yet."

"Mark's doin' his utmost, Gunter. He's still shufflin' around the funds to make it happen."

Heydrich shrugged and said, "Virg, as dedicated as they are, my guys aren't naïve. They know they'll never see any official recognition for their work here, whether we're successful or not. These are smart men, with skills that are currently in high demand. They could be working for North American, Grumman, or any of the other contractors building Apollo. At least they could go home at night and tell their wives what they accomplished that day. As it is, they'll leave here with a five-year blank spot on their resumes, so it's hard to keep them motivated. We can only wave the flag in front of them for so long before their eyes just permanently glaze over."

Heydrich had clearly been waiting for a good opportunity to vent. "Virgil, my manpower problems are just the tip of the iceberg. Look, at the rate we're going, it's just a matter of time before the hardware fails. This *verdammter* simulator wasn't designed to operate around the clock. We barely have time to swab out the sweat and grime from the last crew before it's time to wedge in another pair. We just can't sustain this pace, especially when there's no downtime for maintenance."

"Duly noted, Gunter," replied Wolcott, lightly fanning himself with his Stetson.

"Just get me more men, Virg. That's all I ask."

"I'll do my best, Gunter. I promise. Now, how about this new para-glider simulator? If it arrives in mid-June, how long will take it take you to get it installed and operating?"

Heydrich scratched his forehead. "Probably no earlier than mid-July, assuming that I can get my guys trained in time. But getting it installed and functioning is a lesser issue, Virgil."

"How so, pardner?"

Heydrich waved his hand toward the interior of the crowded hangar. "There's no room at the inn. I don't have adequate floor space to put any more equipment. Unless you can get me another building, we're cubed out."

Wolcott nodded. Heydrich was right. There was a pregnant moment of silence as he scratched his head and closed his eyes, as if pondering the dilemma. Opening his eyes, he looked at the Dyna-Soar mock-up at the far end of the hangar and asked, "Gunter, why do we still keep that danged old monster stashed in here?"

Heydrich chuckled. "Why do we keep it, Virg? Because over a year ago, Mark Tew personally directed me to retain it. He thought it might be a museum piece someday, so we should make sure it wasn't accidently destroyed."

"Museum piece?" Wolcott sniffed. "Yup, I guess it might eventually land a prime spot in the Museum of Grandiose Ideas. Man, I get the willies every time I clamp eyes on that danged thing. It's always felt like a bad omen to me. As soon as I wean Mark from his nostalgic streak, we'll haul that damned piece of junk out of here to make some room."

"Sounds good to me, Virgil."

"So long as I'm over here in your domain," said Wolcott, looking toward the Gemini mock-up, "how about a scoutin' report on our first string? How is Crew One handling the new plays?"

"Carson and Agnew? Drew Carson flies the Box faster than we can throw problems at him," answered Heydrich. "My troops can't keep up with him, and he knows it."

Heydrich coughed, and continued. "But Tim Agnew, there's another matter entirely. This whole game hinges on the right-seater keeping pace with the calculations and feeding them to the left-seater. To his credit, Tim's the most qualified right-seater in the bullpen, but he just can't keep up with Carson, and he can't keep up with us."

A technician looked back over his shoulder and said, "Hey, Gunter, Carson is just sitting there, cooling his heels. We're going to throw a stuck maneuvering thruster on him. Scenario item number Two-Oh-Six, failed circuit breaker. Go on that?"

Heydrich stuck up a thumb and replied, "Go on the kerplunk thruster. Keep 'em guessing."

The technician threw toggle switches and twisted knobs to introduce the problem into the simulation. True to form, Carson diagnosed and resolved the errant thruster in record time, and resumed the ongoing mission without breaking stride.

Wolcott turned back to Heydrich and asked, "So how do we fix this problem with Agnew?"

"*Fix*? Fix? You really want my opinion, Virg?" Heydrich replied in an exasperated voice. He stubbed out the cigarette in an empty paper coffee cup. A small desk fan hummed nearby, riffling papers on the console as it slowly pivoted side to side. "It's not Agnew's fault. We're asking him and the other right-seaters to execute tasks that NASA hands off to big computers on the ground, and it's just more than they can handle. I don't see us flying any missions until the Block Two computer is delivered, regardless of the deadlines we're under."

"How about the current onboard computer, pard? Is it any help at all?" The Gemini spacecraft was equipped with a fifty-nine pound computer custom-made by IBM. During missions, it performed several key roles, assisting the two-men crew from lift-off all the way through to their reentry back into the Earth's atmosphere. It was a powerful tool, but it was designed for NASA's flights, in which it was almost constantly updated with mission data uploaded from the ground. The long-anticipated Block Two was based on the sophisticated new Apollo Guidance Computer, but with one substantial improvement: the Apollo computer relied on "core rope memory," in which an individual mission's data and software instructions were physically woven into a series of wires and washers, while the Block Two could be reprogrammed—using tape cartridges—for different scenarios, even while on orbit.

"Is the existing machine of any help? Yes and no," answered Heydrich. "For our mission profiles, the current machine may handle roughly forty percent of the workload, but that still leaves a lot of stubby pencil work for the right-seater."

Wolcott shrugged and asked, "Okay, Gunter, are you implyin' that it ain't feasible to do these intercept profiles without the Block Two?"

Heydrich yawned, stretched and replied, "No, Virgil. What I'm telling you is that it's just beyond the capacity of the guys we have now. Granted, all of these boys are superb test pilots, and most of them are extremely talented engineers, but we're working in a whole new realm here. Now, if you could stick Buzz Aldrin in the right-hand seat, this would be a wholly different ballgame. But Aldrin's not available to us, because he's busy doing other things. So in my opinion, again, we should wait on the Block Two."

Wolcott grunted. "That's easy for you to say, friend. But I'm concerned that we're going to lose the initiative. If we keep crawfishin' on the schedule, we run the risk of having the whole danged program scrubbed. We're treadin' on thin ice as it is. I would hate to see this Project go the way of the dinosaurs." He looked toward the Dyna-Soar mock-up.

Heydrich nodded in agreement. "Amen. Five years working on that monster, all for naught."

Suddenly there was an explosion of profanity from the simulator, blaring over the speakers mounted on the control consoles. Heydrich cringed at the outburst. "Ouch. That's Drew Carson at his finest, in his natural element. Sorry you had to hear that little tantrum."

Even with the speaker volume turned down, Wolcott could still hear Carson's anger. "TRY…TO…KEEP…UP!" Carson bellowed. "Just this *once*, Agnew, try to keep up with the damned program. At this rate, I'll never have a chance to fly!"

"*Whew.* How long have they been in the Box, pardner?" asked Wolcott.

"Eighteen hours on a twenty-four hour full-up," replied Heydrich. "I'll be surprised if Agnew makes it through the last six hours without a nervous breakdown. And to make matters worse, if we adhere to the current schedule, we commence forty-eight hour full-ups in two weeks."

Wolcott shook his head at the prospects. Even after years of working under Bennie Schriever at AFSC, slogging for days without normal sleep, he could hardly fathom operating at that intensity for two straight days without a break. Unfortunately, according to preliminary computer models generated by the experts, it could theoretically take forty-eight hours of continuous flying to intercept some potential targets. That equated to forty-eight hours of Agnew working calculations and taking star shots, while Carson converted the calculations to operational maneuvers. On orbit, they might snatch ten to fifteen minutes of rest at a time, at best.

"Well, pard, aren't they just about due for a break?" queried Wolcott. "I thought you gave them fifteen minutes every six hours."

Heydrich nodded in assent. "Their next break is at the top of the hour. Two minutes away."

"Well, let's lend them the extra two minutes. Unlock the Box and let 'em out, pardner."

Heydrich noted the deviation on his log and then directed the technicians to pull the pilots out of the simulator. Two technicians clambered up the platform to unlatch and swing open the big hatches. Carson slid out, bounced to his feet, stretched, and made a beeline for a Coca-Cola machine against a wall. He dropped a quarter in the dispenser, selected a bottle, popped off the cap in the door, and then drank deeply before sauntering toward the latrine.

Assisted by the technicians, Agnew slowly emerged from the Box. Obviously in agony, he stood up and then carefully negotiated the four steps from the mock-up platform to the main floor. He took two faltering steps and then crumpled to the painted concrete like an exhausted kitten. He lay there, quietly whimpering, trembling in pain. Agnew's flight suit was drenched with sweat. Wolcott noticed shiny wet spots, one on each shoulder blade and third one just below the small of his back, and suddenly realized the stains were oozing blood.

Concerned, the two technicians rushed toward Agnew to aid him. Shaking his head, Wolcott raised a hand to stop them in their tracks. He walked to Agnew and knelt beside him. Agnew's face was a white masque of anguish. In a low, quiet voice that only Agnew could hear, he spoke. "You are a commissioned officer in the United States Air Force. Get to your feet and act like one." There wasn't the slightest nuance of folksy twang in Wolcott's voice.

Agnew blinked. Slowly gathering his composure, he lifted himself off the floor and stood straight. "Sorry, sir," he mumbled. "It's just…"

Wolcott interrupted him. "You have fourteen minutes left. You've wasted three already. Do what you need to do, and then scramble your ass back in there." Feeling sorry for him, Wolcott watched Agnew slink away.

The man was clearly spent, both physically and emotionally. It would be a miracle if he endured the next six hours. But Wolcott also wouldn't be the least bit surprised if he later learned that Agnew had taken a sudden crash-dive into the mental ward of the base hospital.

Wolcott bounded up the steps to the simulator, and knelt by one of the open hatches. A fetid stench emanated from the crew compartment, an aroma like two men being packed into a broom closet with a space heater and then left for several days to ripen. He watched as a technician sponged small splotches of blood from Agnew's seat; the blots matched the pressure points where the pilot's shoulders and hips habitually rubbed against the hard seat back. Wolcott felt a faint tap on his shoulder; it was Ourecky.

"Sir, may I?" asked Ourecky. He looked as eager as a puppy.

"You want to squeeze in there?" replied Wolcott. "Well, Ourecky, you're nothin' if not brave. Just be careful climbing in, and don't dare bump any of their switch settings."

Ourecky started to slide into the left-hand seat that Carson had just vacated.

"I wouldn't advise that," stated the technician, shaking his head. "If you value your life, you don't ever want to be caught sitting in Major Carson's throne. Here, let me help you slide in on this side."

Assisted by the technician, Ourecky wriggled into the right-hand seat. "That's the computer interface?" he asked, pointing at the gray instrument panel.

The technician leaned into the mock-up. "That's right."

Enthralled, Ourecky continued to identify features in the cabin. "And that's the Encoder Controller? Circuit breakers? Water management? Rate gyros?" Nodding, the technician affirmed the Captain's observations.

Wolcott watched Ourecky chat excitedly with the simulator tech, comparing notes on how information was entered into the computer. On his back inside the capsule, Ourecky acted like a starving kid in a candy store. It was clearly obvious that he was familiar with the various controls and displays. Wolcott thought that was very significant, since the young captain had no previous formal training on the Gemini. The technician knew the instruments and cabin inside and out, probably better than the pilots, yet Ourecky was giving him a run for his money.

Heydrich walked up the steps and stood beside Wolcott. "So I guess we're done talking about the Block Two? Are we still stuck with the current computer?"

"'Fraid so, pard," replied Wolcott. "Sorry, Gunter. I'll stay on it, and circle back up with you after I talk to the eggheads at Cambridge."

Heydrich knelt down and eavesdropped on Ourecky chatting with his technician. He turned to Wolcott and quietly commented, "Man, someone's been hitting the books."

"And then some, pardner," said Wolcott, grinning broadly as he fanned himself with the brim of his Stetson. "I just hope we can peel him out of there before it's time for you to restart."

12

THE ELUSIVE EJECTION POD

Auxiliary Field Ten, Eglin Air Force Base
4:20 a.m., Friday, July 5, 1968

They ran ten miles every Friday. It wasn't a mind-numbing, foot-dragging jog in mass formation, but a timed individual event. Henson thrived at competition, and enjoyed the opportunity to run all out from start to finish. They were being judged according to some standard, but the instructors rarely lent the slightest clue as to what was satisfactory and what was not. In any event, he figured that his safest option was to finish ahead of everyone else; if *they* passed, there should be little doubt that he was also achieving the mysterious goal.

Most of the candidates sprinted off like crazed demons when the PT instructor blew the whistle to start the clock. Invariably, they ended up in a massive clot, elbowing for position, but gradually lost steam in the first mile. In short order, they were strung out along the route, wheezing for breath, running at a much-reduced speed. Henson maintained a steady pace throughout, and breezed past most of his cohorts even before they finished the second mile. Two wiry PJs paced him over halfway, but in time even they sputtered out. By the end of the sixth mile, they had fallen several hundred yards behind him.

The course was a sandy jeep trail that meandered through the pine forests, eventually depositing the runners back at Aux One-Oh. Running alone, Henson enjoyed the sounds of nature awakening: swaying trees rustling in the wind, tree frogs chirping in the swamps, and birds rehearsing the opening notes of their morning songs. He ran fast in the humid gloom, perhaps swifter than he had ever run on any previous Friday. A hard rain had passed through last night, and the damp sand had yet to be disturbed by dozens of churning feet.

It seemed like he had barely started when he saw the red pinpoint glow of an instructor's cigarette before him. The sun had not yet come up, and although the instructor saw him lope across the line, he apparently didn't recognize him. "Name and roster number?" asked the man.

"Henson. Roster thirty-eight." Gasping for oxygen, he bent over and grabbed his knees.

The instructor switched on a penlight, looked at a stopwatch, and then scrawled an entry on the roster. "Sixty-two minutes, ten seconds," he commented, then whistled quietly. "That's mighty good, Henson. Keep walking, or you'll cramp up."

"Thanks," he replied, still catching his breath. He relished this time on Friday morning, when he normally had at least ten minutes to himself before the others began straggling in. The brief solitude was a luxury, much worth the pain.

Arriving at the barracks, he prepared for the day. The past few weeks had taught him to be painstakingly efficient. Even though this morning's clock allowed a more leisurely pace, he refused to squander precious seconds. He showered and shaved in under four minutes, and had donned a clean uniform long before his classmates started filtering in. Combing through his short Afro with a metal pick, he paused at the bulletin board to review the training schedule.

Most of the morning would be consumed by a graded search exercise. By now, the search drills were such a standard part of their grueling routine that they executed them by second nature. For any given mission, a patrol leader was selected from the candidates, and his overall tactics score rested heavily on how well he organized and prosecuted the search. Each man was granted three shots at being a patrol leader, and he was expected to be successful—by the instructors' obscure criteria—on at least two. At this point, Henson was one for two.

On his first graded search, the helicopters had spilled them out at dusk, landing in a tiny clearing adjacent to the Yellow River, to hunt for a pilot who had ejected over the swamps. Henson quickly discovered the search area wasn't a slightly damp chunk of real estate, but a genuine quagmire, heavily inundated by recent monsoonal rains.

All night, they fumbled through dense mazes of thorny "wait-a-minute" vines, fallen trees, and submerged cypress stumps. After eight hours of thrashing through the foul-smelling mire, they found the pilot on a small hummock, desperately clinging to a cypress knee. It was a night to remember, or perhaps one to swiftly forget. Henson received a passing grade for the search, but just barely so.

His second graded search was to locate an ejection pod from an F-111 "Aardvark" fighter-bomber. The F-111, an unusual swept-wing aircraft built by General Dynamics, was unique in that the aircrew did not eject as individuals in an emergency, but the entire cockpit—the ejection pod or "crew capsule"—was blasted free of the airframe and descended under its own parachute. Additionally, unlike most two-seaters, the men sat abreast, as if they were riding in an automobile.

Ironically, most of their searches were for the pods, as if the objects habitually rained out of the sky. Compared to some things they hunted for, an ejection pod wasn't tremendously difficult to find; it was about the size and shape of a compact sports car, and weighed roughly the same, about three thousand pounds.

They had quickly pinpointed the pod in a pine forest, but the exercise held a vile twist: the pod was unoccupied. According to the scenario, the impatient aircrew had departed on foot, and the exercise wouldn't conclude until the two men were found. The search continued long into the night, until the instructors finally called it off. Henson flunked the search, but it was doubtful that any of the candidates would have been successful in that particular fiasco.

In the interim, he had since paid strict attention and was now a master of the art. He was absolutely confident that he would nail his next search, which would give him the two for three he needed to graduate, so he was well on his way to matriculating to an operational squadron.

According to the schedule, after this morning's search concluded, they were scheduled to go to the Safety Range for the remainder of the day. Henson always looked forward to their Safety Range sessions, where the candidates were exposed to a multitude of aircraft and other hardware that might conceivably fall out of the sky and necessitate a search.

While it was logical that most pilots would eject to save their skins in an emergency, the instructors emphasized that it was equally as likely that the candidates might be dispatched to locate a crashed aircraft with the aircrew still aboard. So they learned how to carefully approach each particular type of airframe to extricate its occupants.

Additionally, they learned how to "safe" different types of ejection seats and ordnance that could still be attached to a downed aircraft, and they were taught to locate and destroy the various pieces of equipment—usually classified communications, electronics and cryptographic gear—that could not be left intact if a disabled aircraft was to be abandoned in the field.

Some of the Safety Range training was particularly esoteric. Three weeks ago, the candidates had undergone extensive sessions on the U-2 and SR-71 spy planes. Only last week, they had spent three days learning the particulars of the Gemini two-man spacecraft. Henson had asked the instructor why it was necessary to be familiar with a capsule that was already obsolete, since NASA had already completed their Gemini flights and was transitioning to the Apollo program. The instructor explained the Air Force would use the Gemini as a return vehicle for the Manned Orbiting Laboratory even after NASA shifted its sights to the moon.

Henson thumbed through the rest of the schedule. The last week of Safety Range training was dedicated to "Special Topics," whatever that was. A holdover from a previous cycle claimed that guest instructors from Project Blue Book lectured them on the proper procedures to handle crashed UFOs during the Special Topics week. Additionally, according to the holdover, two days of the Special Topics training focused on "biological isolation" procedures, which included practical sessions on donning protective gear and decontamination protocols.

Even as his sweating classmates were still finishing the run, Henson wolfed down a greasy breakfast of bacon and scrambled eggs at the chow hall and then returned to the barracks to ready his gear. Precisely at seven—0700 on a military clock—the class marched out to a grassy field located about four hundred yards west of the Aux One-Oh compound.

Henson was surprised to see that the Training Flight commander, Captain Lewis, was present this morning. There was also a guest instructor, an Army sergeant from the Ranger School camp located at nearby Aux Field Six. From all appearances, the Ranger wasn't there to interact with the candidates, but rather to impart the finer points of patrolling and small unit tactics to the Air Force instructors.

The Air Force cadre seemed to be in awe of the guest and treated him with great deference. Henson had heard his name—Sergeant First Class Nestor Glades—and had overheard one instructor tell another that Glades had just returned from a classified assignment in Vietnam, where he had run high risk recon missions into Laos and Cambodia.

"Henson, you're up," announced an instructor. "Finn, you're the Assistant Patrol Leader."

Knowing that the clock was ticking, Henson immediately went to the instructor to receive the specifics of the search. He was joined by his deputy for the mission, Ulf Finn, a tall, sandy-haired man in his mid-twenties. Lacking even the slightest semblance of body fat, his sinewy body looked like it had been fashioned from intertwined ropes. With sharp Scandinavian features and blond hair shorn in a taut crew cut, he talked with a Midwestern twang. A five-year Security Police veteran, he had done two stints in Vietnam, guarding highly classified communications facilities, but had spent the past two years staring at the vast plains of Kansas as a member of security patrol protecting a Titan II missile silo.

The instructor handed Henson an index card and noted, "Six Hueys will be here in fifteen minutes. They will be on the ground for two minutes. Any questions?"

Henson shook his head, grabbed Finn by the shoulder, and jogged back to the waiting candidates. Barely pausing to catch his breath, he tersely addressed them. "Gentlemen, I'm the patrol leader for this morning. Finn is the APL. This is your warning order. We're tasked with an immediate search mission. Aircraft will be on station in one-five. Six Hueys. Squad leaders, the APL will coordinate the cross-load plan for the aircraft. Finn, chalk me for the first aircraft. Fourth squad, task organize three hasty intercept teams and cross-level their gear to the other squads. Patrol order in five minutes. Time hack is"—everyone present looked at their watches and stood ready to adjust them if necessary—"Zero Seven One Eight. Any questions?"

Henson scanned the crowd, but saw no hands. As the other candidates scattered, he sat down to refine the plan. Using a transparent coordinate scale and grease pencil to plot the pertinent coordinates on his map, he scrutinized the faint brown contour lines to glean a feel for the lay of the land.

He divided the search into two parts. The first was a straightforward matter of establishing a search grid and banging away on it until they discovered the pod. Henson's previous failure had taught him to assume, at least initially, that the aircrew would abandon the pod. To address that contingency, he applied his knowledge of what the crew had been trained to do in an "E and E"—Escape and Evasion—situation. Analyzing the terrain, he also had to intuitively factor in human nature. Evading aircrews were trained to avoid "natural lines of drift," like ridgelines and creek beds that allowed for easier walking, but the further they travelled, the more likely it was that they would become tired and sloppy. Exhausted people also inherently tend to walk downhill, almost like water flowing with gravity, and subsequently find themselves trapped in low areas where moisture and thick vegetation are more prevalent.

Henson assayed the limited resources available to him—thirty candidates—and applied them to the task at hand. To save time, he had already directed one of the squads to break into three hasty intercept teams. The hasty teams would lighten their loads by handing off their heavy gear to "mules" in the other squads. Less encumbered—carrying just weapons, ammunition, water and basic gear—they could travel fast to strategic points that Henson would designate, ideally to head off the survivors before they managed to venture too far.

Henson had another weapon in his arsenal. Three weeks ago, the candidates had learned the fundamentals of mantracking. If the aircrew had hastily departed the scene on foot, trackers would look for the subtle clues of their passage—the slightest fragments of footprints, scuff marks on the ground, disturbed earth—and trace them through the woods.

If the plan worked, they would be done well before lunchtime, and Henson would have the two successful searches that he needed to graduate. If it didn't, the candidates would likely still be searching into the wee hours of tomorrow morning, and Henson would soon depart for a less challenging assignment in the American Midwest.

Brushing a tick from his neck, Henson made the final touches to his plan. He looked at his watch: he had over a minute to spare. "Mission order in one minute!" he announced. He noticed that the tactics instructors, including Lewis and Glades, had taken a seat to listen.

Looking at his watch, he started the order immediately as the second hand finished its sweep of the minute. "Gentlemen, this is your patrol order. Hold all questions until the end. This order will be in abbreviated format. All SOPs and other standing orders remain in effect. Situation remains the same, as stated in previous orders."

Leaning back against their rucksacks, seated on the ground with their weapons across their laps, the men listened intently to Henson. The sun had not climbed very high over the trees, but the temperature was already starting to soar. Cicadas buzzed in a raucous symphony while a woodpecker chiseled intently at a pine's scaly bark. The bird's staccato rhythm was drowned out by a pair of F-4 Phantoms roaring overhead, on their way to an early morning of practice bombing runs. The jets disappeared in the distance, and their noise faded with them.

"Mission. The flight mission is to conduct an immediate search for an F-111 ejection pod…"

"*A pod?*" barked half the class in unison. "*An ejection pod!*" howled the other half. It was a running joke with the class, even tolerated by the dour instructors, some of whom actually laughed at the passing moment of levity. In their time at Eglin, the candidates had searched for and located far more ejection pods than the total number of F-111s in the Air Force's inventory.

As the candidates settled down, Henson continued. "Mission. The Flight's mission is to conduct an immediate search for an F-111 ejection pod located in the vicinity of Echo Juliet 336905, in order to rescue or recover the aircrew. Execution. Coordinating instructions. Pick up LZ is here. Primary infil LZ is 344889. Alternate infil LZ is 305924. Third squad, task organize into three hasty intercept teams. Hasty One will cover an intercept point located at 31179202. Hasty Two cover 32479273. Hasty Three 35029277. Center mass for the initial search grid is Echo Juliet 32388855, oriented north, fifty-meter spacing between men."

Pausing momentarily to draw a deep breath, Henson knelt down and used the sharp point of a pine needle to indicate the pertinent points on his map to the three squad leaders.

"First squad, be prepared to deploy a tracking team if the survivors are not at the pod. Second squad, provide a detail to destroy sensitive items on the pod. Service and support remain the same. Command and Signal. I will be at the center mass of the search grid, and will displace to the pod when it's found. Order of succession is APL, then first squad leader, second squad leader, third squad leader, and then by order of seniority. Signal. Primary FM freq is 37.25, alternate is 30.75, air freq is 40.50. Current signals plan and SOP code words remain in effect. Any questions?" Henson looked over the men and saw no indication of confusion or questions. "Good. Choppers in eight. Squad leaders, give me an up when you're ready."

The squads rallied in tight clusters to examine their maps and collaborate on their parts of the plan. Only a few minutes passed before each of the three squad leaders indicated their readiness to execute. Despite any discord that they might have in their off hours, the Flight consistently came together in the field and worked in unison. Henson respected them for that. After his conversation with Lewis, most of the signs of racial animosity had disappeared; the undertone was still there, like it always was, but he felt like he was an integral part of this group.

He folded his map and jammed it in his thigh cargo pocket. He took the bulky PRC-25 tactical radio from the radio operator who normally carried it, made sure that the primary and alternate frequencies were pre-set, then stowed it in his rucksack. It was an extra twenty pounds on his back, but carrying the cumbersome radio enabled him to monitor the situation and communicate without going through an intermediary.

Besides that, it freed up an extra man to fill out the search grid, so they could cover more ground on each successive pass.

He heard the faint but unmistakable sound of the helicopter rotors beating the thick morning air. Finn had already organized the men in a cross-load plan that dispersed leaders and other key personnel among the six aircraft, so that the loss of a single helicopter would not cause the mission to immediately grind to a screeching halt. Making their final checks, the men knelt in two lines on either side of the landing zone, facing out from where the helicopters would land.

The aircraft arrived. Their doors were painted orange, indicating that they were flown by student pilots from the Army's aviation school at nearby Fort Rucker in Alabama. Finn gave a hand signal, and the men quickly boarded the aircraft. Henson clambered into the lead chopper and knelt down just behind the communications console between the two pilots. The crew chief handed him a radio headset. Henson spread his map out on the console and quickly conferred with the fledgling aviators, verifying the primary and alternate landing zones near the search area, and what actions the aircraft should take if they took fire coming into the LZ.

As he chatted with the pilots on the intercom, Henson noticed the crew chief open a laundry bag stuffed with food; from the bag, the aviator furtively passed out candy bars to the candidates seated in the back. He leaned forward and offered a king-sized Baby Ruth. Shaking his head, Henson politely declined. The crew chief appeared confused by the apparent snub.

Tracing the aircraft's progress on his map, Henson smiled as he realized that it was a case of mistaken identity. The "orange door" student pilots normally flew in support of Ranger School students undergoing Florida Phase, the third and final segment of the grueling eight-week course. Apparently, thought Henson, this student aircrew wasn't completely sure of who they were ferrying and had just automatically assumed that they were Ranger students.

At Eglin, when the Ranger students executed their final twelve-day patrol in the swamps, their food consumption was limited to one C-ration meal per day. Since the neophyte Rangers moved almost nonstop all day long, one meager C-ration provided less than a third of the calories that their bodies were actually burning in a twenty-four hour period. Consequently, the Ranger students were always starving, and there was a longstanding tradition that the orange door crews snuck extra food and goodies to them.

Skimming the treetops, the pilots flew low and fast. The pilot gave Henson the two-minute warning. He affirmed, flashed two fingers to the other candidates, and then handed the radio headset to the crew chief. The LZ was barely large enough to accommodate two aircraft at a time, so the candidates were dropped off in three serials.

Henson was impressed with the pilot on his aircraft; the man skillfully maneuvered the helicopter into the small clearing, flaring just in time to land with a slight bump. Almost certainly the pilot would be in Vietnam in a few weeks, flying a similar Huey in combat. In seconds, the candidates tumbled out of the two aircraft and they were gone, with the next two already on final approach. Without hesitation or any instructions, the first hasty team moved out on the run, scrambling toward their intercept point.

The second serial of helicopters landed, dropping off their charges, followed rapidly by the third serial. All three hasty teams were moving, as Henson guided the remainder of the men to the starting point for the search grid.

They had been searching for several minutes when a booming clatter of gunfire erupted to the northeast, far off in the distance. Henson suspected that one of the hasty teams had encountered an "enemy patrol." He knew better than to jump on the radio and immediately demand information; the patrol was busy and would report as soon as they were able.

The radio crackled, and Henson pressed the black headset against his ear. "Sierra Six, Hasty Two is in contact, enemy patrol, five personnel, over," was the team's initial report. Henson spoke briefly, acknowledging the message.

Moments later, there was a second report. "Sierra Six, Hasty Two has engaged a five-man enemy patrol at Echo Juliet 336899. All enemy Kilo-India-Alpha. No friendly casualties. Conducting hasty search. Will Charlie Mike in two, over." The voice was calm and businesslike, declaring that the hasty team was unharmed, that they would quickly search the bodies for maps and documents before continuing with their mission.

"Hasty Two, Sierra Six, copy Charlie Mike in two, out," replied Henson, writing down the coordinates for the skirmish and quickly checking his map. Unfazed, the search line proceeded through the pine woods and oak thickets, looking for the elusive ejection pod, covering a thousand-meter swathe with each successive pass through the search area.

The radio crackled again. "Sierra Six, Hasty One is set, over." The brief message informed Henson that the first hasty team had arrived at their intercept point. "Hasty One, Sierra Six, roger set, out," replied Henson, checking his watch. *Not bad,* he thought, *They must have really been trucking to hit their mark that quickly.* Within just a few minutes, the other two hasty teams called in to report that they were also at their assigned locations.

He pulled off his bush hat and wiped sweat from his brow with his sleeve. With the backstops in place, the rest of the mission was essentially just a monotonous slog through the pines and thick undergrowth, running the grid, then re-setting it and running it again, over and over, until the pod was located. As the sun progressively crept up in the sky, with the heat becoming relentlessly oppressive, he knew it was easy to lose focus and become complacent.

After two hours of relentless searching, he heard a clicking noise to his right; the next man was passing a signal that the target was located. Henson signaled the men to halt and then went to investigate the finding. The ejection pod was located in a grassy clearing approximately the size of a baseball diamond amid short clumps of saw grass and sumac bushes. Dented and scraped, with several patches of bare metal glinting through its dull paint, the unusual object had obviously seen better days.

The first squad had already formed a security perimeter around the pod. The squad leader signaled Henson that he had already spotted two men through the Perspex windscreen and that one appeared to be conscious and communicative. In a hushed voice, Henson spoke into the radio handset, informing the far-flung hasty teams that the pod had been found. "All Hasty elements, Sierra Six, Bingo at Echo Juliet 33649206. Two pax. Over."

Henson motioned an approach team to move forward. Weapons at ready, the four men advanced on the pod from the rear on either side. Two of the men crept forward with crash axes and other tools; their job was to extricate the crew. Finding the pod intact, with no need to bash their way into the cockpit, the pair unlatched the clamshell cockpit and pried the sides open.

"We are an Air Force rescue team," Henson declared to the conscious pilot, waving an American flag in one hand while holding his GAU-5 at the ready with the other. "Follow our instructions and you will not be harmed. Put your hands out where we can see them."

Pale-faced and sweating profusely, the stand-in pilot nodded weakly, playing his role to the hilt. Leaning forward in his seat and gurgling quietly, the co-pilot was unresponsive; the cadre had apparently directed him to feign unconsciousness. Two taciturn instructors stood by quietly, scribbling down notes.

Henson directed the Flight's PJ to focus on the co-pilot. "We're here to rescue you," he said to the pilot. "But we need to authenticate you before we can move you. Are you injured?"

The pilot slowly shook his head.

"Do you have a blood chit?" Printed on cloth or waterproof paper, a blood chit had an American flag on one side and phrases in different languages on the other, requesting assistance in exchange for a reward. Most importantly, it bore a unique serial number to identify the bearer. The instructors had constantly stressed the importance of positively authenticating aircrew members. Until the survivors could be authenticated, which might take hours even if their information was transmitted immediately, they had to be handled with the utmost care and guarded as if they were enemy combatants.

Although the candidates moved slowly and deliberately, the scene was quietly abuzz with activity. Three men examined the pod to identify all the equipment that had to be salvaged or destroyed before the Flight departed the scene. The HF radio operator assembled his radio, switched it on, and made contact with their headquarters. With a competent operator, the HF—high frequency—radio could literally transmit a low-power signal around the world.

Only a few seconds passed before they received word that the aircrew was legitimate. Henson breathed a sigh of relief and then contacted the hasty teams to pass on the good news. "All Hasty elements, this is Sierra Six. Papa Charlie verified. Fall back to my position. Over." The three hasty teams responded and let Henson know that they were moving to his location.

"Outstanding job, Henson," declared an instructor. "You've just set the class record. The mission is complete, so everyone can stand down, grab some chow and relax." The instructor pointed at the PJ, who was already re-stowing his medical gear. "Baker, Captain Lewis wants to see you over at the pod. Now."

Henson was elated. The candidates clapped and cheered, and several came up to slap him on the back and offer congratulations. Now that he was two for three, he felt as if a ponderous yoke had been hoisted off his shoulders.

It had been an exceedingly long and stressful morning, and he was famished. He found a tree to lean against and flopped down to eat. Using a tiny P-38 can opener, he sliced into a can of C-ration spaghetti. As he spooned up the cold glop, he couldn't help but notice that there was a heated argument ensuing at the pod. As best as he could tell, Lewis had some issue with the medical treatment that the PJ, Steve Baker, had bestowed on the co-pilot. The discrepancy appeared to be the focus of a disagreement between Captain Lewis and the other instructors. From the looks of things, most sided with Baker and argued on his behalf.

Henson watched as Lewis berated Baker and the instructors backing him. The squabble went on for several minutes, and then Baker returned to join the other candidates. He sat down quietly, not making eye contact with Henson, and finished stowing his kit. As he pondered what might have happened, Henson watched the trucks arrive to take them to the Safety Range.

The candidates relaxed for a few minutes until a sergeant called them to assemble at the trucks. "The captain has some observations on this morning's search," he declared. "We'll load up afterwards." Henson noticed that the sergeant, who had been the primary grader for the exercise, seemed uncomfortable and made every effort not to look in Henson's direction.

Lewis leaned against a hand-carved oak walking stick as he spoke. His comments reminded Henson that this was not a training regimen for the fainthearted or shy. Criticism was harsh, immediate, and brutally public. "Henson. Shrewd plan, tactically sound. Excellent ground search, well executed. But your PJ failed to identify internal bleeding in one of your survivors, so you're coming off the field with one for two. You're an overall failure for this mission."

The harsh words stunned Henson like a mallet. With one brief sentence, Lewis had dashed his triumphant redemption into the mud. Several candidates shook their heads, and a murmur grew in the ranks. Lewis's summary judgment was blatantly unfair. Moreover, since it was his second failure, it warranted Henson's immediate dismissal from the course.

To Henson's left rear, an angry voice erupted. "That's not fair! He shouldn't be held accountable for Baker's mistake."

Recovering gradually from his shock, Henson glanced back to see his defender. It was one of the Five White Crackers he had earlier suspected of leaving the racial threats, a skinny kid with buckteeth from Phenix City, Alabama.

"And that's a *negative* spot report for you, Mister Green," answered Lewis calmly, twirling the walking stick between his palms. "Minus twenty-five points. Anyone else care to lend their opinion?" He paused, looking over the mass of nearly exhausted men, and added, "I didn't think so. Remember that as a leader, you're responsible for everything your men do or *fail* to do. Now, fall in on your gear. There's still a long day ahead."

As they loaded the trucks, Baker walked up to Henson. "Sorry, Matt," he said, heaving his rucksack onto the truck. "We're moving fast, and there's just no way to catch everything."

"Water under the bridge, baby," replied Henson. "It's done, already forgotten." He stuck out his hand, and they shook. It was unlike Baker to make even the slightest of medical errors. He had come here after four years in the pararescue field, including a combat tour in Vietnam, where he had earned a Silver Star and a Distinguished Flying Cross. He also wore an Airman's Medal on his dress blues; he had received it for selflessly jumping into the crashing seas of the Bering Straits to rescue an interceptor pilot who had had ejected from a disabled F-102.

The instructor walked up. "Henson, you won't go to the Safety Range with the class," he said quietly. "Jump in the bed of that pickup over there. We'll haul you back to Aux One-Oh."

Henson nodded glumly.

"Sergeant, is there any chance that we can appeal this thing?" asked Baker. "It was my mistake, not his. It doesn't seem right that he should be booted for my blunder."

The instructor shrugged and shook his head. "I don't think Captain Lewis is too inclined to change his mind." He turned to Henson and added, "Leave your ammo and mission gear with Finn." The sergeant glanced around to check that Lewis had departed. "Henson, you did good. Best search I've ever seen. Sincerely, I'm really sorry about this."

"Thanks, Sergeant. That means a lot," Henson said. Turning away, still quietly overwhelmed with shock that was increasingly turning to anger, he found Finn and relinquished his ammunition, radio, and other equipment. Then he climbed into the back of the pick-up truck and made himself as comfortable as he could, given the circumstances.

As he waited, he closed his eyes, fanned himself with his hat, and pondered his future. He hadn't thought too long before he heard a voice. "What's your name, boy?"

Prepared to be angry, he looked up and saw Glades, the Army instructor, standing next to the truck. In contrast to everyone else who had walked all morning through the sunbaked pine barrens, the unruffled Ranger hadn't even broken a sweat. He was almost exactly six feet tall and looked deceptively thin. His crew-cut hair was dark brown, and his eyes brown as well. Deeply tanned, his features were square-cut, almost severely so, like he had been hewn from wood, a lethal sort of cigar store Indian. "What's your name, boy?" he repeated.

As Henson realized that Glades wasn't calling him "boy" as a racial epithet, his welling rage quickly dissipated. The word just seemed disarmingly natural coming from Glades, not threatening or derogatory. Glades spoke with a strange sort of accent, like a cross between sharecropper Southern and Midwestern twang, with the slightest hint of Irish inflection, a tongue that Henson could not readily place. "My name's Henson, Sergeant," he answered, sitting up straight and donning his hat. "Airman Matthew Henson."

"You done well out there, Henson," said Glades. "Very well. I just wanted you to know that."

"Thank you, Sergeant."

Without speaking another word, Glades turned and left. Henson watched as a new leader was pulled out of the pack of candidates, and he mused whether the new guy would suffer a similar fate. On the way back to the compound, as the truck bumped along the sand trails and gravel roads, he reflected on his dismal circumstances. He had seen enough guys leave the training to understand that there was scarce chance for recourse. Once the instructors decided you were gone, you were gone and quickly so. Since he had been progressing so well in the course, he strongly suspected that his expulsion was racially motivated. He recalled the brother he had met at the slop chute. If the operational squadrons were a club reserved for whites only, perhaps they already had their token Negro to sit by the door for the quota checkers to see.

Arriving at the headquarters thirty minutes later, Henson forced himself to keep his head up and maintain his composure; regardless of what they told him, he would not lose his cool. If they sent him to guard penguins for the next three years, fine. At the end, there would be the GI Bill money that was his ticket back to school.

A sergeant ushered him into the Wing commander's office. Although Colonel Fels had spoken to them at the beginning of the cycle, the candidates had not seen him since. The office was filled with awards and paraphernalia accumulated over the course of a long and distinguished career. A large framed picture of an F-105 Thunderchief fighter-bomber adorned the paneled wall behind his desk. The remainder of the wall space was almost completely covered by squadron photographs, plaques, and awards.

"Airman Henson reporting as ordered, sir!"

"Sit down, Henson," answered Fels, returning Henson's salute.

Henson took a seat as instructed. "Sir, I would like to request an appeal for…"

"We're not here to discuss appeals," Fels said abruptly, holding up a hand. "I know that you haven't been in the Air Force very long, Henson, but I'm sure that you're aware this is not a normal Air Force organization. We have a critical mission, and we operate worldwide. We can't afford to have any personnel operating in the field unless we have *absolute* trust and confidence in them and their abilities. You understand me, son?"

Trying to restrain his anger, Henson nodded. At this point, if he was powerless to change the situation, he would've preferred that Fels shut up and send him on his way, wherever that may be.

"And that's why we put everyone through a very stringent selection process. It might not be readily apparent, but this process is carefully designed to separate the wheat from the chaff."

What is the point of this lecture? Right now, all Henson wanted to do was return to the billets so he could pack his stuff and slink away before the class returned from the Safety Range. He hadn't made any close friends since he'd been here, so there was no one to say goodbye to. If he was going to leave in disgrace, he would just as soon be gone before the rest of the candidates showed up to offer their shallow attempts at condolence.

Fels smiled. "Henson, to alleviate any confusion here, you've been selected. We've been watching you closely since you've been here, and you're just the kind of man we're looking for."

Henson was taken aback. Several seconds passed before the colonel's words sunk in and he was again able to speak. "Sir, how about the rest of the course?" he asked. "Don't I have to finish the course before I go to an operational squadron?"

Fels resolutely shook his head. "Henson, the rest of these candidates will continue through the wringer, and if they pass, they'll be assigned to the squadrons. As for you, you've been selected for a higher calling, a more crucial assignment."

"Sir, may I ask—"

"Henson, when you leave here, you'll turn in your gear, and then you'll be reassigned to the Logistics Support Office. Pending final approval of your paperwork, you'll undergo some further interviews and psychological testing, and then you'll start into your LSO training. Most of the technical training will be here at Aux One-Oh, but you'll go to some special courses elsewhere."

Logistics Support Office? Henson sighed in disgust; he wanted *nothing* to do with the LSO. Consisting primarily of washouts from the assessment course, the guys stuck in LSO looked as if they had been sequestered in hell. They plodded around the compound like sullen wretches who had been ripped from a life of constant excitement—tactical assault drills, range firing, explosives work, rappelling from helicopters—to a pitifully mundane existence of shuffling paperwork and filling in vouchers.

The LSO lepers were housed in their own billets, separate from the Training Flight and operational squadron, and they kept strictly to themselves. Henson could envision few fates worse than being assigned to the LSO. Even a five-year tour guarding sand piles in the molten heat of Saudi Arabia was preferable to life as a miserable outcast who couldn't make the cut.

"Sir, begging your pardon, I didn't volunteer to come here to work in Logistics Support. If there's any way possible, I would prefer to finish out my cycle and then go to a squadron."

"Son, maybe I'm not being clear enough. You've been selected for something *very* special. I know that duty with the squadrons looks appealing, but the fact is that I can fill those slots with virtually anyone who shows up at our doorstep, provided we give them sufficient training. I can promise you that your assignment will be considerably more rewarding and challenging than working in a squadron."

"But, sir, I…"

"Henson, I'll make you a deal," offered Fels. "If you finish your LSO training and you're not happy with your assignment, then come back to me and I will personally stick you right back into a training cycle at the same place you left, and you can go to an operational squadron if you pass the course. But trust me, Henson, once you've got a feel for what the LSO really does, you won't be back to finish the course."

13

THE WRIGHT STUFF

Atlanta Municipal Airport, Atlanta, Georgia
9:05 a.m., Monday, July 15, 1968

Seated by the window, Ourecky was engrossed in a crossword puzzle. He was mulling over a clue, chewing on the eraser end of his pencil, when a black man took the seat next to him. The handsome man appeared to be in his mid-twenties, with an extremely athletic physique. His skin was so dark that it took Ourecky aback; he could not help but stare.

"Something wrong, mister?" asked the stranger, fastening his lap belt. "Am I in the wrong seat? This ain't like the bus in Montgomery; I'm obligated to sit up here with you white folks."

Ourecky shook his head and replied, "No, nothing's wrong. Please don't take offense, but you're the blackest man I've ever seen."

The newcomer laughed; his bright white teeth were a startling contrast to his ebony flesh. "No offense taken, baby. Yeah, I guess I'm just about as dark as they come. They shipped my ancestors out of Africa to chop sugarcane on a plantation in Jamaica, and my grandparents made it to the States later on. So I guess you could you could say I'm as close to pureblooded as you can get."

"Well, I was raised in Nebraska. I never even saw a Negro until I went into the Air Force."

Latching his seat belt, the man laughed so hard that he seemed on the verge of convulsing. "Negro? *Negro?* Oh, man, you just don't hear *that* word much anymore. So you're from Nebraska? Did they somehow beam you up out of the fifties? Well, I suppose you wouldn't see too many brothers up that way, unless they were *really* lost. At least you're honest, though. And you're in the Air Force? I was in the Air Force. Are you still in?"

"I am. My permanent station is Eglin, but I'm working on a temporary duty project at Wright-Patterson. I'm Scott Ourecky, by the way."

"Matt Henson. Yeah, man, I just got out. I was also at Eglin, at least for a little while."

"So where are you headed now? Back home?"

"No, my home is New Orleans. I'm on my way to Dayton to interview for a job."

"Really? Dayton? What kind of job?" Ourecky folded the crossword puzzle, and slid it into the seat pocket in front of him.

"I'm interviewing with a company called Apex Minerals Exploration. Mostly, they scout places where mining claims have been filed. They track down equipment, trucks for rent, local people to hire, stuff like that. Scut work, really. Nothing too exciting. They've been awarded some contracts in Africa and South America, so they need new people."

"Sounds exciting to me," Ourecky said. "So, how did you find out about this company?"

"I answered a classified ad. Apex hires people who speak different languages. From what I've heard, they don't much care about the color of your skin so long as you can do the job."

"Interesting. How many languages do you speak?"

"Besides English, three: French, Portuguese, and some Spanish. English was really my second language. I grew up speaking mostly French at home and didn't start into English until I went to first grade. My mother owns a little café in the French Quarter, and I picked up Portuguese from her cook. I guess I just naturally soak up languages.

"It's really sort of funny," continued Henson. "When I first hear a new language, I don't even listen to the words. Initially, I just listen to the rhythm, and it seems like the words and context just sort of fill themselves in later."

Ourecky looked up and saw Bea. She was slowly making her way through the plane, verifying that the passengers were buckled in. She stopped, smiled at Ourecky, and held out a magazine. "Here's that *Life* magazine you wanted, Scott. I saved it for you." The glossy cover showed President Johnson kissing his new granddaughter.

"Whew. *Pretty* girl," commented Henson, watching attentively as Bea sashayed up the aisle toward the first class compartment. "She's obviously got a thing for you, man."

Frowning, Ourecky dismissively shook his head. "No. I just fly this route every month, so we've gotten acquainted a little bit. I'm sure she has a boyfriend somewhere."

Henson chuckled. "You believe what you want to believe, baby, but you learn things growing up among voodoo people in New Orleans. I can *assure* you that girl is sweet on you."

"Uh, I really don't think so." Ourecky opened the magazine in his lap and thumbed through the pages. Three rows ahead of them, a young couple tried to soothe their squawling baby.

"So you like *Life*?"

"Yeah. It's okay, I suppose. Mostly, I like their coverage of the astronauts. Lately, though, it seems like all their stories have been about the Middle East."

"Arabs and Israelis? Sounds to me like the Arabs bit off more than they could chew in that situation. You would think all those people would just learn to live in peace, since they're jammed in there together so close."

"Just like we get along here in this country?" asked Ourecky, scanning an article about racial unrest and riots in Atlanta, Boston, Buffalo, Cincinnati, and Tampa.

"Touché," answered Henson, looking over his shoulder. "You got me on that one, babe."

Aerospace Support Project
Wright-Patterson Air Force Base, Ohio
10:00 a.m., Monday, July 15, 1968

Unless there were more pressing matters to attend to, Tew's key staff convened on Monday morning to compare notes. Out of Tew's earshot, his staff jokingly referred to the weekly gathering as the "Joe Friday" briefing. Tew wasn't a fan of small talk or conjecture; his dry style was more in line with Jack Webb's painfully dour "just the facts" detective on *Dragnet*.

"Gentlemen, let me call this meeting to order," said Tew. "First order of business is to introduce Major Ed Russo, our new liaison officer from the MOL office out in California. Russo, welcome aboard. I know a lot of our business is going to be new to you, so if you have any questions, don't be bashful about throwing up a hand."

"Thank you, General." With broad shoulders and a medium build, Russo was a couple of inches shy of six feet. His black hair, flecked with premature gray, was cut in a short flattop. He was dark, almost swarthy, with aquiline nose and dark eyes reminiscent of a Mediterranean heritage. Unlike the others at the table, Russo was in uniform. His chest was adorned with three rows of ribbons, most earned during two tours in Vietnam.

"By the way, gentlemen, congratulations are in order," stated Tew. "Major Russo was tentatively selected for the next group of MOL astronauts. That's close hold information; it won't be formally announced until next year. Until then, Major Russo will work here with us."

"So, pard, I guess that makes you our resident Can Man," said Wolcott.

"Can Man?" replied Russo. "What's that, General?"

Wolcott chuckled. "Can Men is our little term of endearment for the MOL crews. After you've hung your hat here a while, pard, you'll probably see it's a completely different culture. To accomplish their mission, our pilots are obligated to *fly* their spacecraft. On the other hand, to do their job, the MOL crews have to *sit* in a Can. "

"I beg to differ, General," said Russo. "The MOL…"

Tew curtly interjected, "Okay, let's go around the table. Thompson? Personnel?"

Lieutenant Colonel Byron Thompson was Blue Gemini's personnel officer. Short, thin, and balding, he was the only member of Tew's staff who had not come up through the flying ranks. As a non-pilot, he had a tendency to overreach in his efforts to fit in with the rest of the staff. He inserted a cardboard-mounted chart into a large opaque projector. "General, these are our current personnel numbers, tabulated as of Monday last week. As you can see, our largest personnel investment is in the 116th Aerospace Operations Support Wing at Eglin. They have a total of 433 personnel assigned at Eglin and other locations."

Thompson pulled out the chart and replaced it with another. He turned the projector back on. It buzzed loudly before its bulb blew out with a loud pop. The staff officers groaned. "So much for that, General," he said. "May I continue, or do we wait for another bulb?"

"Continue," said Tew. "Otherwise we could be here all day."

"Yes, sir. The 116th currently has thirty-six personnel on temporary duty in Vietnam. The Logistics Support Office also has sixteen personnel forward-deployed in seven countries."

Thompson continued. "Sir, as for the rest of your numbers, we have ninety-seven personnel currently working here at Wright-Patterson. There are thirty-two men presently committed to the PDF construction project. Additionally, we are due to receive twenty-eight launch support and flight control personnel from the 6555th Aerospace Test Wing. General, your total headcount—excluding personnel on loan from other organizations—is 590. Any questions, sir?"

"No. Good job. Russo, did you get all that?"

"I did, sir, but what's the PDF project he mentioned?"

"Pacific Departure Facility," answered Thompson. "It's our launch site on Johnston Island, near Hawaii. It's currently under construction."

"Anything else, Thompson?" asked Tew.

"One item, sir." Thompson opened a folder, removed a form, and handed it to Tew. "Major Agnew submitted *another* transfer request."

"Really?" asked Wolcott, picking up the request. "Shucks, I just talked to him about that a couple of months ago." He crumpled the form into a ball and added, "You can tell Mr. Agnew that we contemplated his request, but it ain't happenin'."

"Intelligence?" asked Tew.

The intelligence officer, Ted Seibert, was a studious lieutenant colonel who dressed and acted like an Ivy League college professor. Handsome, with a thick mane of blond hair, he wore a suede-elbowed tweed jacket over a blue oxford shirt. Cupping a briarwood pipe in his hand, he cleared his throat and said, "Sir, things are not looking very good in Southeast Asia. All indications are that Westmoreland will request another 200,000 troops. On another note, the Soviets are very focused on the Chinese right now, since the Chinese detonated that hydrogen bomb last month. There are still fears of a nuclear confrontation between those two."

"And not a minute too soon," noted Wolcott, slapping the table. "Good riddance to both."

Seibert drew on his pipe and then continued. "Concerning the Soviets, there's also a new wrinkle, sir. We're hearing murmurs that the Soviets are aggressively developing a new anti-satellite system, and they intend to start live testing as early as next year. If our information is accurate, theirs is a co-orbital system, not unlike our old SAINT satellite interceptor concept."

Tew nodded, made a note, and asked, "Anything new concerning orbital bombardment systems? Any developments with their UR-500 heavy lift booster?"

"No news on either, sir. They're obviously holding those cards very close to the vest."

Frowning, Tew nodded. "Anything else?"

"That's all. That concludes my briefing, General. Nothing else significant to report."

"Thanks," said Tew. "By the way, I want your Special Security operatives to keep tabs on someone for me. One of our temporary workers." Tew wrote a name on a slip of paper, and then held it out for Wolcott to read.

Wolcott chuckled. "Jimmy Hara's boys will have an easy time with that one," he observed. "As far as I know, he ain't even set foot off the base."

Tew handed the note to Seibert. He read it and said, "General, we'll cover this for you."

"Thanks. Virgil? Operations and Training?"

Wolcott parked his cigarette in the ashtray and then leaned back in his chair. "I want to remind all present that our first unmanned mission is still scheduled to launch in February of next year and our first manned shot is still in June. Unless there's some kind of miracle and the PDF is ready, pardner, we'll fire the first shot from the Slick Four-W pad at Vandenberg."

"The PDF *will* be ready, General," insisted the logistics officer, Grady Rhodes, a portly Colonel. Rhodes was obviously peeved at Wolcott's persistent implications that the new launch facility would not be finished on schedule.

"If you insist, pard," commented Wolcott. "Beyond that, our simulator facility is still runnin' full bore. Crew Two is in the Box today. Afterwards, they head to Johnsville for reentry runs on the Big Wheel centrifuge. Crew One goes into the Box tomorrow for a twenty-four hour full-up. Crew Three is in Connecticut, at Hamilton Standard, getting measured for their new suits. One item for your approval, Mark: we want to move that old Dyna-Soar mock-up out of Gunter's shop to make room for the new Paraglider Landing Simulator."

"Approved," said Tew. "I don't understand why we were still holding on to that damned old thing, anyway. Logistics?"

Frowning at Wolcott, the corpulent Rhodes stood up. He tried to fasten the buttons of his snugly fitting blue sport coat, but swiftly abandoned the futile attempt. He cleared his throat and spoke. "General, first up, I have a progress report on the Critical Target Upgrade program."

"Go ahead."

"Sir, we've successfully de-fueled and removed the first two Titan II's from their silos, and they're en route by rail to the contractor for the Critical Target Upgrade modifications."

"General, I'm not familiar with this upgrade he's referring to," said Russo.

"Colonel Rhodes, would you mind giving Russo some background on this program?"

"As you wish, sir," said Rhodes, turning to face Russo. "Major, the Air Force has about sixty Titan II's buried in silos in California, Arizona, Kansas, and Arkansas. As the new Minutemen missiles are fielded, the Titan II's will be kept in their silos to provide a redundant nuclear strike capability. They will be phased out, starting in 1971. I'm sure that you're aware the Titan II has proven to be a reliable launch vehicle for orbital payloads. So, our plan is to quietly pull some of them out of their silos early and transition them from ICBMs to launch vehicles."

"Certainly, we want our Titan IIs to meet the same standards as NASA's man-rated Titan IIs, so we established a program to designate certain Titan II ICBMs for priority targets. We used this as justification to establish the Critical Target Upgrade program to overhaul and retrofit the designated launch vehicles to higher standards, which just happen to closely mirror NASA's standards for man-rating the Titan II. After they're modified, the missiles are returned to their silos. Theoretically, of course."

"So this is essentially just a big shell game?" asked Russo, raising his eyebrows.

"Shell game? Essentially, yes," answered Rhodes smugly. "Sir, with your permission, I'll address General Wolcott's concerns about the PDF work."

"Go ahead," said Tew.

"We've run into some significant setbacks, but none that have completely shut down the construction. The main problem has been with concrete curing on some of the major structures. Our construction specs anticipated a temperate climate, not the tropical environment that we're working in. To make a long story short, we demolished the structures that weren't cured to spec and we'll start pouring the replacement structures next week, provided all our materials are shipped on schedule. The only loose end is disposing of the scrap concrete."

"How much scrap concrete are we talking about?" asked Tew.

"About two hundred cubic yards, General."

"And what does that weigh, pard?" asked Wolcott.

"Approximately three hundred tons," replied the chagrined colonel, wiping his brow and sipping from a glass of water.

"*Really?*" asked Wolcott. "That much? Speakin' of shell games, Grady, where exactly will all this busted up concrete go? Are you plannin' to shoot it into orbit?"

"Right now, we're planning to dump it offshore to build a breakwater and artificial reef."

"*Whew.* That should make an awfully fancy reef," noted Wolcott. "I guess I had better tote some fishing tackle with me when and if we finally start launchin' from there."

"Anyway, as I said, we'll begin pouring the new structures starting next week," said Rhodes, ignoring Wolcott. "Besides that, the Navy is assisting us with dredging the ship channel necessary to bring in the LST transports."

Tew turned to a Navy liaison officer and asked, "Is Admiral Tarbox going to abide by his commitment to support the PDF?"

Nodding, the Navy officer answered, "We have two LSTs undergoing modifications at Norfolk. Those are the only vessels specifically committed to Blue Gemini. All other resources will be shifted as necessary to support each launch."

"Good," replied Tew. "Anything else concerning logistics, Rhodes?"

Rhodes shook his head. "Nothing significant, sir."

"Fine," said Tew. "Unless anyone has something else, let's return to work."

James M. Cox Municipal Airport, Dayton, Ohio
2:10 p.m., Monday, July 15, 1968

Pushed by a favorable tail wind, they had arrived in Dayton a full twenty minutes ahead of schedule. Ourecky chatted with Henson as they waited in the queue to disembark. He was fascinated by Henson's stories of growing up in New Orleans and had made up his mind to visit the Louisiana city when and if he had some free time.

They were almost off the plane when Bea met him at the door. "Scott, if you're not in a huge rush, could you sit down for a moment?" she asked, gesturing towards a first class seat. "There was something I wanted to ask you."

Perplexed, Ourecky nodded.

Henson winked and said, "See, man? I told you so. Hey, I enjoyed chatting with you."

"You, too, Matt. I hope your interview goes well. Maybe I'll see you around."

"Maybe," replied Henson, stepping out into the sunlight.

Ourecky waited patiently. Finally, Bea joined him, sitting in the adjacent seat. Sighing, she slipped her shoes off and wriggled her toes. "Oh, these new pumps are just killing me," she declared. "So, Scott Ourecky, we've known each other for three months, but you've never once asked me out. Is there a reason for that? You don't like me?"

Ourecky was dumbfounded. *Maybe that guy Henson was actually right.* "Well, I, uh, just assumed that you had a boyfriend or fiancée, or that you dated one of the pilots."

"I don't date pilots," she answered emphatically, shaking her head. "*Ever.*"

"Well, I've been really busy," he stammered. "I just…"

"Excuses, excuses. Here, sweetie, I'll make this easy for you." She handed Ourecky a folded slip of paper. "That's my number. You'll still be in town Friday night, right?"

Ourecky nodded.

"Good. Ring me on Thursday night, right after *Bewitched*, and we'll make plans for Friday."

"*Bewitched?*"

"The television show, silly. The one where that pretty witch is married to the cute advertising guy. It's on from eight to eight-thirty. Just call me after that, please."

Ourecky made a note on the slip of paper, and said, "But I don't have a car, Bea."

"Well, I do. I'll drive, unless you're offended to be driven around by a woman."

Ourecky shook his head. He gathered his things and stood up.

She smiled. "Don't forget to call."

Simulator Facility, Aerospace Support Project
3:12 p.m., Wednesday, July 17, 1968

Gunter Heydrich took a swig from a bottle of Pepto-Bismol. *Just a few more hours to go, he thought, and then a night off and then back into the barrel again.* To Heydrich, this was beginning to feel considerably less like a job and much more like an endurance contest. He was tired, and his men were tired. At least things were going smoothly today; Carson and Agnew, the crew locked in the Box, were relentlessly closing on their intercept target.

Suddenly there was a commotion. A blistering torrent of profanity blasted from the intercom, except this time it was Agnew's voice and not Carson's. An observer standing next to the simulator yelled down, "Gunter, we have a problem in the Box! A *big* problem!"

3:15 p.m.

Wolcott was immersed in paperwork when there was a knock at the door. A captain popped in and calmly announced, "General Wolcott, the Simulator Facility is reporting an emergency, sir."

Emergency? thought Wolcott. *An emergency in the Simulator Facility?* He immediately recalled last year's tragic Apollo 1 fire, when Gus Grissom and his two crewmates were incinerated during a pre-flight test. His thoughts were interrupted by the wail of a siren. He sprang out of his chair and headed toward the simulator hangar.

With red lights flashing, the ambulance was just departing as Wolcott arrived. Gunter Heydrich, nervously smoking a cigarette, met him outside. He was unshaven, and his white shirt was rumpled and dotted with coffee stains. His black hair was awry, and he looked distraught.

"What happened?" demanded Wolcott, almost out of breath. "Who went to the hospital?"

"You'll just have to see this one for yourself, Virgil," replied Heydrich, shaking his head. "I *warned* you. You can only stew people in a pressure cooker for so long before the stress finally takes over." The two men walked into the hangar, breezing right past the well-armed guard without showing any identification.

Carson was seated on the stairs leading to the simulator. His right eye was purple and the knuckles on his right hand were scraped and bloody, clearly evidence of some sort of a scuffle. Several technicians examined the interior of the simulator. The remainder of the simulator faculty sat idly at their consoles. One man had his head down on his desk, snoring loudly; he wasn't loafing, but apparently had succumbed to sheer exhaustion.

Rushing toward Carson, Wolcott erupted in anger. "Major, can you possibly fathom how inappropriate your actions were? Officers do not come to blows, regardless of the circumstances. Trust me, pard, I fully intend to file disciplinary paperwork on you after I collect some witness statements, and I can assure you that some negative action will be forthcoming. Do you understand me, Mister Carson? Am I making myself clear?"

"Virgil, I didn't start the fight," avowed Carson quietly, furtively rubbing his discolored eye. "I just defended myself." The collar of his Nomex flight suit was torn, and he smelled like he was badly in need of a long hot shower.

Heydrich corroborated Carson's story. "He's right, Virg. They were twenty hours into a twenty-four hour full-up. Agnew just detonated without warning. My platform guy saw it all through his observation window. One minute, Agnew is planted in his seat working calculations, and the next moment he's piling over onto Carson's side with his fists flying. Carson just defended himself. He didn't do anything I wouldn't have done in the same circumstances."

Wolcott turned back to Carson. "Pardner, go pack some ice on that eye before it swells up. Your duty day ain't over yet, not by a long shot." He turned back to Heydrich and asked, "Gunter, how long before you can have the Box back up and operational?"

"Maybe an hour," replied Heydrich, watching Carson walk away. "My guys need to slide in there and clean it up, plus there are some bent switches and broken bulbs that have to be replaced. Just for future reference, a spacecraft cabin is not an appropriate venue for a fistfight."

"Well, pard, we can thank our lucky stars that it happened in the Box, and not upstairs." Wolcott bounded up the four steps and looked inside the mock spacecraft. Simulator time was extremely valuable, and he couldn't bear to see it wasted. "How bad is Agnew? Can we put him back up on the bronco? The faster he saddles up, the better."

Heydrich shook his head. "Virg, that isn't happening. Agnew's beat up pretty bad."

Wolcott frowned. He knew that Carson had been a varsity boxer at West Point, and that he had also been a Golden Gloves champ before he went to the Academy. Agnew obviously wasn't cognizant of that; otherwise he wouldn't have been in a hurry to pick a fight.

"Carson apparently packs a mean punch, even in close quarters," said Heydrich. "To his credit, he held back. He could have really pulverized Agnew if he wanted to. Anyway, Agnew made it abundantly clear that he would resign his commission before he climbed back in the Box with Carson or anyone else."

Cringing, Wolcott closed his eyes. Losing a pilot this way was so incomprehensible that they had never considered the potential consequences. It could jeopardize the entire Project. But Wolcott was a practical man, and he had immediate issues to contend with. "Gunter, I'll deal with Agnew later. We still have Carson. Can we load one of your techs in there with him?"

Heydrich shook his head and took a long pull on his cigarette. His hands were trembling. Exhaling a pall of smoke, he replied, "*Ja*, Virg, I can stick one of my guys in the Box to run basic procedures with Carson, but that would be futile."

"Why?"

"Carson has the basics down cold," answered Heydrich. "And my guys aren't up to speed on the intercept procedures, so they wouldn't be able to do the calculations in a timely manner. If Carson was frustrated with Agnew, you could just imagine what he's going to be like with someone who's learning all this on the fly. I strongly recommend that we keep my guys out of the Box. I can't afford to lose anyone."

"This ain't good, pard," observed Wolcott. "It ain't good at all."

Heydrich continued. "Virgil, I think you need to just accept that you're going to lose a day of simulator time while we wait for the next crew to come in. Why don't we call it a day, and just pack it in? Everyone could use the break."

Wolcott scratched his head and asked, "If they were twenty hours in, how close were they to completin' the intercept? Anywhere near close to puttin' the horse in the barn?"

Nodding, Heydrich used a finger to push his heavy glasses back on his nose. "They were probably two hours out from final closure when Agnew snapped. Unless they made some really ugly maneuvering errors in the last phases, they would have closed the deal. We had already thrown them an intermittent power failure on the new radar, but they resolved that in no time. I was really hoping that this run would boost Agnew's confidence."

Wolcott watched a technician use a screwdriver to remove a damaged toggle switch from the left side instrument panel. The technician leaned out of the hatch and held out the bent switch for their perusal; there was blood and a small fleck of skin on the steel post. "So once they rendezvoused, Gunter, what was next in this profile?"

Heydrich referred to his notes. "The usual. Nothing exciting. They do a fly-around inspection of the target to make sure that it was safe to execute a close approach. We're still working on the close approach procedures, so normally we jump the scenario forward at that point."

"They go through a power-down sequence and then park the spacecraft in loiter mode. We generally leave them in loiter mode for about twenty minutes to let them grab a little shut-eye. Then we wind the clock forward before they do a power-up, and then reentry. We're not doing the landing yet, because the paraglider simulator isn't ready yet."

Wolcott stuck a finger in his mouth, extracted a worn lump of chewing tobacco, and flicked the brown wad into a trashcan. Pulling an envelope of Red Man from his pocket, he stuffed a replacement into his mouth. "Okay, Gunter. Let's assume the worst case. At least for today, Carson is the only hand left in the bunkhouse. What can we do with just him by his lonesome?"

"Not much," replied Heydrich. "We can do an immediate abort-to-reentry scenario, with the assumption that the co-pilot is disabled. Unfortunately, Carson's been through that drill so many times that he could probably fly it in his sleep. Virg, I know that you don't want to waste any simulator time, but sticking him back in there by himself would just be negative training."

A worker walked up and announced that the simulator would be ready to go again in thirty minutes. As the technician recounted the damages to Heydrich, Wolcott walked to the second row of desks and made a phone call.

Thirty minutes later, Ourecky walked into the hangar. Wolcott instructed one of the technicians to fit him with a communications headset. Then he waved Carson over.

"Sir?" asked Carson. He had gone to the locker room to change into a clean flight suit but emerged still smelling fairly rank. He was obviously sore and very tired. He held a cloth-wrapped icepack against his injured eye.

Wolcott pointed at Ourecky. "That's your simulator buddy for the rest of the day, Carson. You had better treat him appropriately, pard."

Obviously rankled at Wolcott's decision, Carson vigorously shook his head. "Virgil, I have to lodge a protest. I don't—"

Wolcott cut him off. "Pardner, you *will* finish the simulation with Ourecky, or you will suddenly discover yourself permanently on the low end of the pecking order for any potential mission assignments. You understand me, bub? Savvy?"

It was a bluff, and a shaky one at best; Wolcott knew that if they launched in accordance with the current schedule, he had no option but to slot Carson on the first mission since he was the only pilot even remotely ready to fly into space. Of course, although Carson was a shoo-in, that plan hinged on whether they could find someone sufficiently compatible to fly with him. Ideally, it should be someone who could make the trip without being sedated or wearing boxing gloves.

Placing his hand on Carson's shoulder, Wolcott gently nudged him toward the simulator, where the two met Heydrich and Ourecky at the base of the stairs leading up to the mock-up.

Wolcott leaned toward Heydrich and said something quietly. Heydrich nodded his head and then called for one of the technicians. "Chris, help the captain into his seat and answer any questions he might have on the controls or where we are with this scenario."

The technician nodded. Ourecky was obviously thrilled with the opportunity to go through an actual run on the Box. When he was out of earshot, Wolcott spoke quietly to Heydrich and Carson. "No slips, gentlemen. Gunter, be patient with our young captain, and adjust the pace as necessary. Same for you, Carson. He might not be up to full speed on the instruments, but he knows the intercept calcs inside and out, so let's take full advantage of that. Are we all ridin' in the same direction?"

"We are," answered Heydrich. Rubbing his uninjured eye, Carson nodded in affirmation.

"Splendid. Gunter, I'm going to mosey on back to my office to wrangle some paperwork. Do me a favor, pard, and call me before they start into their reentry sequence. If Mark doesn't have me tied up, I'll come back over here to watch."

Carson climbed the stairs and slowly edged into the cockpit. He nodded his head, and the technicians closed the hatches. He slipped on his headset and adjusted the microphone.

"Gunter, switching off VOX," announced Carson brusquely. He threw a switch that cut off the voice-activated intercom circuit to the simulator controllers. "Don't get comfortable over there, Captain, because you aren't going to be sitting in here for very long. Once they figure out that you can't keep up, they'll shut down this charade. You understand me?"

"Understood, sir," replied Ourecky.

"Good. When you're ready to quit, just say the word. But let me warn you: If you feel a sudden urge to freak out, you had better restrain yourself and stay on your side of the Box. If you pull an Agnew on me, I'll send you to the hospital. Understand, Captain?"

"Uh, I understand, sir."

"Good." Carson reached out and switched the intercom circuit back to the voice-activated mode. He spoke into the microphone. "Toggling back to VOX. We are ready to restart in here. Call the clock, please."

Heydrich's tired voice came over the intercom circuit. "The clock was stopped at Ground Elapsed Time 20:12:25."

"I copy clock stopped at GET 20:12:25," answered Carson. "Ready to resume."

"Clock will resume at GET 20:12:25, at my count," declared Heydrich. "Five…four…three…two…one…Mark. Clock started. Good luck, gentlemen."

8:45 p.m.

Wolcott returned and took a seat next to Heydrich. "How goes it?" he asked. "Are they still in orbit, Gunter? No bloodshed?"

Heydrich strained to hear a piece of the intercom dialogue and then answered, "Virg, they're just a few minutes from lighting their retros. This whole shebang will be wrapped up in less than thirty minutes."

Wolcott was astounded. "Ourecky made it all the way through to reentry?"

Heydrich nodded. "He did. I don't know where you found that kid, but he knows the procedures book inside and out, at least as good as Agnew. It's kind of scary. He lagged some during the intercept and power-down, but while Carson was snoozing during the loiter phase, he studied the checklist. He seemed to have a pretty good grasp of it during the power-up."

Wolcott scratched his chin. He picked up a clipboard and reviewed the reentry procedures checklist. "Pard, do you s'pose Ourecky could fill in until we find someone to replace Agnew?"

"Maybe," replied Heydrich. "Virgil, if you gave me sufficient time, I could train anyone to fly the Box when things go right. The real test starts when things go wrong. Ourecky *almost* kept up today. When I say he *almost* kept up, you need to remember that we didn't throw any curveballs at him. No glitches. No malfunctions."

"Duly noted, pard." Wolcott sat down and turned up the volume on the intercom speaker.

Inside the simulator, Carson announced, "Retro Attitude indicator is amber." He pointed at a series of lighted switches running along the left side of the center instrument panel and added, "Captain, unless I dictate otherwise, keep your eyes glued on these sequencing lights."

"I confirm Retro Attitude," noted Ourecky. Sweat beaded on his face. A small fan did little to dissipate the heat. The cabin was considerably more cramped than he had previously imagined. Shifting slightly in his seat to relieve some pressure spots, he found it difficult to move without inadvertently bumping into protruding switches and circuit breakers. He and Carson were lying virtually shoulder to shoulder, separated by mere inches. With the hatches closed and dogged down, the cabin fit like an uncomfortably tight glove. It was definitely not a working environment conducive to anyone even mildly claustrophobic. To make matters even worse, the cockpit's cross-section was funnel-shaped, so there was scarcely any room for their feet in the narrow footwells at the truncated lower end of the cone.

As he tried to make himself comfortable, Ourecky recalled a story he had heard about the Gemini's development. After he had made his suborbital spaceflight aboard the Mercury "Liberty Bell 7" space capsule, NASA astronaut Virgil "Gus" Grissom had become intimately involved in the design of the Gemini spacecraft. Much of the cabin's interior layout, to include positioning of critical controls and instruments, were the result of his personal input to McDonnell Douglas engineers, even as the two-man spacecraft was still on the drawing boards.

While Grissom's contributions were obviously instrumental in creating an agile and versatile spacecraft that flew almost like a fighter plane, there was also a negative aspect of his legacy: the engineers built a spacecraft perfectly suited to accommodate Gus, who at five feet-six inches was the shortest of the original seven astronauts. The other spacemen, not quite as diminutive as Grissom, referred to the fledging spacecraft with a humorous but fairly accurate moniker: "The Gusmobile." After realizing that some of the taller astronauts—like five-eleven Ed White, who would later become America's first spacewalker—could not possibly fit into the small cockpit, the engineers were forced to alter the specifications to allow the larger men to squeeze—barely—inside. Ironically, Ed White later burned to death alongside Gus Grissom aboard Apollo One, in preparation for the maiden flight of the new three-man spacecraft.

"O2 High Rate switch is amber," stated Carson. He pushed that switch and it changed to green. "Battery Power telelight is amber. Switching off the adapter power supply." He waited a moment and then said, "Captain, switch on the main batteries. *Now.*"

"Huh? What? Main batteries?" asked Ourecky, scanning the instrument panel immediately to his front. His head reeled, trying to remember the locations of the battery switches. He thought he had a good grasp of the reentry checklist, but he was struggling hard to stay abreast.

"The main batteries," snapped Carson, reaching over Ourecky and pointing at a small panel. "There. Look to your *right*, on the wall. Those four switches right there. Switch them on." Ourecky did as he was instructed and toggled the switches. "Main battery switches all to On," he announced.

"Now, try your best to keep up, Captain. Battery Power light is green. RCS light is amber." Carson reached out and pushed in the switch. "Squibs fired. RCS is activated. RCS is green."

"RCS is *green*," confirmed Ourecky.

"SEP OAMS LINES light is amber." Carson yawned audibly, flexed his fingers, and then pushed in the switch. "OAMS lines are separated. SEP OAMS LINES light is now green."

"SEP OAMS is green," stated Ourecky, verifying that the light had indeed turned green.

Carson continued with the checklist. "Separate electrical light is amber. Pyrotechnics activated. Separate electrical light is green."

"SEP ELEC is showing green," said Ourecky.

"SEP ADAPT light is amber. Pyro activated. Sensors confirm adapter separation. SEP ADAPT light is now green."

"I confirm SEP ADAPT."

"ARM AUTO RETRO is now amber."

Ourecky wiped his face with his sleeve. "Confirm ARM AUTO RETRO is amber."

"Watch your hands. Watch your hands. Don't flail around so much. Keep your hands away from the instruments," snapped Carson. "Counting down to retrofire. 4…3…2…1…Retrofire."

Ourecky heard a loud thump behind him, followed by a dull roaring sound, which was followed by three more thumps and an even louder racket.

"Four retros," said Carson, counting the rockets as they fired in turn. "All four retros firing."

"I verify that all four retros firing."

"IVI 328 aft, 102 right, four up," noted Carson. "Attitude looks good."

Within a few minutes, the simulated mission was complete. Exhilarated, Ourecky all but bounded out of his seat. After Carson gingerly slid out of the mock-up, two simulation technicians climbed in. Armed with diagrams of the control panels, they painstakingly reset the instruments to the way they would appear when the spacecraft was on the launch pad.

Two folding chairs waited at the base of the stairs. Carson stiffly sat down as a technician offered him some crushed ice wrapped in a white terry washcloth. Wincing, he pressed the ice against his bruised eye. A medical technician assisted him in unzipping his flight suit and pulling it down to his waist. Carson was an impressive physical specimen; his body had changed little since his days as an athlete at West Point. The technician took his pulse, then wrapped a blood pressure cuff around his large bicep. Carson's eyes were red from strain and sleep deprivation.

Ourecky took out a notepad and jotted down some observations. "Hey, Major, can I ask you something? It's kind of personal."

"*What?*" asked Carson impatiently, examining his Omega wristwatch. He angrily shook his head; the face of the expensive timepiece had been scratched during the earlier altercation.

"Well, uh, I kind of have a date for Friday night, but I'm not from around here, so I have no idea where to take her. Can you offer any suggestions?"

"You're kidding me, aren't you? A date? *You* have a date for Friday night? My God, don't we live in a world just filled with irony." Massaging his swollen eye, Carson yawned. Smiling broadly, he said, "Well, Captain, here's what I would do. Friday night? Probably your best bet is to take her out for dinner at a nice place and then maybe go to a nightclub afterwards."

"I haven't been off base yet, so I don't know any clubs."

"Really? There's a place called the Falcon Club, just outside the main gate. It's where me and the guys hang out when we're in town. Trust me, if she's from around here, she knows where the Falcon Club is."

"Thanks, sir," replied Ourecky. "Uh, I think I'll go with your suggestions."

Carson grinned, turned to the medical technician and said, "Are you done with me, sawbones? If you don't mind, I would really like to head back to my billets and hit the sack."

14

THE FALCON CLUB

Aerospace Support Project
9:35 a.m., Thursday, July 18, 1968

Ourecky was hunched over his makeshift desk, meticulously penciling an arcane equation onto graph paper, when he heard footsteps in the hall. He glanced up to see Wolcott entering his miniscule office.

"Pardner, I ain't had a chance to chat with you," drawled Wolcott. "But I really appreciate you fillin' in yesterday. You did a bang-up job in the Box, and I ain't goin' to forget that anytime too soon."

"Uh, my pleasure, General. I was honored that you would even ask me."

Fanning himself with his hat, Wolcott smiled. "Well, hombre, I just wanted you to know that we might be a little shorthanded for a while to come, and we may need you to help out with the simulator more often. Any problem with that?"

Ourecky grinned. "None whatsoever, sir. Working in there gave me a lot of ideas on how to improve the worksheets to make the whole process flow even smoother. I think the more time I can spend in the simulator, the better I'll understand the details of the intercept missions."

Wolcott frowned, and then said, "Pardner, you should know that if you go back into the Box, you'll be climbin' back in there with Carson again. Are you sure you two can geehaw?"

Setting aside his slide rule and pencil, Ourecky nodded. "Yes, sir. I'm confident I can work with Major Carson, sir."

"Good. *Very* good. Hey, we're due to receive our Paraglider Landing Simulator next month. It's brand spankin' new. I sure would appreciate your help in workin' the bugs out of it."

"Sounds like fun, sir."

"Look, I have another favor to ask of you, Ourecky." Wolcott plopped a thick stack of paperwork on the drafting table. "I want you to haul this stuff over to Colonel Walters at the base hospital. He's a flight surgeon who works with us. I want you to schedule a flight physical with him."

"Flight physical, sir? What for?"

"Two reasons, pard. First, the simulator rules require that you have a full-blown flight physical before you can participate in extended mission scenarios. Second, I don't want your head to swell bigger than your hat, but it's becoming pretty danged apparent how important you are to this Project. That means your time is precious to us. As it is, we're burnin' up a lot of that time when you fly commercial. Our crews fly T-38s to stay current, so I reckon we could whack two birds with one stone. One of the guys can swoop down to Eglin, drop you in the back seat, and you can be here in a flash instead of sittin' on an airliner all day."

Leafing through the medical forms, Ourecky's first thought was of Bea, and that he would no longer see her on the Monday mornings that he flew from Atlanta to Dayton. "You don't want me to fly commercial anymore, sir?"

Wolcott shook his head. "It's a waste of time and resources, pard. Call me selfish, but so long as you're punched in on our clock, I want you up here working. There are a few other things that you'll need to accomplish before we can strap you into a back seat, though. We'll make an appointment at the altitude chamber to have you checked out, and you'll go through egress trainin' over at the base parachute loft."

"Egress training, sir?"

"The survival gear techs will run you through a day's worth of ejection seat trainin' and show you how to work a parachute if the need should ever arise." Wolcott looked around the tiny room where Ourecky worked; he always seemed claustrophobic when he came in here.

"I'll do my best, sir."

"One more thing, hoss, and this is kind of a sensitive topic. Doc Walters will subject you to the same work-up as our flight crews. It's a lot more extensive than a normal flight physical, but it's the only type of physical that's authorized in our budget. I would greatly appreciate it if you didn't share this little nugget of information with Carson and the other crew guys. And the same applies to Mark Tew. He's always tryin' to squeeze more copper out of every penny, and I don't need him crossways with me over spending a little extra money to move you around the country a mite faster. So, just to be safe, let's just keep this secret twixt you, me, and Doc Walters. Fair enough, buckaroo?"

Setting aside his pencil, Ourecky nodded.

"Splendid," said Wolcott. He turned to leave the room, then hesitated. "By the way, is there any reason you decided not to be a pilot when you joined the Air Force?"

"I wanted to, sir, but I flunked the aptitude test. I've applied for a waiver to attend flight school, but was turned down five times. After all that, I just figured that flying just wasn't in the stars for me."

"Interesting," observed Wolcott, departing the room.

Ourecky placed the stack of medical forms on a bookshelf. *Bea. I'm supposed to call Bea tonight.* He pulled out his wallet and found the folded scrap of paper with her number. Looking at it, he suddenly realized that he didn't even know her last name.

Parking Lot 20, Wright-Patterson Air Force Base, Ohio
9:00 p.m., Thursday, July 18, 1968

With his hands jammed into his pockets, Air Force Staff Sergeant Eric Yost strolled out of the warehouse where he worked. He was roughly midway through his daily stint—the graveyard shift that ran from four in the afternoon until midnight—so he paused for his allotted hour "lunch" break. As he crossed the darkened parking lot, he watched the lights of a Ford Mustang fade in the distance; the car was occupied by his three co-workers, on their way to an all-night diner in Dayton. He wasn't at all hungry, so he would do as he customarily did every night at this time: shuffle out to his 1957 Chevrolet panel van, smoke a couple of cigarettes, contemplate his miserable life, and wish that he could sneak a drink without getting caught. It had been over a month since he had tasted any liquor, and Yost was having a tough time with sobriety.

Alcohol had wreaked a harsh toll on his military career; summarily busted two pay grades after showing up for work drunk twice in the past year, he was now on the verge of losing his security clearance. After over eighteen years spent tracking critical electronic components around the globe, he was now relegated to driving a forklift in this warehouse. The menial job was Yost's last chance for redemption. If he didn't stay sober and make good, he would be booted from the Air Force right at the cusp of twenty years' service, with no chance of drawing his retirement pay.

He had serious doubts that he could squeak through the next twenty months in his current assignment. The hours were horrible and the working conditions left much to be desired. His workplace—Contingency Stocks Commodity Warehouse #2—was an old brick aircraft hangar that pre-dated World War II. It was poorly lit, drafty, musty, and infested with mice. As best as Yost could determine, none of their "contingency commodity stocks" ever actually departed the warehouse. Mostly, he and his co-workers spent their time shifting the pallets from one rack to another, when they weren't playing poker or telling war stories.

If work wasn't agonizing enough, he dreaded going home, since his wife was driving him absolutely crazy. Gretchen constantly harangued him about his recent demotion and griped that they didn't have enough money to make ends meet. Thankfully, they owned their house free and clear—courtesy of a windfall left by his mother after she died last year—or they would really be under water with their bills. Of course, the house wasn't much to speak of; located in a crappy neighborhood, it was in constant need of costly repairs.

Out of school for the summer, their three kids screamed and squawked all day long. Gretchen used the kids as an excuse not to get a job to help with their finances, even though she ignored the little monsters most of the day as she watched soap operas and stuffed her face with candy and potato chips. Since it was futile to obtain any rest at home, he had taken to driving onto the base several hours before the start of his shift. He would habitually park in a back space, jam cigarette butts in his ears to dull the noise of planes taking off from the adjacent runway, and doze for a few hours in the back of the van before going on shift.

Dozing now, Yost was awakened by a dull rumbling noise. He opened his eyes and watched a large flatbed truck park in front in of a nearby hangar. The hangar doors gradually clanked open, and a strange object slowly materialized. It appeared to be some sort of aircraft, mounted on a roller-mounted platform, being pushed by roughly twenty men.

In the course of his career, Yost had seen virtually every aircraft in the Air Force inventory, but this was unlike anything that he had ever witnessed. Its black fuselage was roughly thirty feet long, with short delta-shaped wings that curled up sharply at the ends. Its sleek nose was shaped like a dolphin's.

Suddenly he realized that the stubby vehicle lacked any air inlets that would be associated with jet engines. Obviously, it had to be powered by something, perhaps some sort of anti-gravity propulsion system or other mysterious technology. As he pondered the strange vehicle, the men struggled to manhandle it from the platform onto the bed of the trailer. As they strained to load it, several other men worked to cover it with canvas tarps. Yost saw an obese man standing off to the side, shouting instructions. Periodically, the fat man swore loudly in German; Yost had met Gretchen while stationed near Frankfurt, so he clearly recognized German curses when he heard them, since he had been on the receiving end so many times.

In a few minutes, the truck and its strange cargo were gone. As the hangar's doors were closing, Yost realized that the hangar's interior lights had been switched off the whole time, obviously to prevent anyone from seeing what else was within the structure.

Since starting this job, Yost had been extremely curious about what occurred inside that particular hangar. Hangar Three wasn't used as a warehouse, like the three other decrepit hangars in the same row. From his observations, approximately forty men worked inside the facility. With the exception of some machine gun-toting guards who periodically emerged to walk around the building's exterior, none of the workers wore uniforms. Most of the men appeared to be civilians, possibly engineers or scientists, and they obviously were busy night and day, even on weekends.

Rumors abounded that Wright-Patterson was home to dozens of Soviet bloc aircraft that were being examined and reverse-engineered, and there were pervasive tales—constantly denied by the base's leadership—that a secret hangar housed the wreckage of crashed UFOs—and possibly the remains of their alien pilots—recovered at Roswell, New Mexico, and other locations. Now, he was convinced that he had stumbled upon a highly classified facility—perhaps even the fabled UFO hangar—that the facility's operators had elected to hide in plain sight rather than burying it behind barbed wire. It all looked rather innocuous. In front of the big brick hangar, an almost unobtrusive sign vaguely declared "Aerospace Support Project – United States Air Force."

Yost's head spun with the possibilities. *Could he have just witnessed an alien spacecraft?* All things considered, it certainly seemed that way. Now he was even more curious and decided that he would keep a close eye on Hangar Three, particularly since he was already spending so much idle time in this parking lot.

Aside from the otherworldly craft, there were other strange occurrences that aroused his suspicions. Just yesterday afternoon, some sort of accident had occurred in the hangar; sleeping soundly, Yost had been jarred awake by an ambulance's siren. An old friend, a former poker buddy from Germany, was now assigned to the base hospital. Yost thought he might ring him up to perhaps glean some juicy details about what happened yesterday. At this point, there was no way of knowing where his curiosity might lead him, but certainly it had to guide him to a better place than where he was now.

The Falcon Club, Dayton, Ohio; 8:30 p.m., Friday, July 19, 1968

They drove to the Falcon Club after a dinner of spaghetti, meatballs, and small talk. Bea found a parking place in a secluded area of the lot, next to a light pole. This red VW Kharmann Ghia convertible was her baby, and she didn't want anybody swinging their car doors into it or leaning against it. Apparently uncertain of the correct protocol, Ourecky climbed out and awkwardly ran around to her side to open her door.

"Thank you, Scott. You're such a gentleman," she said, climbing out. She smoothed her blue dress and took his arm. "I hope you don't mind about the driving. I've only owned this car a few months, and I don't dare let anyone else drive it."

"It's okay," he replied. "I really don't mind."

Gravel crunched under their feet as they made their way to the entrance. Car horns blared in the distance, and a pair of moths fluttered madly in the flickering glow of a buzzing streetlight. The humid night air smelled of stale beer and a nearby factory. Several people milled around in the parking lot; more than a few men openly ogled Bea as she passed by.

"Are you *sure* about this club, Scott?" she asked. "This just doesn't seem to be the kind of place where you would hang out. Have you ever been here?"

"No," answered Ourecky. "One of the pilots recommended it, so I'm sure it's okay."

"Don't take it wrong. It's fine, but I just couldn't picture you here. Let's go in."

They walked inside; the interior was only slightly better lit than the parking lot. They went to the bar and ordered drinks. The club mostly catered to Air Force men from the base and the sort of women who gravitated to such men. Black-framed pictures of obsolete airplanes and long-dead pilots filled the paneled wall behind the bar. At the far end of the club, a four-piece house band played Elvis Presley's "Suspicious Minds."

Fumbling with his wallet, Ourecky counted out bills to a frumpy barmaid and then handed Bea her drink. As he sipped his Schlitz from the bottle, he noticed Carson walk up. "Bea, this is Major Carson," he said, introducing the pilot. "Uh, we work together on the base."

"*Bea*? Please call me Drew," said Carson suavely. His right eye was still purple and slightly swollen. "*Wow*. When my friend Ourecky here told me he had a date for tonight, he *sure* didn't let on that it was with a gorgeous fashion model. How in the *world* did you two meet?"

Bea felt more than slightly uncomfortable as Carson scanned her up and down, as if he were a butcher appraising a hanging slab of beef. "We met on a plane," she explained, forcing a polite smile. "I'm a stewardess for Delta Airlines."

"I should have known," observed Carson, rolling his eyes.

Bea motioned to a vacant table and two chairs on the other side of the room, next to the dance floor. "Scott, I think that table's open. Would you mind? I would dearly love to sit down. These new shoes are still hurting my feet."

Ourecky escorted her to the table and held her seat. "Uh, Bea, I need to visit the little boy's room," he mumbled. "I'll be right back."

"Go ahead. I'll be right here, waiting for you." She pulled a compact from her purse and checked her makeup in the mirror. She clicked the compact closed, slipped it back into her purse, and took a sip from her drink. She noticed that Carson had drifted away from the bar and was now standing with two other men. He kept gazing in her direction, obviously trying to make prolonged eye contact.

Curious, Bea watched the three men. She was surprised that Scott knew them and was even more surprised that he worked with them. She had been in this club before and had been conscious of their clique for several weeks. As the daughter of a pilot and stepdaughter of another, she had been around all manner of aviators her entire life and was familiar with their idiosyncrasies.

But even among pilots, this close-knit group seemed somewhat different; although they shared virtually nothing about what they did or the planes they flew, they seemed to carry a certain mystique that set them apart from the multitude of other pilots who came and went at Wright-Patterson. To her knowledge, none of them were married or had even the slightest inclination toward matrimony or any other semblance of permanent attachment to another human being. Rather, based on the comments of some of her girlfriends, they used and discarded women like so many Kleenex.

From her observations, if there was a dominant leader among their little pack, it was Drew Carson. As Ourecky sauntered toward the restroom, Carson swiftly descended upon her, like a raptor swooping—talons outstretched—to scoop up an unsuspecting prey. "Gosh, baby, you look ravishing tonight. Simply delectable," he declared, swinging into Ourecky's chair and casually pushing his beer to the side.

She smiled politely. "You're in my friend's seat," she observed. "He's coming right back."

"Oh yeah, babe," he replied. He made a show of checking the time, pushing up his paisley shirtsleeve to reveal a grossly oversized chronograph. "I know him. He works for us, but only in an engineering support role. Me, I'm a pilot."

"I know *precisely* who you are, Drew Carson. My friend Jill went out with you a couple of times. It would have been about a month ago."

"Jill?" He looked up at the tiled ceiling, as if looking for Jill's face there. Bea half-expected him to pull out a little black book to check his notes. Of course, his little black book probably comprised many volumes, far too many to carry.

"Jill Osborn. She's in a steno pool at the base. About my height, long black hair." Bea looked down and added, "Much bigger on top than me. Sound familiar?"

"No. Sorry, honey. I still can't place her. I guess she just wasn't that memorable, huh?"

Bea sampled her drink. She pursed her lips and wrinkled her nose; the bartender had made it much too sweet. "Well, Drew Carson, *your* memory may not be that great, but Jill sure remembers *you*. In spades."

He grinned. "I suppose she wants another date?"

"Not really. She didn't have anything particularly nice to say about you. By the way, what happened to your eye? Run into a doorknob?"

Self-conscious, Carson massaged his black eye. "Accident at work. Speaking of eyes, did I tell you that yours are absolutely dazzling?"

She grinned. "Funny you should say that," she crooned, fluttering her eyelashes. "Since you don't seem to be looking anywhere near my face very often, I wouldn't think you would be much dazzled by my eyes…"

Carson glanced up and spied Ourecky approaching from the restroom. "Hey, baby, life is short. Wouldn't you much rather be spending your precious time with a real aviator instead of some egghead?" he asked. "My Corvette is parked out front. Why don't you jettison Ourecky so you and

I can go for a ride? It's a perfect night. I'll put the top down; maybe later you could return the favor."

"Oh, that's okay. I think you would probably be a lot more comfortable by yourself. I've ridden in a Corvette before, and I don't think there would be enough room for me to squeeze in there with you and your ego."

Ourecky returned from the restroom. He stood beside the table, trying his best to act nonchalant. "Uh, Major Carson," he muttered. "Uh, thanks for keeping Bea company."

Carson ignored him, focusing all of his attention on Bea. "A good-looking woman like you shouldn't be hanging around a guy like this. An *engineer*," he said, sniffing. "You should be with a *pilot*. I'll go start the car. Black Corvette, right outside."

Bea laughed softly. "Oh, that sounds so tempting! I have my own car, though. Now, Drew Carson, why don't you just run along and leave us be?"

Undaunted by her rebuff, Carson stood out of the chair and leaned over the table. "Last chance," he declared. "Pilot or loser?"

"Pilot?" sneered Bea, dismissing him with a casual wave of her hand. "I've known plenty of pilots, men who could out-fly you any day of the week. Why don't you just jump into your fancy little Corvette and take a drive with yourself? I'm quite content right here, thank you."

Chagrined, Carson turned and left.

Ourecky sat down and they chatted. Fidgeting, he was obviously uncomfortable. He looked like he didn't know what to do with his hands, like he would have been more at ease if they were occupied with a slide rule or a compass.

"Are you okay?" implored Bea, leaning toward him and gently touching his forearm. "You seem a bit distant."

"Just work," he answered. "It seems like I can't ever leave it completely behind. There's always one or another thing that has to be done, and it's always running through my mind."

"Oh. I thought it was *me*. To be honest, I was waiting for the '*I don't think this is going anywhere*' speech. I was curious to see which version you were going to recite. I think I've heard them all."

His eyes opened wide, accompanied by a surprised look spreading over his face. "Oh no, that's not it at all, Bea. Look, I'll be honest. I really haven't been out with any girls since college, and even then I didn't date too much. I've just been too busy. I'm still sort of stunned from Monday. I thought you were playing some sort of trick on me when you gave me your number."

Bea looked at the table and smiled modestly. "So you're not used to being around girls then? Now, be honest, Scott Ourecky."

"Honestly? Not really. I work long hours, usually seven days a week, and it's hard to meet anyone," he answered. He took a sip from his beer and then gently swirled it in the bottle.

"Well, isn't *this* a horse of a different color," she said. She sat quietly and looked at him. He seemed so sincere and innocent, but she had been around plenty of guys who were masters at feigning sincerity and innocence. *But maybe there was a chance.* "Dance with me, Scott. I'm really tired of sitting here."

Ourecky took her arm and escorted her to the dance floor, joining four other couples. He faced her, slipped his hand into the small of her back, and held her close. He was a little clumsy, but at least he was polite enough not to step on her feet. The house band performed a reasonably good cover of Aaron Neville's "Tell It Like It Is."

The song wasn't even halfway done when yet another interloper arrived. He looked to be in his late thirties, with a puffy face and sunken eyes. He was dressed in a wrinkled white shirt and gray slacks, with a plaid tie loosened at his neck, as if he had come to the club straight from his office. He obviously had been drinking for a quite a while and looked to be well into his second drink past too many. He tapped Ourecky on the shoulder and mumbled, "Mind if I cut in, buddy?"

"Rather you didn't," said Ourecky.

"Oh, c'mon, baby," insisted the loathsome stranger, pawing Bea's shoulder. "Just one dance. What can that hurt?"

Bea whirled out of his grasp and faced the man. "I think you heard what my friend said. I only want to dance with him, so why don't you find someone else and leave us be?"

The man stared at her, and then suddenly a flash of recognition came to his face. "Hey, I thought you looked familiar," he ranted. "Aren't you on TV? No, no, *wait*...I know. You're a *stewardess*, aren't you? Delta, right? You fly the Atlanta to Dayton route, right? Am I right? I'm right, aren't I?"

Stepping in front of Bea, Ourecky confronted the man. "Uh, I think the lady asked you to leave her alone."

"I wasn't talking to *you*, pal," slurred the man, now getting irate. "Why don't you mind your own beeswax?" Wobbling side to side, he leaned toward Ourecky and balled his fists.

Almost in a blur, a man appeared next to Ourecky. He was short, stocky and wore a denim jacket over a blue T-shirt. He looked slightly foreign, maybe a Caucasian and Oriental mix. "Wilson!" he declared, subtly clutching the drunk's shoulder with his right hand. "Baby, it's been a while! How are things?"

Bea couldn't help but notice that the obnoxious drunk winced like he had suddenly been slammed with a baseball bat. She was sure she had seen the stocky man just a few minutes before, lingering in the shadows by the bar.

The stranger turned his eyes to Ourecky and said, "Hey, man, Wilson and me have some catching up to do. You don't mind, do you?" Ourecky shrugged his shoulders; then he and Bea drifted away to dance again.

"Who was that guy?" asked Bea. "The short one? Do you know him? Is he on the base?"

Ourecky couldn't help but notice that the stranger faded away just as quickly as he had appeared—and had taken the drunk with him. "Not that I know of. I think I've seen him around some, but I'm really not sure where. Did you know the other guy from somewhere?"

"The drunk? Yeah, but I don't really know him. He's a salesman, I think. Flies in here about every two weeks. He's asked me for my phone number a few times, but I've kept him at bay. I'm pretty sure he's married, not that I would be interested anyway. I'm just grateful that his friend took him away."

10:05 p.m.

Ourecky hadn't said anything during the drive back to his quarters. She cut off the car, and switched off the headlights. "You've been so quiet," she said. "Cat got your tongue again?"

"Uh, Bea, I have some bad news," he confided. "There's something I need to tell you."

"Let me guess. You're *married*, right?" She groaned softly and shook her head in dismay. "I suppose I should have known. Why am I always so damned stupid?"

Ourecky looked startled. He hadn't expected her response. "No, no, nothing like that, Bea. I'm just not going to be seeing you much anymore. I won't be on your flight."

"*Really*? So this means you're not coming up to Dayton anymore? Is your project over?"

"No, I'll still be coming up here. They've just worked out a way for me to fly directly from Eglin to here without going commercial. I'll be flying on military aircraft instead. Sorry."

"Well, I'll sure miss seeing you, Scott Ourecky."

"Me, too."

"But you still have my number, right?"

"I do."

"Then call me," she said, kissing him lightly on the cheek. "Anytime."

15

DOUBLE NOUGHT SPY SCHOOL

Apex Minerals Exploration Inc., Dayton, Ohio
8:15 a.m., Monday, July 22, 1968

O n paper, Matthew Henson and the other nine men were the latest employees of Apex Minerals Exploration. In reality, except for a rented building in a rundown industrial park near the Dayton airport and a phone picked up by an answering service, Apex Minerals Exploration did not exist. Henson recognized four of the other men from his stint at Aux One-Oh; two had voluntarily dropped from the course while the other two were academic failures like him. He assumed that the five others came here from previous training cycles.

The ten men were no longer members of the US military. Their records had been painstakingly expunged of any reference to military service or government affiliation. According to a briefing that Henson had received at Eglin, once he was released from this effort, his military status would be reinstated; at such time, he would be credited with time spent on the Project, and he would receive back pay. Until then, he and the others were employees of Apex.

They were here to be tutored by a Mister Grau. They had been warned not to ask Grau about his background or previous employment, but knew that he had recently retired from an agency where he had been deeply immersed in decades of clandestine work around the globe.

"In here. Take a seat, eh, any seat. Don't touch anything until I tell you to," said Grau, as he stood beside a door and ushered them into a storage space that had been converted into a makeshift classroom. He didn't appear capable of standing entirely still, but slowly wobbled back and forth like a derelict building on the verge of imminent collapse.

In the cramped room, there were two long tables facing an unpainted cinderblock wall. Each table was set with five metal folding chairs. A single fluorescent light fixture sputtered and hummed, casting more gloomy shadows than light. The room smelled of mildew and dust. A rusted coffee can sat in a corner, catching a steady drip trickling from the ceiling.

On the table, in front of each seat, was a bag or satchel of some kind. Henson took a place at the back table, behind an ancient attaché case. Its cordovan leather was faded, deeply cracked, and smelled like an old pair of work boots. He thought to open it and look inside but remembered Grau's admonition and thought better.

In the course of his covert career, Grau had obviously endured considerably more wear and tear than the average government bureaucrat. He was missing the middle and ring fingers from his left hand and the little finger from his right hand. He walked with a pronounced limp in his right leg, as if his knee was partially fused. A black patch covered a vacant socket where his right eye had been. His remaining hair was white and his skin was sickly pale.

Using an old wooden cane, Grau slowly hobbled to the front of the room. He hooked the cane's crook on the edge of the front table and lowered himself into a folding chair. He sighed, closed his remaining eye, and then softly announced, "I'm Mr. Grau. Welcome to Double Nought Spy School." His hands trembled almost constantly, as if he was stricken with palsy, but his voice was steady and calm.

He opened his eye, blinked, looked over his charges, and then continued. "Hah! Just kidding. First of all, for you fans of James Bond or Jethro Bodine, if you have notions of running around the world wearing a tuxedo, sipping martinis and shooting enemy agents, then you need to dispel those thoughts from your head.

"After you're done here, and we shove you fledglings out of the nest and into the world, you will have no official status that affords you any special protection. In most cases, you'll be working entirely by yourself. You'll carry only documents and credentials that identify you as employees of Apex Minerals Exploration. We'll give you sufficient training to pass yourselves off in this capacity, but otherwise you're on your own.

"Let me make something emphatically clear: there is no safety net or back-up. If you're stranded in some godforsaken rat hole of a country and have to call the US Embassy or consulate for help, you've already screwed up. The Embassy people won't have a clue that you're in the country as anything other than someone just trying to make a fast buck. Consequently, they're not going to be inclined to throw you a rope and drag you out of trouble.

"I will train you how to *not* act like a spy. In most places you'll visit, there'll be some form of secret police. They've been trained to not trust foreigners. They will naturally assume that you're visiting their domain in an intelligence-gathering or illicit capacity. Your behavior has to convince them otherwise. If you act like a spook, they're going to assume that you're a spook." Grau looked up at Henson. "Eh, spook's just another word for spy. No offense."

"None taken," answered Henson. He smiled. Most of the other men laughed. He noticed that some of the men seemed uncomfortable. He wasn't sure if it was because of Grau's spook reference or whether they were just uncomfortable with him being here in the first place.

"Good. Roughly ninety percent of what you'll be doing is logistics work," stated Grau. "It's not glamorous or exciting, but it's very necessary. All of this effort is to support a contingency mission, so mostly you'll be locating and contracting equipment that we'll probably never use.

"When you go into a location, you'll usually establish one of two different types of contingency sites. The first type is an emergency landing site. That's a remote airstrip where a strategic asset can land in an emergency. In the coming weeks, we'll teach you what to look for and how to survey such an airstrip without drawing attention to yourself.

"Since things can drop out of the sky in the middle of the night, we'll teach you how to light an emergency strip so a pilot can land safely. The pilot might also need an electronic beacon for the pilot to home in on, so we'll show you how to assemble a beacon and operate it.

"The second type is a staging site, where we can stash people from the operational squadrons. Normally, we position them in relatively friendly countries so we can be ready to conduct rescue and recovery operations in adjacent countries that might not be so friendly."

Grau reached into a pocket of his rumpled poplin jacket and tugged out a plastic pillbox. He selected a white pill from a multicolored assortment, and casually popped it into his mouth. He swallowed with some difficulty and then continued. "Setting up a staging site is normally more involved than establishing an emergency landing site, since you'll usually have to coordinate for the trucks and airplanes required to ferry our folks from the first country to the second."

"I don't understand," said one of the men. He appeared to be in his mid-thirties, older than the rest of the group. Henson didn't recognize him from Aux One-Oh, and he apparently had been out of training long enough to cultivate long hair, a shaggy beard, and a paunch around his middle. "Instead of all this coordinating and contracting, wouldn't it be a lot simpler just to bring in a C-130 full of troops? They could just sit there on strip alert until they're needed." A few of the men nodded in agreement and talked quietly among themselves. Henson had to agree; all of this clandestine Apex business sure seemed like a complex and unnecessary undertaking.

"Settle down, settle down," admonished Grau, as if he were a schoolteacher contending with a roomful of unruly schoolchildren. He flipped up his eye patch and gently scratched the interior of his empty eye socket. "Well, let's ponder that question. Let's assume that we have advance word that a highly classified reconnaissance aircraft will be flying from Point A to Point B. It's imperative that this aircraft not fall into enemy hands—or anyone's hands—if it has to make an emergency landing. So, we place emergency landing sites in friendly or relatively friendly countries, as I've told you, and we also develop staging sites so that we can launch into less-than-friendly countries if need be. Make sense so far?"

The scruffy man nodded.

"Now, concerning those staging sites, we have at least two alternatives which we'll call Option A and Option B. In Option A, we land a C-130 full of troops at a staging site and leave them there for the duration of the mission, however long that might be. The troops come loaded for bear, and they're ready to spring into action at a moment's notice. Eh, how's that sound?"

"Sounds mighty damned good to me," declared the man enthusiastically.

"I thought you might agree, but there's a problem. Landing a C-130 full of troops somewhere lets everyone know that something big is about to occur. Besides that, it usually requires a stack of diplomatic paperwork and a lot of cash on and under the table to secure landing rights and such. So Option A might not always be the best way to go, agreed?"

"I guess not."

"Hah! Let's not *guess*," replied Grau, briskly slapping his thigh. "Now, let's consider Option B. Since we have advance notice, let's find a place to hide a bunch of men and their equipment, and then we'll slip them into the country quietly, one or two at a time, from different directions."

Grau continued. "Let's assume that we'll need an airplane to fly them into the second country. Granted, it would be nice to have a C-130, but an old DC-3 will suffice for most of our requirements. They're everywhere. There might not be a single flush toilet in the entire country, but it's almost a certainty that you'll find a DC-3 for hire, or something of similar capabilities. Anyway, our little pocket army stays hidden there for the duration of the mission, and if nothing happens, they just dissolve away the same way they came. So, Option A or Option B?"

"I guess I would go with Option B," conceded the man finally.

"Good. If you remain with us, you'll eventually see that there's a method to our madness." Suddenly, Grau was beset with a coughing fit that bent him over double. It took over a minute for his hacking to subside. Regaining his composure, he pulled out a worn handkerchief and wiped his mouth. Henson could see that the faded handkerchief was spotted with fresh blood.

Grau slipped a silver flask from the inside pocket of his jacket and took a sip from it. "Now, since I'm training you to be double-nought non-spies, let's discuss cover stories. All of you have probably read a spy novel, so you probably know the phrase 'cover story.' A cover story is simply a plausible reason for being somewhere and doing the things you're doing.

"We strive to keep things simple," declared Grau. "As long as you toil under the Apex umbrella, you will use just one cover story. One size fits all. So long as you stick to the story, you shouldn't have any problems. If you deviate from it, or decide to fabricate your own embellishments, then you're on your own.

"Here's the story. You're visiting their country because preliminary geological studies indicate the presence of a potentially valuable mineral. It won't be something of immediate value, like gold or platinum or diamonds, but a substance that will require considerable processing to render it into something of worth. That is, if it's found, and if it's found in sufficient volume.

"Now, we'll impart you with just a smidgen of geological training, at least enough to be reasonably conversant, but we don't intend for you to pass yourselves off as geologists. As far as the locals are concerned, your job with Apex is more that of a scout, to poke around to find traces of the mineral. And you also establish contacts and locate resources to move people and equipment around to extract the mineral should it be found in large enough quantities."

The scarce color in Grau's sallow face suddenly faded, like he was about to keel over. Sitting down, he said, "If you do your job well, local officials will be cooperative because they'll assume the project's success will bring money for the local economy and—more importantly—cash that will flow directly into their pockets. But always remember that this is a long-term effort. It's important that you forge lasting relationships but also that you don't make promises."

"I'm not sure I like this," said Henson. "If we know that the mineral is not there, aren't we just building up people's hopes? Aren't we just exploiting them?"

Grau slowly stood up and consulted a class roster. "Eh…Henson, right?"

"Yes, sir."

"Well, Mr. Henson, ultimately you're correct. But here are some points to consider. First, you don't *ever* make any promises. When you communicate with the locals, you always portray the situation as a long shot and consistently express doubts that the mineral will ever be found. Let's face it, if we were confident that the mineral was going to be found there—theoretically—then Apex would send in a *real* geologist and not a second-string player like you. As far as the locals are concerned, you're just another working stiff who's just doing what you're told to do."

Placing his hands over his head, Grau stretched slowly, like a bear emerging from months of hibernation. His body literally creaked as he stretched. "So we'll never find the mineral. But while you're there, you're spending money, so there *really* is money flowing into the local economy. So I don't agree with you that we're *exploiting* anybody. Do you agree, Henson?"

Henson nodded reluctantly.

"Now, this story lends you the latitude to move around the countryside with relative freedom. You can purchase maps, be inquisitive, take lots of pictures, and write copious notes. And although this is not an intelligence mission, you still need to pay close attention to details like road networks, military forces, police forces, airport security procedures, and the like. Over the course of the next few weeks, I'll show you what to look for when you visit a country."

Then Grau frowned and added, "But let me warn you, it's almost a certainty that you'll be hauled in for questioning at some point. As I implied earlier, some local officials will assume you're a spook, no matter how innocent you appear to be.

"Once you're arrested, your first instinct will be to buy your way out of trouble. That's a huge error. When you're swift to offer up a juicy bribe, you're causing the locals to think two things. First, they'll assume that you're connected to an organization with a lot of dough, so now your options are limited to slinging cash around instead of sticking to the story and being frugal. Second, and more importantly, resorting to a bribe verifies that you have something to hide.

"Since we're on the topic of money, open those bags," said Grau.

Henson opened the attaché. Peering inside, he saw that it was jammed with loose US currency of different denominations, as well as rolls and packets of other currencies. He remained quiet, as did most of the other men, but some of the new students whooped like they had struck the big jackpot at a Vegas casino.

"Once you're overseas, you'll be managing a lot of cash," explained Grau. "The objective of this exercise is to make you comfortable with handling and accounting for large sums of money. We could talk about this in theory, but the best way we can condition you to handle money is to place it in your hands. So, that money will stay with you for the duration of our training here.

"Each of those bags contains two thousand dollars in US notes and a thousand dollars in other currencies. I want you to count and organize it, and then you'll sign a voucher for it. While you're here, you'll pay for all your living expenses out of that money. Hotel rooms, meals, transportation, supplies, film, whatever you require. There's also a ledger in there. You are to account for every penny that you spend."

One of the men asked, "What about this foreign money? We can't spend this here, can we?"

"No," replied Grau. "You'll turn that in at the end of the training, but we want you to become accustomed to handling it. When you travel, you'll not always have the luxury of paying for everything in US greenbacks. As an example, if you're in a country where the most commonly accepted currency is the franc, you don't want to accidently hand a taxi driver fifty dollars worth of francs for a three-dollar cab ride. So we'll run through some simple drills to keep you on your toes. If I ask you to produce the equivalent of, say, ten dollars in German marks, then you'll have thirty seconds to dig into your bag and plop that amount on the table. *Verstehen Sie?*"

The man smiled slightly, nodded and replied, "*Jawohl, Herr Grau.*"

"*Sehr gut.* Very good." Grau smiled, and said, "Now, here's the big news. Every operational site is budgeted for a certain amount of money. The amount varies according to the requirements and where it's located. Before we send you on an assignment, we'll front you the budgeted amount in cash, or we'll make arrangements to wire it to you if it's not safe to travel there with a big bankroll stashed in your suitcase.

"Now, here's your incentive to stick to the story, be frugal, and not pay for everything with bribes," said Grau. "If you come in under budget for the site, you keep the remainder. But there's a caveat: we'll reconcile your books and double-check your work to make sure that you did all the coordination and contracting necessary to execute the contingency, should it be necessary, but you gentlemen stand to accrue a sizeable profit if you play your cards right."

This was a dream come true, thought Henson. Not only would he be able to work on his own, but there was actually a chance to make a profit. *What could be better than this?* Looking to the future, Henson mused on how he could use the surplus cash. Certainly, between the GI Bill and the extra money he stood to garner, he could finish his degree, probably at a school more prestigious than LSU. Maybe there would be enough to help his mother expand her café.

A voice interrupted Henson's thoughts. "But what if we go over budget?" asked the pudgy man with long hair and unkempt beard. "What if we need more money to work a site?"

Grau shook his head. "If you need more funds, then we will disburse more funds. But let me explain something. Provided you're doing your job right, we want you to make some money for yourselves. If you have a stake in the game, it's far more likely that you'll play your part and stick to the story."

Henson was intrigued with the merits of the ingenious scheme. Conscious that the money was effectively coming out of their pockets, they were more likely to act responsibly, instead of running amok like drunken salesmen on an unlimited expense account. They would be more motivated to stay in a reasonable hotel instead of a lavish five-star palace, and wouldn't be as inclined to squander a lot of cash on wine, women, song or fancy dining. They were more likely to conduct themselves in a sensible manner, and subsequently draw a lot less scrutiny to their actions. It also made sense from a business standpoint; instead of writing checks on an endless account, they would be encouraged to actively negotiate.

"So if you need more cash, we'll front you more cash," explained Grau. "But be aware that your opportunity to make a profit vanishes as soon as your ledger goes into the red. And one more thing: unless you have a really good excuse, we only allow you to overdraw on two assignments, and then you leave Apex and go back to the Air Force.

"Okay. We'll talk in greater detail about the accounting process at a later date. Now, I want to pass on some fundamental rules of conduct. You should engrain these in your head and adhere to them always. When you're travelling, unless we instruct you otherwise, stay away from border areas. Stay away from kids. Avoid contact with young women, at all costs."

"How young is young?" asked one man, smiling and winking. The man had a flamboyant Fu Manchu moustache, like Joe Namath's, and an impish smirk like he considered himself a latter-day Don Juan.

"How young?" asked Grau. "If you enjoy being on this side of the grass, you should avoid any females between six and sixty. No matter how innocent your intentions may be, if you approach a female if only to ask directions, it's a safe bet that at least one local guy already has an eye on her, so you're setting yourself up for a very ugly confrontation."

Obviously ignoring Grau's guidance, the man grinned. He leaned toward the chubby bearded man and whispered a crude comment. The two shared a quiet laugh.

Grau cleared his throat. "As an old man who has seen my share of scrapes and close calls, let me offer some advice," he said, sharply focusing his one-eyed attention on the mustachioed man. "If you want to be successful in this job, you need to be mindful of human nature. Moreover, you should always strive to treat people with dignity and respect. Let me reiterate: When we send you out in the world, you'll be on your own with no safety net. If you go out there and insist on being stupid, then you'll be lucky if you *just* end up in jail. If you do end up in jail, you'll be cooling your heels until I come to yank you out, and I probably won't be in a rush. Understood?"

The man nodded soberly.

"So it's clear, the rules are entirely different out there. Some infractions will get you expelled from a country, some will land you in jail, but other gaffes will land you in a shallow grave. If you can't focus on the job and keep your pants zipped, it will be the latter."

16

PICNIC

Kapustin Yar Cosmodrome, Astrakhan Oblast, USSR
11:25 a.m., Monday, July 29, 1968

Puffing nervously on a cigarette, Gregor Mikhailovich Yohzin waited in his *Moskvitch* sedan as his driver—a very reliable sergeant from the Ukraine—spread a gray wool blanket in the grass. The sergeant knelt down, smoothed the blanket, and then anchored its corners with three large stones and a picnic basket. He then neatly arranged the basket's contents—a half-loaf of dark bread, two pans, five bowls, a wedge of cheese and a vacuum bottle of hot tea—next to the blanket. Glancing at the rear-view mirror, Yohzin stubbed out his cigarette in the ashtray. He glanced to his left, where a sleeping dog nestled beside him in the narrow back seat. Magnus, the handsome black-and-tan canine, was his constant companion.

A major general in the RVSN—Soviet Strategic Rocket Forces—Yohzin oversaw initial testing of medium-range ballistic missile prototypes at the sprawling Kapustin Yar cosmodrome. While his posting was certainly a plum assignment by Soviet military standards, he perceived it as menial make-work. He didn't participate in the actual design process, contributing his creativity and perspectives, but merely evaluated the works of others. He deplored his job; his facilities were abysmal, the working environment was blatantly hazardous, and he was constantly compelled to beg and plead for every single kopek of his modest budget.

Ironically, although he was one of the most accomplished aerospace engineers in the Soviet Union, he was relegated to mundane tasks simply because his extensive training and prowess were not formally recognized by any of the premier aerospace design bureaus. In a fair world, he should have been working shoulder to shoulder with the likes of Sergei Korolev or Vladimir Chelomei, envisioning the rockets and spacecraft that would send explorers to the distant reaches of the solar system and beyond, but his destiny was denied. Instead, he was trapped here on the desolate steppes of Purgatory, frittering away his precious time, laboring in obscurity as lesser men lined up to assume the role that rightfully belonged to him.

He swiveled his head to look at the blanket stretched out in the stubbly grass and compelled himself to focus on more immediate matters, rather than dwell on the countless indignities heaped upon him by his superiors. This picnic—a weekly rendezvous with a long-time friend—was one of his favorite diversions. In fact, it was one of the few distractions that Yohzin allowed himself. Unlike others who held elevated ranks in the Soviet military, he didn't partake in the excesses and amenities that his contemporaries viewed as hard-won perquisites of the office. His lifestyle was modest, if not downright austere.

He wasn't a bloated glutton who gorged on expensive foods and premium liquor. His palate leaned more toward common staples like buckwheat *kasha*, onion-stuffed *pelimnin*, and cabbage-filled *pirozhki*. Caviar and other extravagant delicacies rarely adorned his plate, unless he was served them at some official function, and even then he ate them grudgingly, strictly out of polite deference to his hosts.

He didn't keep a jaded mistress, plucked from the enlisted ranks, ready to service his every physical desire. If the truth be known, he had been intimate with only two women—his first wife, who had died of breast cancer nineteen years prior, and his current spouse—in his entire life.

Yohzin, his wife, and their two sons made their home in a simple apartment. It was his sole abode; there was no dacha of rough-hewn logs in a nearby forest where he could seek refuge and solitude.

But despite his otherwise stoic existence, Yohzin did occasionally yield to one vice. He enjoyed the performing arts—particularly orchestras, opera, and the ballet—and so when he visited Moscow on official business, usually once a quarter, he made a point to take in a show or two. Additionally, whenever practical, he brought his family with him to Moscow and the theater, in hopes that his two sons might also eventually acquire his appreciation for the finer things in life.

The sergeant tapped lightly on the window before opening Yohzin's door. "All ready, sir," he announced quietly. "The usual, sir? Bring the car back around in an hour?"

Yohzin looked out, pivoted in his seat and nodded. He stepped out the car, stretched, and whistled over his shoulder. Magnus awoke, bounded out of the car behind him, obediently fell into position behind his left heel, and kept pace as he strolled toward the blanket.

An Alsatian, Magnus was descended from a pair presented to Yohzin over two decades ago. Yohzin smiled as he reminisced about the events that led up to that. In the early thirties, as a promising and dependable engineering undergraduate, he had been dispatched to study at the acclaimed *Technische Hochschule* in Berlin. He closed his eyes and sighed; those were heady days. In that distant era, he had met the esteemed Wernher von Braun, now the fair-haired hero of the American space program. He had even attended meetings of the fabled *Verein für Raumschiffahrt*—the Society for Space Travel—and had been present when von Braun assisted visionaries Willy Ley and Hermann Oberth as they launched their experimental liquid-fueled rockets from a nearby Army base. Shortly thereafter, von Braun was pressed into service with the German military, designing the *Vergeltungswaffen* rockets that would eventually wreak terror on London, and the next decade was dark history.

In the aftermath of the war, as the Allied conquerors scrambled to consolidate their plunder, von Braun and his colleagues were valuable commodities. Although he could never openly express such a thought, Yohzin was immensely thankful that von Braun and most of his brighter acolytes had escaped to the West after the inevitable fall of Hitler's Third Reich. Scooped up by a clandestine OSS intelligence mission called Operation Paperclip, the best and brightest of the German rocket scientists were whisked to the United States. Most of them went to work for the US Army, at least initially, and later matriculated to NASA where they developed the massive Saturn boosters that would soon propel men to the Moon.

As America reaped the best of the German rocket scientists, the Soviet Union gleaned what was left. Even as the war was grinding to a halt and German forces raced westward to avoid the wrath of the Soviet onslaught, Yohzin and hundreds of his contemporaries were ordered to the former German rocket test facilities in *Lehesten*, in what would eventually become the *Deutsche Demokratische Republik*—East Germany. There, they were set to work analyzing the A-4 rockets left intact at the end of hostilities.

Yohzin and other Soviet officers were assigned to test-fire and appraise the A-4s. Surprisingly, a substantial number of German engineers and scientists also participated in the tests. Some were essentially forced to toil for their former enemies, but many joined the effort voluntarily. The luring prospects of steady wages and better living conditions were powerful inducements in a country shattered by war. Yohzin had been selected to handpick and oversee a group of them, not only because he was fluent in their language, but also because he actually studied with many of them before the War.

But his was not the only team of Germans. Most of his fellow officers triumphantly lorded over their own German subordinates as if they were barely more than indentured servants. In contrast, Yohzin treated his charges with dignity and respect. In return, his Germans consistently rewarded him with productivity and—more importantly—reliability. While the rockets of his counterparts frequently fizzled or obliterated themselves on their launching pads, Yohzin's rockets—granted extra attention by their Teutonic engineers—flew straight and true. In time, the burgeoning research effort was shifted from East Germany to Kapustin Yar, then a newly established rocket testing range near the Volga River, approximately a hundred kilometers west of Stalingrad. Later, as the program gained momentum and the Soviet engineers developed new rockets that were superior to those of their predecessors, the German scientists—no longer trusted to work on the new designs—were allowed to return home.

As they departed the Soviet Union in 1950, Yohzin's Germans, greatly appreciative of his kindness, presented him with the breeding pair of Alsatians. Like von Braun and his cohorts, Yohzin vehemently believed that rocketry's true value was for mankind's peaceful exploration of outer space. In the years that followed, he progressively rose up the ranks within the fledgling RVSN, but despite his successes, he also found himself ostracized for his previous relationships with the Germans. Now, snubbed by most of his contemporaries, disillusioned and bitter, he was just a few years away from the meager promises of military retirement.

Yohzin sat down on the blanket and loosened his stiff collar. The sky was clear and the sun shone brightly. The grass and clover were ruffled by a gentle breeze. He nibbled on bread smeared with yogurt as he awaited the arrival of his lunch companion, Lieutenant General Rustam Abdirov, also of the RVSN. The two had been close friends since their first meeting in Lehesten at the conclusion of the war; a former artillery officer, Abdirov had been Yohzin's boss during the years of testing the German A-4 rockets, and much of his early career had been built upon the string of Yohzin's successes with his German team.

Now that Abdirov had returned to Kapustin Yar, the two men had resumed their close acquaintance. While all of the work at the cosmodrome was secret, Abdirov's endeavors were considerably more secret than most. Although he could only hint of his labors to Yohzin, Abdirov was now apparently at the forefront of the Soviet military's efforts to exploit space. Literally fenced off from outsiders, an isolated section—the old Burya launch site—of the cosmodrome had been allocated for his efforts. The elaborate Burya facility had been built to test a ramjet-powered intercontinental cruise missile, but closed in 1961 when the project was cancelled.

A large blue sedan pulled up. Two aides disembarked, assisted Abdirov from the vehicle, climbed back in, and the car puttered away in the distance. The general was tall and terribly thin; from a distance, he might easily be mistaken for a scarecrow or an emaciated concentration camp survivor. Leaning on a cane, he shuffled with a stiff-legged gait, but made good headway nonetheless.

As Abdirov drew near, Yohzin resisted the urge to gasp; even though he now saw his friend on at least a weekly basis, he could never become entirely accustomed to his ghastly appearance.

Four years older than Yohzin, Abdirov was something of a mentor. He was one of a handful of senior officers who appreciated Yohzin's training and abilities. Over the years, he had repeatedly approached their superiors within the RVSN, making pleas on behalf of his protégé, striving to land Yohzin a posting with his organization or one of the premier aerospace design bureaus. Unfortunately, his petitions fell on deaf ears, and Yohzin remained stuck on the same rung of the ladder.

They embraced like brothers, and Abdirov loudly kissed Yohzin on both cheeks, in the custom of his nomadic forebears. Never fond of Magnus—and vocally so—the elder man cast a baleful gaze at the canine. Magnus, obviously unsettled by the general's gruesome appearance, tucked his thick tail between his legs and scampered in retreat. He eventually plopped down in the grass several meters from the blanket and kept a wary eye on Abdirov.

"Help an old invalid?" asked Abdirov. Owing to his extensive injuries, Abdirov's knees had only a few degrees range of motion, so sitting down and getting up were enormous chores for him. Obliging, Yohzin clasped his mentor's wrists and gently lowered him to the blanket. Abdirov winced as his desiccated skin literally crackled like old cellophane; Yohzin could only imagine the pain he endured on a daily basis. Making himself as comfortable as he could, Abdirov reclined on his left flank like a shepherd, propping himself up on his left elbow. Yohzin opened the vacuum bottle, poured a cup half-full of steaming tea heavily laced with sugar, and handed it to his friend.

"Some *lapsha*, Rustam?" said Yohzin, holding out a ceramic bowl of noodles. "Or perhaps some mutton? Luba made it especially for you."

Abdirov grinned. Only the left corner of his damaged lips turned up when he smiled; the right side was permanently frozen in a drooping smirk. "Then I must eat it, for Luba's sake!" he declared. Because of the damage to his mouth, his words came out in somewhat of an effeminate lisp. He reached into a bowl, popped a chunk of boiled mutton—*besbarmack*—between his teeth, and chewed with gusto. "Delicious! Just like my dear *Ana* used to make. Your Luba is really some cook. I am *so* smitten! If she wasn't yours, Gregor Mikhailovich, I swear that I would pull her up on my horse and whisk her away."

As they shared the lunch that Yohzin's wife had lovingly prepared, the two men exchanged gossip and speculation about their contemporaries. They conversed about advances in rocketry and space exploration, but some subjects were taboo, even between friends. Yohzin was intensely curious about Abdirov's activities, for various reasons, but he had learned not to be overly inquisitive. He had long ago learned that Abdirov answered probing questions with silence and was quick to curtail a conversation if Yohzin seemed too eager to tread on forbidden ground.

When they had finished lunch, Yohzin packed the remnants in the wicker picnic basket. Only a small scrap of dark brown bread was left, which he tossed toward Magnus. The dog snapped the morsel out of the air and quickly devoured it, even as Yohzin realized that he had offended his friend. "*Prastite*," he mumbled, apologizing for his insensitive gaffe. "Sorry."

Wagging one of his few fingers, Abdirov scowled as he chastised him. "I've warned you, brother. I know that you're fond of that cur, but you should leave it at home. As I've told you repeatedly, your past affiliations are holding you back."

"I'm sure that you're correct, Rustam," admitted Yohzin, nodding glumly.

"You're damned right I am! If you're serious about improving your station, trotting that damned dog around doesn't help matters much. It's as if you insist on rubbing it in their faces. If you would accept my counsel and amend your ways, maybe I can convince the powers that be to allow you to come to work for me."

"I would like that, Rustam," replied Yohzin.

"Then *listen* to me, Gregor Mikhailovich," declared Abdirov, proffering an aluminum flask. "Trust me, *bratanik*, I am working on phenomenal things, and I could really use someone with your abilities."

Yohzin drank deeply from the flask and handed it back. In moments, his belly was warmed by the *Stolichnaya*. Grasping the flask with the three fingers of his right hand, Abdirov drew a long swig. The two men were silent for a few minutes, passing the flask back and forth until the vodka was depleted.

"So tell me, Gregor Mikhailovich, are you are still staying current with technology developments in the West?" asked Abdirov, shielding his eye from the sun's glare.

"I am," answered Yohzin. In one of the few instances where he was recognized for his true abilities, Yohzin was widely considered as an authority on the American space program. Six years ago, because of his intimate knowledge of many of the German rocket scientists now working for the Americans, he had been seconded from the RVSN to the GRU—*Glavnoye Razvedyvatel'noye Upravleniye*—foreign military intelligence directorate of the Soviet military. He had worked at the GRU's headquarters—nicknamed the Aquarium—at Khodinka Airfield near Moscow. In this capacity, granted unprecedented access to intelligence materials concerning American rocket development, he became familiar with NASA and its manned spaceflight programs. Periodically, he was called back to Moscow on follow-up assignments.

"Then let me ask you a hypothetical question," said Abdirov. "Let's say—theoretically, of course—that I had a requirement to return a vehicle from orbit to a specified landing point, with a high degree of precision."

"How precise?"

"Within two or three kilometers of the desired landing point," answered Abdirov, swishing away a fly.

"That's quite some degree of accuracy," answered Yohzin. "Perhaps unattainable."

"Well, for the sake of discussion, let's at least entertain the thought."

Leaning toward a patch of bare ground, Yohzin sketched out a concept with his finger. "Perhaps your solution might be some sort of beacon system, slaved to a steerable parachute, where the vehicle is automatically guided to a transponder."

"Good idea, but let me expand my requirements," said Abdirov. "The system should not require positioning any equipment, like a beacon, at the desired landing point. At any point during orbital flight, it should be possible to continually update the desired landing point so that the reentry vehicle can arrive at any point on earth, at least within its orbital path. Lastly, the entire control system to accomplish these tasks should be carried on the orbital vehicle itself, so that there is no need for ground-based control stations."

Yohzin chuckled. "If I didn't know any better, Rustam, I would suspect that you are trying to drop bombs from orbit."

"Perhaps it's best that you not speculate about such things," growled Abdirov, glaring at him. A pair of MiGs rushed by overhead.

When the noise abated, Yohzin swallowed and said, "I surely didn't mean to offend you."

Abdirov nodded. "I know that, but you need to be much more cautious about expressing your thoughts." He paused and added, "But for the sake of discussion, let's explore *that* idea, since it's really irrelevant what we're trying to return from orbit. Imagine that we *were* tasked to drop bombs from space, and we had to deliver them with at least some reasonable degree of accuracy. Do you know if the Americans have any system that might accomplish this task, as I've defined it?"

Yohzin closed his eyes and thought for a few moments. He started to shake his head, but then realized that the Americans *did* possess something that might be a suitable solution for the task. He opened his eyes and smiled broadly. "You know, the Americans developed a very powerful but small computer for their Gemini spacecraft. It weighed less than thirty kilograms."

"Thirty kilograms? *Hah*! Impossible!" sniffed Abdirov. Yohzin's sedan pulled up and parked about a hundred meters away. Abdirov gestured at the car and observed "If our illustrious Soviet computer experts built an equivalent machine, it would probably be as big as your *Moskvitch* there. It's not feasible that the Americans built something that small."

"Honest, Rustam, they did. It was quite a feat of engineering. It could accomplish almost all of your requirements. It was designed so that their astronauts would enter the latitude and longitude for their desired point of impact, and the computer would automatically generate the guidance instructions to deliver them there."

"Amazing."

"Sincerely, it is, but on the negative side, since the Gemini was a manned spacecraft, such a computer would require a man aboard to update it. I stringently doubt that an entirely autonomous control system could be made that small."

"Agreed," said Abdirov.

"Moreover, I doubt that it would work effectively with our reentry vehicles. The Gemini reentry vehicle is aerodynamically shaped and its weight distributed so that it generates lift. Part of the their reentry process involved rolling the spacecraft during reentry, in a controlled manner, so that it remained oriented on the correct trajectory. Conversely, our reentry vehicles are predominately round, so they don't generate lift. Consequently, like with the *Vostoks* and *Zenits*, the reentry is more of a ballistic trajectory, which cannot be appreciably altered."

"Interesting points," noted Abdirov. "Certainly food for thought. If only I could lay hands on one of those machines…"

"Perhaps you might speak with the GRU," replied Yohzin, signaling his driver. "They have developed quite a knack for acquiring technology from the Americans."

17

THE FORTY-EIGHT HOUR QUESTION

Flight Operations, Eglin Air Force Base, Florida
9:35 a.m., Monday, August 12, 1968

Drew Carson hung up the phone, grinned, and tucked his little black book into the sleeve pocket of his Nomex flight suit. Today's schedule would be tight, but they could still rendezvous at their usual place and time. He just had to convince Ourecky to keep his yap shut about the arrangement, but it really shouldn't be that difficult for him to intimidate the engineer into silence.

He picked the phone up again to file his flight plan. Now, there was little to do but wait for Ourecky to jock up. Leaning over the operations counter, he studied the weather charts as he finished the stale remnants of a cheese sandwich bought from a vending machine in the pilot's lounge. He interpreted the meteorological symbols, precisely drawn in grease pencil on a Plexiglas sheet covering a map of the eastern United States, and saw that the weather couldn't be any more favorable. It looked like uninterrupted smooth air between here and Ohio. "Is this your most current forecast?" he asked the airman behind the counter.

"It is, sir," replied the airman, reading a paper printout as it spooled from a clattering teletype machine. "Do you want a quick briefing, sir?"

"No need," answered Carson, noting the projected winds for the Gulf of Mexico. Wishing that he had thought to bring along his cut-down toothbrush, he swigged down the last of his coffee, crumpled the paper cup, and tossed it into a trash can next to the airman's desk.

He consulted his watch—a massive Omega Flightmaster that he had bought last week—and cursed Ourecky for being slow. Behind him, he heard the distinctive squeak of boot soles on linoleum and turned to see the engineer exiting the aircrew locker room.

Outfitted in spanking new flight gear, smiling like a raccoon in a henhouse, Ourecky carried a "B-4" suit bag in one hand and a pristine white flight helmet in the other. Seeing him, Carson was immediately filled with disgust. Scarcely a year ago, he had graduated at the top of his class at ARPS—Aerospace Research Pilot School—and now he was relegated to being a glorified taxi driver. *Was there any way that he could win back Wolcott's favor? At this point, was it even worth it?*

"Wipe that stupid grin off your snout, Captain," snarled Carson. "Quit dawdling and bring your dumb ass over here. Who draped that parachute harness on you?"

"Uh, one of the riggers helped me with it, sir," answered Ourecky, tugging at the unfamiliar webbing. "Something wrong?"

"Yeah, there's something wrong. *Terribly* wrong. For starters, you don't rate wearing that flight gear. Besides that, it's too loose. Here, let me lend you a hand."

Ourecky obediently stood still as Carson yanked several straps.

"This is a little uncomfortable," noted Ourecky. He appeared to be in considerable pain.

"It's a parachute harness, Captain. It's not meant to be comfortable. It's designed to save your stupid ass. Grab your junk and let's move." Walking comfortably upright, Carson confidently strolled out the door toward the flight line. Hunched over and struggling to walk, Ourecky trudged close behind, dragging his suitcase, grunting with every painful step. To a distant observer, the two probably looked like an organ grinder and his faithful but overworked monkey.

A gentle breeze blew from the south, lofting the faint mixed scents of salt air, jet fuel, and sunbaked asphalt. The busy field was awash with the sounds of planes taxiing, taking off, and landing. As they approached the sleek T-38, an airman threw Carson a brisk salute. Carson casually returned it, as if the honor rendered was a passing nuisance. "All tanked up and ready, Major," said the airman, handing Carson a clipboard with the service sheet.

Carson glanced at the numbers on the sheet, signed it, and handed the clipboard back to the airman. He jerked his thumb towards the cockpit and said, "See to it that this captain gets buttoned up in the back seat while I pre-flight. Got any air sickness bags?"

"I do, sir," answered the airman.

"Then lend him some extras. And make damned sure he knows how to use them."

Carson turned to Ourecky. "This is your pre-flight briefing," he said curtly. "Strap in and shut up. Don't you dare touch any controls, or I'll chop your hands off. Any questions?"

Donning his helmet, Ourecky shook his head.

"Didn't think so," noted Carson brusquely. "Pass me that B-4."

Ourecky handed the fabric suitcase to Carson, clambered up the ladder, and awkwardly wormed into the rear seat. Carson jammed the suitcase into a luggage pod mounted under the starboard wing, latched the pod closed, and immediately rolled into his pre-flight check.

Minutes later, they roared off into the blue skies of a perfect Florida morning. Carson retracted the landing gear before thumbing the microphone switch on the stick. "Eglin Tower, this is Reaper Four Four. Request divert from filed flight plan to Area Two Charlie."

The tower answered immediately: "Reaper Four Four, your divert is approved. Continue climbing to five thousand and turn left to One-Eight-Zero. Maintain five thousand until feet wet."

"Eglin Tower, Reaper Four Four, climbing to five thousand and will turn left to One-Eight-Zero." Carson completed the climb and banked the aircraft south. Almost immediately, they crossed the narrow white strand of Gulf beaches and were cruising over azure open water.

Over the intercom, Ourecky asked, "Uh, Major, shouldn't we be headed *north*?"

"I'm going to link up with some Navy buddies," answered Carson, checking the map clipped to his kneeboard and dialing in the TACAN beacon channel for the USS Lexington. The Lex, an aircraft carrier used to train fledgling Navy aviators, was virtually a permanent fixture on the Gulf. The rendezvous plan was to meet at ten thousand feet over the ship.

His Navy friends were Pensacola-based instructor pilots. Eager for an opportunity to go head-to-head with the Air Force pilot, they had cadged a pair of Douglas A-4 Skyhawks for a "proficiency" flight. Both were experienced combat vets with three Vietnam cruises apiece.

As much as he despised the notion of ferrying Ourecky around, Carson relished the opportunity for some aerial combat practice. He lived to fly and never spent enough time in the cockpit, particularly since he was now logging more time in the Box than in the air.

Although Blue Gemini seemed so intriguing at the outset—and it still was, to a certain extent—it was really wearing on him. Few good pilots would forgo even the slightest opportunity to fly into space. Carson had volunteered, knowing that he would likely never receive public recognition, particularly like the endless adulation lavished upon the NASA astronauts, and that his actions would never be annotated on his official records.

He also knew he would never wear silver astronaut wings on his Air Force uniform. To him, that was probably one of the most cruelly ironic aspects of this invisible space program; if he ever rode a rocket, he would soar more than twice as high as the eight Air Force pilots bestowed with astronaut wings for flying the X-15 above fifty miles. Aside from the X-15 pilots, the MOL "Can Men"—the Blue Gemini pilots' derisive moniker for the Manned Orbiting Laboratory crew members—would also be awarded astronaut wings upon their return to earth.

Carson was becoming ever less confident that he would ever ascend to orbit, even though he was by far the best pilot of the six—no, make that *five*—assigned to Blue Gemini. There were persistent rumors that the MOL program was on the verge of being cancelled. Even though many aspects of the MOL mission were highly classified, that program was at least visible in the public eye. If a publically recognized effort like the MOL could be eliminated—despite the legions of congressmen who would fight to keep it alive because it was so important to their districts—then certainly their quiet little space program could also suddenly fade into oblivion.

Although he still desired to fly in space, what Carson wanted *now*, more than anything else that he had ever desired, was an opportunity to fly in Vietnam. He could be the greatest aviator since Rickenbacker, but that meant absolutely nothing if he didn't prove his mettle against an enemy in the air. His uncle, a fighter ace who had downed eight German aircraft, constantly reminded him: *There is no substitute for experience in combat.*

But national sentiment was quickly turning against the war, so Carson's chance to fly in combat was swiftly fading. Vietnam was likely to be over soon—at least for the United States—and he saw no good shooting wars looming on the horizon.

He had long yearned to be a general, but he was painfully aware that his decision to aim for the stars—in secret—now made it increasingly less likely that he would ever wear stars on his epaulettes. Despite a glowing record—academic and athletic excellence at West Point, top marks in flight school, superb ratings in his first squadron assignments, early selection to test pilot school—his Air Force career could plummet into obscurity if combat experience was not prominently reflected on his records. Without exception, every single one of his flight school classmates had already flown in Southeast Asia. Granted, some of them had been killed and others were languishing in POW camps, but they had seen *combat*, and he had not.

And while Blue Gemini might be of great strategic significance, this interlude would appear as a glaring blank spot on his military resume. Carson had some insight into how promotion boards worked; he knew that *some* future board members would glimpse that void and understand what it really meant, but most would interpret it as cowardly shirking of the combat duty shared by his contemporaries. No, just as his uncle asserted, there was only one legitimate way to earn his spurs: *There is no substitute for experience in combat.*

Shielding his eyes from the sun, Carson saw the Lex's churning wake in the distance; the rendezvous point was near. "Listen to me, Ourecky," he said over the intercom. "We're making a detour to log some combat training. Yank out an airsick bag and be prepared to use it. You know how to unhook your oxygen mask, right? You remember that from egress training?"

"I do, sir."

"Good. Whatever you do, don't barf in your mask, or you'll have big problems. When I start maneuvering, keep your head and eyes locked to the front. If you turn your head while I'm making a fast turn, you'll never recover and you'll be puking for the rest of the flight. Ready?"

"I am, sir."

Carson threw back his head, laughed like a man possessed, and then thumbed the microphone switch. "This is *Reaper* Four Four. I am five miles north of the RV at Angels Ten."

On the bootleg "bump" radio frequency, an exceedingly calm voice answered: "Reaper Four Four, this is Badger. Snuffy and I are at your ten o'clock, Angels Twelve."

Carson looked up and to the left. He saw the two small A-4 Skyhawks. "Badger, I have a visual. Tally Ho, Navy pig."

"Reaper, this is Badger. Tally Ho. Prepare to die." The two A-4s immediately broke to the left and began maneuvering for the merge. The fight was on.

Most of Ourecky's forward view was obstructed by Carson's ejection seat, so he saw very little of the close-in action. Still, he was mesmerized by the speed of the engagement. After a few brief minutes of twisting and turning through the sky, Carson had maneuvered in behind one of the nimble little A-4s. But the Navy pilot definitely wasn't throwing in the towel and tried everything in his arsenal to escape Carson's closing grasp.

As the T-38 snapped and pitched through the sky, with his stomach sickeningly following just a step or two behind, Ourecky remembered his childhood trips to the state fairgrounds in Omaha. From his perspective, the dogfight was like a giant rollercoaster, but without the rickety steel rails and the tacit certainty that he would safely return to earth when the ride was over.

Ourecky had initially suspected that Carson conjured up the mock dogfight to make him violently ill and reluctant to ever hitch aboard the T-38 in the future. But now it was obvious that Carson was intently focused on the engagement. His maneuvers were smooth and methodical; it was as if the plane was being steered by a machine. Like rendezvous and orbital mechanics, aerial combat was an exercise in energy management, except in a far more violent and dynamic form, and Carson was obviously a consummate master of its nuances. It seemed as if he was always think- ing at least a blink ahead of his adversary, deftly handling the controls to subtly dissipate energy or add it on to maintain the advantage.

Keeping his eyes locked forward, Ourecky followed Carson's instruc- tions. Soon he discovered that if he focused on his "eight-ball," the flight attitude indicator, it was easier to mesh his ear's vestibular system with what his eyes saw and what his sloshing stomach felt. After a while, he was even comfortable looking around in the cockpit.

"Got you, Snuffy!" boasted Carson, closing the deal on the third engagement. "Splash one!"

"Where's Badger? Where is he?" asked Carson aloud, banking the plane into gentle S-turns and craning his head to scan the stark blue sky.

"I see him," exclaimed Ourecky, turning his head back sharply to look to the left rear.

There was a moment of silence, and then Carson exploded over the intercom: "Did I not *tell* you to keep your head and eyes to the damned front? Can you not do what I tell you to do?"

Seconds later, Carson furtively asked, "So where is he? Are you *positive* you can see him?"

"Yes, sir. He's a couple of miles back behind you. He drifts out slightly to your seven o'clock when you turn and then pops back in behind you. He's trying to close the gap."

"Okay. Hold on. We're going steep." Carson waited a few seconds and then racked the agile trainer into a hard left break. The Navy pilot reacted just a half-second late, and Carson quickly assumed the upper hand, seal- ing the deal in less than a minute.

The skirmishes went on for another thirty minutes. Carson didn't win every engagement, but he more than held his own, considering that he was flying against two seasoned Naval Aviators with two MiG kills apiece. The three planes flew in tight formation until they neared the coastline, and then the Navy pilots waved, broke off, and headed for Pensacola.

"Hope you enjoyed that little jaunt, Ourecky. Maybe you should inform your stewardess girlfriend that I do know how to fly after all," said Carson, chuckling. "And make sure you've got your puke bags sealed up good."

"Uh, I didn't need to use any bags, sir," replied Ourecky.

"None?" asked Carson incredulously. *That's not possible. A newbie like Ourecky couldn't possibly still be in possession of his breakfast.* He unlatched his oxygen mask, drew in a deep breath from the cool flow and then swung the rubber mask to the side. Taking a slight sniff, he expected to gag; it was impossible to conceal the acrid smell of vomit in the close confines of the T-38. But there was nothing but the usual odors of a fighter cockpit. He sniffed again. *Nothing.* Then he quickly sucked in a whole breath. *Still nothing. It just wasn't possible.*

"Everything okay, Major?" asked Ourecky.

"Yeah, it's okay," replied Carson, swinging his oxygen mask back into place. He dialed in the next checkpoint, quickly scanned his instruments, and verified the radio settings. "Look, Ourecky, we keep this dogfighting business to ourselves. Understood? Absolutely no mention to Virgil Wolcott or anyone else, or I crunch you. Got it?"

"Our secret, sir," replied Ourecky from the back seat.

Wright Arms Apartments, Dayton, Ohio
9:00 p.m., Thursday, August 15, 1968

"Did you like the movie?" he asked.

"I guess," she said, fumbling with her keys. "But I didn't completely understand it. One minute, a monkey is throwing a bone in the air and the next thing you know, astronauts are walking on the moon. I have no idea what that black slab was supposed to represent, and why they kept finding it everywhere. And that spaceship just looked like a big sperm cell to me."

Ourecky blushed. "Uh, I didn't notice that," he said.

"Really? I suppose it was some kind of metaphor or symbol," she commented. "And that last part was just a little too crazy for me. It reminded me of an LSD trip, not that I've ever dropped any acid. And I didn't understand about the baby at the end. Do you suppose that was meant to be that Dave Bowman guy? I guess it was all a little over my head."

Bea's apartment was tiny. The furnishings were sparse; besides a well-worn couch, there was a phonograph and a small television. To enhance reception, the television antenna's rabbit ears were adorned with shiny clumps of aluminum foil. A coffee table held a small collection of fashion and Delta in-flight magazines. The other end of the room was a miniscule kitchenette.

"I need to wash up," he said. He held out his hands and spread his fingers, as if presenting them for close order inspection; they were still greasy with popcorn butter.

She nodded, and pointed at her bedroom door. "Through there, on the right, just past my dresser," she replied, sitting down on the couch and slipping out of her shoes. "Excuse the mess. It's been a hectic week, and I haven't had a chance to pick up."

He navigated his way through the clutter of her bedroom and found the bathroom. It wasn't much larger than a cramped lavatory on an airliner. There were nickel-sized spots of rust in the sink. On the glass shelf above the sink, a blue toothbrush with flayed bristles stood upright in a Delta coffee mug, next to a half-empty tube of Pepsodent wintergreen toothpaste. The shelf also was home to her make-up, a small atomizer of Houbigant Chantilly perfume, and a round plastic dispenser of birth control pills.

He turned on the tap, let the water run until it was warm, and washed his hands with Ivory soap. Shutting off the flow, he heard her muffled voice from the living room: "There's a clean towel in that little cabinet over the toilet."

He found the towel, dried his hands, and retraced his short route back to the living room. As he walked through the door, she struck a match and lit a pair of candles on an end table next to the couch. "Wine?" she asked, switching off the table lamp. "I think I still have some."

"Uh, sure, Bea. That would be nice."

Rummaging through the refrigerator, she found a bottle. Carrying a green bottle in one hand and two mismatched wine glasses in the other, she floated back to the couch on bare feet. "I'm going to put on a record. You like the Beatles? I think they're really groovy." She turned, knelt down, and flipped through a collection of LP albums in a wooden milk crate.

"They're okay," he said, admiring the small of her back as he poured the wine.

Bea placed the album on the spindle and switched on the phonograph. She watched as the black disk dropped and the arm swung over. "I'm saving my pennies to buy a new record player," she observed, sitting down next to him on the couch. "I've just about worn out this one."

He tried not to stare at her. She seemed even more beautiful in the flickering light of the candles. Her perfume smelled vaguely like flowers. He was physically drawn to her, but beyond that, she was an enigma. He had yet to comprehend why she would be attracted to him. "I hardly know anything about you, Bea," he commented.

"There's not much to know," she replied. She took a sip of the wine and set the glass on the coffee table. "You've seen enough of my life to know what it's like. I fly five days a week. Same route, mostly the same people every time. It used to be exciting, but now it's just another job."

"But that's just what you do for a living. There has to be more. How about your family? Do you have any brothers or sisters?"

Closing her eyes, she slowly shook her head and murmured, "No. Just little old me."

"How about your parents? Do they live here in Dayton?"

"No, they're gone now. My dad was killed in Korea. My mum remarried after that, but then my stepfather died in '62. Mum passed away three years ago." She sniffed and reached for a Kleenex, but the cardboard box on the coffee table was empty. She reached into her purse and found a tissue. "I really miss my mum and dad."

"Tell me about them," he said. "Please."

"My parents met in England during the War," she said, daubing her eyes before wiping her nose. "He was in the Army Air Corps. He met my mum at a dance on the base, and they were married a few weeks later. He was twenty and she was seventeen."

"Sounds nice. Very romantic."

The record skipped. The words "*Roll up for the mystery tour…*" droned over and over. Bea stood up to fix it. "Yeah, I suppose it would have been endless days of bliss and rose petals if he hadn't been shot down a few weeks before I was born. He ended up in a German POW camp."

Bea continued. "He came back to Molesworth after he was released, and then he rotated back here, to Wright Field. My mum and me weren't able to ship over for another year. He stayed in the Army for a couple of years after the war and then went to college on the GI Bill."

"What did he study?" asked Ourecky, gently swirling the wine in his glass.

"Engineering. Same as you."

"So your dad was an engineer? That's ironic."

She closed her eyes, sighed, and leaned her head against his shoulder. "That's what he went to college to be, anyway. Scott, those four years were the best time in my life. My dad would come home from class and spread out his books on the kitchen table. Sometimes, I would sit on the floor and play with my dolls, but mostly I would just watch him study. I was very young, but I remember that he had a slide rule, just like yours, because I used to believe it was a magic ruler. And he always had this really serious, focused look on his face while he was studying, but every once in a while he would look up at me and smile. I *really* miss that smile."

"Where was your mother?" asked Ourecky.

"She took in ironing and kept me during the day until my dad came home from classes, and then she went to her job taking tickets at the theater. I used to struggle to stay awake until Mum came home, and then the three of us would spend time together before they put me to bed. We were a real family, at least for a while. That's what I miss most. Does that make any sense?"

Ourecky nodded. "Yeah, Bea, I think I understand." Actually, he was having a difficult time picturing her childhood; his own had been spent in a bustling farmhouse with his parents, three brothers, two sisters, and his paternal grandparents.

She sat up and ran her fingers through her hair. "But then Dad went back into the military right after he graduated. Not the Army; it was the

Air Force by then. He went to OCS, then straight to flight school, and right after that he went to Korea."

"What happened to him?"

"He was shot down on my eighth birthday, 1952. That's why I don't celebrate it anymore."

"So you stayed here in Ohio?"

"I wish. My mum didn't have her citizenship yet, and we were packing up to move back to Molesworth when Dad's wingman showed up." Bea closed her eyes and gritted her teeth. "Captain Ted Andersen. Actually came to the door with flowers, no less. He acted like he was on some sort of mission, like he was duty-bound to take care of me and Mum. My mum was in a fragile state emotionally, so it didn't take much effort for Ted to sweep her off her feet."

"So you think he felt obligated to marry your mother?"

"Oh, maybe. But I think it's likely that he saw it more as an opportunity than an obligation. He roomed with my father in Korea. I'm sure that my dad had that picture up somewhere," she said, gesturing at a sepia-toned photograph hanging on the wall behind the couch. "It was my dad's favorite picture of her. And I'm sure Ted probably drooled over it every day."

Ourecky turned to look at the picture of Bea's parents, obviously taken on their wedding day. "She's beautiful," he declared. "I mean, she *was* beautiful. You look exactly like her."

Bea smiled at him. "Thank you. Anyway, dear old Ted Andersen, my dad's best buddy, married my mother on Christmas day that year, and then our lives left on a conveyor belt. We were like nomads. We lived everywhere: Germany twice, Okinawa, Italy, Hawaii, the Azores, and Spain. We were only in the States twice, one year in California and one year in Alabama."

Ourecky could hardly imagine growing up in such exotic locales. In his Nebraska childhood, the only changes in the scenery came with the seasons. "Wow. That must have been exciting."

"Exciting?" she asked, laughing. "Oh yeah, Scott, it was just *groovy*. Hunky dory wonderful. We rarely spent much more than a year anywhere, and somehow the Air Force *always* managed to transfer my stepdad right in the middle of the school year."

"School? I hadn't even considered that. Where did you go to school?"

"There was always a government issue school for Air Force brats wherever we happened to land. I thought it was a great life, right up until I was in the sixth grade in Montgomery, and I went to a regular school. I was mixed in with normal kids, and I realized that I didn't have any real friends like they had. I guess that with all the moving, I had just never learned how to make friends. And I think as I started getting older, I learned to intentionally *not* make friends, because it was always so painful to say goodbye and leave them behind.

"Anyway, the whole time we were moving around, it was just my mum and me," she said. "It seemed like Ted was always off on a temporary duty assignment somewhere. If he was home, he spent every waking hour at work, flying, or at the officer's club. My mum used to go to the club with him, but she got tired of that and stayed home with me instead. And to make matters worse, he would chase anything with a skirt. It was pretty obvious, and he and my mum used to have such horrible fights about it, but she wouldn't leave him."

Bea took a sip of her wine. "We were stationed in Spain when he died. I remember it like it was yesterday. It was prom night. Boy, was that pathetic. Nine girls and five boys in our senior class. Definitely the offspring of fighter pilots."

"Huh?"

Bea grinned. "For some reason, fighter pilots seem to have more girls than boys. I'm sure that it's never been scientifically proven and no one seems to know why, but it's definitely what I saw growing up. *Always* more girls than boys. What a pain. I didn't have a date for the prom, but my mum insisted that I go anyway. She was pinning on my corsage when the squadron commander and chaplain showed up at the door."

"What happened?"

"Dear old Ted suffered an engine flame-out out over the Med, near Gibraltar. He ejected, but his parachute didn't open. It was really hard for me to act sad when they gave us the news. I had to hold back the urge to break out laughing. There I was, with my hair all done up, standing there in my hand-me-down pink chiffon dress and my mum was in her old flannel housecoat. We gripped each other tight, and I pretended to be overcome with grief, but I could tell that she wanted to laugh also. Isn't that sad, that we could be so overjoyed at something so tragic?"

"Well, if he was that bad a guy and was so hurtful to your mother, then maybe it wasn't that tragic after all," he observed.

"I suppose. I just wish that he had never shown up on our doorstep."

"So you came back here after your stepfather was killed?"

She nodded. "Dayton was the closest thing we ever had to a home. My grandfather was killed in the War, and my grandmother passed away in '57, so there wasn't much sense in me and my mum going back to England. So we came back here."

"Bea, what happened to your mother?"

"Breast cancer," she replied, staring at the picture of her parents. "What a horrible way to die. It took a over a year, and she suffered so." The record ended, and the room was filled with uncomfortable silence. One of the candles sputtered and then slowly flickered out. Minutes passed before she quietly spoke. "And here we are, Scott. That's my life."

"Well, here we are," he echoed. He summoned his courage and kissed her lightly on the lips. He sensed that she was on the verge of crying and held her close until the tears came.

Aerospace Support Project; 10:00 a.m., Monday, August 19, 1968

"We're short on time," said Mark Tew. "So we'll just stick to what's really pertinent. Personnel?"

Thompson, the personnel officer, stood out of his chair. "All of our head counts remain the same as last briefed. Sir, I've also inquired about a replacement for Major Agnew. The bottom line is that we won't be allocated any pilot replacements for at least another year."

Tew grimaced, slumped over in his chair, and grabbed his stomach.

"You okay, Mark?" asked Wolcott. "Do we need to stop now?" He had never seen Tew look so bad; the general's eyes were bloodshot, and he was sweating profusely.

Tew painfully shook his head. "I'm okay," he croaked. Looking up, he gestured at a man standing by the bookshelves behind his desk. "Sergeant, look in my top right drawer and fetch that bottle to me. There's a glass, also. Bring that as well."

The sergeant retrieved a blue medicine bottle and a shot glass. Tew filled the glass with milk of magnesia, brought it to his lips and quickly gulped down the white liquid. He wiped his mouth with the back of his hand. "Intelligence?" he asked. "Colonel Seibert?"

"Not much to report," replied Seibert. "We're seeing clues that the Soviets are preparing to launch a Zond probe to the moon. All indications are that it's a lead-up to a manned flight, possibly early next year. I really don't think that the Russians are going to lie down and let us beat them to the moon. That's all, sir."

Tew nodded. "Operations and training?"

"On track," declared Wolcott. "Nothing significant to report."

"Logistics?"

Rhodes, the portly logistics officer, leaned forward and spoke. "Construction and preparation of the Pacific Departure Facility is now back on schedule. The dredging is complete for the approach channel. As a fortuitous development, we were able to use our scrap concrete to complete a breakwater and pier shoring for the channel. The Navy is bringing in their LSTs for a test run next week."

"Good," noted Tew. He held his stomach and quietly groaned.

"Are you *sure* you're going to be all right, Mark?" asked Wolcott. "There ain't nothin' so pressin' that we can't hold off until tomorrow."

Tew shook his head. "Anything else?" he asked.

Russo cleared his throat and said, "General, we're hearing rumblings that the Space Task Group is considering shutting down the MOL." The Space Task Group was an interagency oversight committee that ensured the nation's money was effectively spent on space efforts.

"I've heard," said Tew. "Gentlemen, I can tell you that I don't feel overly confident about the future. If Hubert Humphrey grabs the Democratic nomination and wins in November, I don't think that we'll ever fly. I think we stood a better chance if Bobby Kennedy had lived to be on the Democratic ticket, but Humphrey will likely let NASA go to the moon a few times and then all but shut down the space program after that. I'm sure he won't be supportive of what *we're* doing."

"And Nixon, sir? What if he runs on the GOP ticket?" asked Seibert.

"Dick Nixon will fall in with us, pardner. He's a smart guy, and he knows what's at stake," observed Wolcott. "Now, hombres, unless someone has anything really earth-shattering to discuss, why don't we lend General Tew a break?"

"Okay, any other pressing issues?" asked Tew. No one replied. "Good. Gunter, Virgil, stay behind for a minute. The rest of you are released."

Wolcott waited for the others to disperse before lighting a cigarette. Tew weakly shook his head. Wolcott snuffed out the cigarette in the ashtray and asked, "What's on your mind, Mark?"

Tew filled another shot glass with milk of magnesia. Closing his eyes and gritting his teeth, he slugged down the chalky liquid. "In January, we go to Washington to present a decision brief to Hugh Kittredge and his key staff," he said. "If we haven't met our key milestones by then, it's almost a certainty that they will shut us down. Between the three of us, I need to know whether we'll be ready or not. I can't tell Kittredge that we're confident of success if we haven't proven the concept. Without the Block Two computer, we're asking these boys to fly up to forty-eight hours unassisted, and we're not even sure we can break the twenty-four hour mark."

Wolcott shrugged. "Mark, we have no control over the Block Two, if that's the steer you're lookin' to rope. Right now, the boys at MIT are totally focused on getting Apollo to the moon, and they don't have much time left over for us. I can't push them to deliver."

"Okay," said Tew. "Let's look at the worst case scenario. Realistically, can we even fly without the Block Two?"

"Yes and no," answered Heydrich. "*Ja*, I think that we can prepare at least two crews to execute by June, with just the existing hardware. But one of the critical milestones is that we validate a full-up forty-eight-hour simulation by January. That's one of the items you'll be briefing at the next In-Progress Review, Mark."

"Gunter, I am *painfully* aware of that. How close are we to clocking forty-eight hours?"

"We ain't, boss." Wolcott sniffed. "Not by a long shot. None of the crews have crossed the twenty-four-hour line yet, so it ain't extremely likely that they'll make forty-eight hours before we brief Hugh Kittredge. But in any event, if anyone can pull it off, I would lay my money on Carson."

"So, Gunter, what's *your* assessment?" asked Tew. "Can we make it by January?"

"At this point, I agree with Virgil that Carson is our strongest contender to make the forty-eight-hour mark by January," answered Heydrich. "Of course, Carson can't do it by himself, so all of this is contingent on finding someone to fill that right-hand seat."

"Could you excuse us a minute, Gunter?" asked Wolcott.

Heydrich nodded, stood up, picked up his notebooks, and headed for the door. "I'll be outside," he said over his shoulder.

"Virgil, we need to make this quick," said Tew, glancing at his Timex watch. "I have a doctor's appointment in twenty minutes."

"Ulcers acting up again?" asked Wolcott.

Tew nodded solemnly. "Heart, too. Virgil, this damned Project is killing me. Literally." He screwed the lid back on the blue bottle. "So what's on your mind?"

Wolcott leaned back and folded his arms across his chest. "Look, after everything we've been through over the years, this is as close as we've ever come to puttin' *our* guys in orbit. Mark, we are *so* close. Ain't you willin' to stretch the truth just a hair?"

Tew shook his head. "No, Virgil. As much as I want to put someone up there, I am not willing to deceive anyone, including Hugh Kittredge, just to make it happen. Either we make the milestone or we don't. It's yes or no, black or white. No wiggle room, friend."

Flustered, Wolcott put his elbows on the table and rested his chin in his hands. "Mark, pardner, I'm beggin' you to reconsider. We've been killin' ourselves for the past two years, and it all boils down to whether you're willing to fudge just one little bullet on a briefing chart. You know full well we won't launch a crew in June if we're not ready, but we *will* be ready by then. Please, Mark, just this once, please be willing to bend a little."

"No."

Wolcott looked at the ceiling, slowly shook his head, and groaned. "Okay, buddy, okay. I ain't wantin' to play this card, but there's another option."

"Virgil, there are *no* other options. It's very cut and dry. We either make it to forty-eight hours by January or we don't. That simple."

Wolcott stood up and walked to bookshelves lining the wall behind their desks. He opened a black binder and placed it in front of Tew. The page detailed the various milestones that had to be completed at different intervals during the Project. "Right there, pardner," he exclaimed, holding his finger on a line of text. "Right there it is, in black and white. That's our savin' grace."

Tew cited the line aloud: "*Demonstrate the capability to complete an unassisted non-cooperative rendezvous by successfully executing a full-profile mission simulation of forty-eight hours duration.*" He sighed and added, "Virgil, we've covered this ground a thousand times."

Grinning, Wolcott said, "Mark, that line says we have to successfully execute a forty-eight-hour full-up simulation, but it doesn't explicitly specify that we complete it with a *flight* crew."

"I'm obviously still missing something here, Virgil. Care to elucidate?"

"Look, Mark, Gunter is confident that Carson can make forty-eight hours. I concur. The trick is wranglin' someone to stick in the right seat."

"And we've covered that ground, too," declared Tew. His voice had been calm, but now it was showing the first traces of anger. "We're *out* of options, Virgil. There is *no* one else."

"That ain't entirely true, Mark. I think that this engineer Ourecky could stick it out for forty-eight hours. Look, pard, he's a danged robot. The kid doesn't sleep. He just works and works and works. So we pair him up with Carson in the Box."

"*Ourecky?* Pair Carson with Ourecky? Ourecky's not a pilot. He's on loan to us, no less. What the hell would this prove?" Tew demanded, slamming the binder shut.

"Accordin' to the rules as they are written, it doesn't matter *who* we stick in the Box, so long as they make the forty-eight hours. Amigo, as it is, Carson and Ourecky look to be the most promising shot to clear this hurdle before January. If they're successful, then you can go to your briefing, tell the bosses we can deliver the mail on a forty-eight-hour run, and not have to lie. That will buy us sufficient time to bring the other two crews up to speed, and it also buys us time for the Block Two to be delivered."

Saying nothing, Tew slowly shook his head. "But what happens to Carson afterwards?"

Wolcott put his hands behind his neck and loudly cracked his knuckles, a habit that Tew found especially annoying. "Look, pard, here's my game plan. We team Ourecky with Carson. We prove the concepts, refine the procedures, and keep Carson sharp. You'll have your chart bullet by January. The other crews continue training. At the opportune time, we look at the other right-seaters and snatch the pick of the litter to replace Ourecky. That simple, *jefe*."

"How about Ourecky? After what happened with Agnew, we would be asking a lot of him."

"Ourecky? Pardner, Ourecky considers it a privilege just to sit in the Box. You ain't goin' to hear any griping from him. And besides, if we lose him, it ain't like we're losin' one of our pilots."

"I'm still not convinced, Virgil," Tew said. "It's a mighty convoluted scheme." He started to reach for the blue bottle, looked at his watch, and thought better of it.

"Come on, Mark, give it a chance. There's nothing to lose and much to be gained. Let's not let all of our work get flushed down the tube. Give me this. No, give *us* this. *All of us*."

"Okay, okay," conceded Tew. "I'll go along with this, provided that you can assure me that we can be ready in June."

"We'll be ready, but we'll have to shift the line-up on the Box to give Carson and Ourecky top priority, at least until January. I'll need you to yank some strings to arrange for Ourecky to be brought up here full-time. We can't pull this off if we're still time-sharin' him with Eglin."

"Done," replied Tew, picking up the phone and dialing a number.

2:30 p.m.

"Busy, Virgil?" asked Jimmy Hara.

"No, come on in, pardner," drawled Wolcott. He was seated at the conference table, trying to make sense of several stacks of paperwork before him. "What's on your mind?

"Surveillance report, Virg. General Tew asked me to keep an eye on one of your temporary workers," said Hara. "Captain Ourecky?"

"Yup. I do recall Mark asking you to do that," said Wolcott, lighting a cigarette. "Look, Mark is sort of indisposed right now. Why don't you just give it to me, hoss?"

Hara nodded and sat down across the table from Wolcott. "Virg, I recall you saying that Ourecky never leaves the base. That's not exactly true. He's apparently met a girl here. She's quite a looker, too." Hara handed Wolcott a surveillance photo taken at the Falcon Club.

"*Woo doggy,*" Wolcott said, looking at the photo. "I have to agree with you there, Jimmy. What an armful. She's a looker, all right. Got a name on her?"

"So far, just a first name: *Bea*. She's apparently a stewardess."

"Hmm," said Wolcott. "Stewardess? You know, it's funny. She looks a lot like that girl on TV. You know, that show with the astronaut and the…"

"Genie," interjected Hara. "I thought so, too. She's just about a dead ringer, except for the hair. Anyway, as you can see, she's kind of hard to miss. I've seen her around town before Ourecky met her. I think every one of your pilots has hit on her. I watched Carson make a pass at her a few weeks ago, and he ended up going down in flames. She's probably the only woman in Dayton who's ever been to the Falcon Club who didn't want to be picked up by a pilot."

"And yet she goes out with Ourecky?" asked Wolcott, shaking his head. He sure hadn't foreseen this situation, but he could certainly see potential problems if the relationship got more serious and Ourecky lost focus on his work.

"Sir, I really doubt that she'll remain interested in him very long. She doesn't strike me as someone who has an extremely long attention span with men." Hara slid the envelope of pictures across the table to Wolcott. "Will there be anything else, Virg?" he asked.

"Yeah, Jimmy. Stay on this for a couple more weeks. Develop a more detailed background on this woman and then circle back to me."

Hara nodded. "Virgil, if you don't mind me asking, is there any particular reason that you're focusing on Ourecky? He doesn't strike me as a key player in the game, but you're sure asking me to devote a lot of time on him."

"Hombre, he's about to become considerably more involved in the very near future. I s'pose you're right about this woman, that it's goin' to be over as quick as it started, but we can't afford for him to become too distracted. We also don't want anyone paying too much attention to his comings and goings."

"So, Virgil, you're going to put Ourecky back in the Box with Major Carson?"

Wolcott was too surprised to respond. He had *just* had that conversation with Tew. Finally, he said, "So you know about that?"

Hara nodded. "With Major Carson's track record, maybe I could give Ourecky some jiujutsu lessons," he said, smiling. "It might level the playing field a little."

"No, pard, I think Carson's going to be on his best behavior from here on out," said Wolcott. He paused and added, "Jimmy, is there *anything* you don't know?"

Hara thought for a moment before replying. "Honestly, Virgil, I guess I don't know."

Parking Lot 20, Wright-Patterson Air Force Base, Ohio
2:30 p.m., Monday, August 19, 1968

As he watched the activities in front of Hangar Three, where he suspected that alien flying saucers were stored and analyzed, Eric Yost sipped from a pint bottle of Old Crow. He had slightly more than an hour before he was due to report for his night shift assignment, so he knew that he had to taper off on the booze lest he be discovered by the snot-nosed lieutenant who lorded over the warehouse.

Yost had monitored Hangar Three for the past several weeks and was certain that something extremely odd was going on within its brick facade. His buddy, Bob Carr, at the hospital, had provided some very interesting details concerning the apparent accident back in July. At Yost's request, Carr checked the hospital's records and determined that the incident involved a pilot—Major Tim Agnew—who was admitted briefly to the emergency room before being placed in strict isolation on the psychiatric ward. Carr claimed that Agnew was gone two days later, with no record of transfer or other documentation concerning his disposition. He had not been discharged back to duty, nor had he been transferred to another hospital; he had just vanished. If that wasn't strange enough, Carr swore that the official admissions register had been altered to eliminate any trace that Agnew was even there in the first place.

Presently, something fairly momentous was obviously happening at Hangar Three. It appeared as if they were making ready for a significant delivery of something. Although the hangar's doors were opened wide, temporary traffic barriers and guards were carefully positioned to prevent any passersby from inadvertently looking inside. Moreover, curtains of white muslin "target cloth" were strategically placed inside the hangar as well. Yost saw that two massive Hyster H-100 diesel forklifts, each capable of handling up to ten thousand pounds, had been staged in the parking lot near the hangar's entrance.

Yost couldn't believe that they would be so audacious as to move another alien spacecraft in broad daylight, but he was curious all the same. He was on the verge of falling asleep when three semi-trailer tractors arrived, each towing a wide-body flatbed trailer loaded with variously shaped crates. As he watched men emerge from the hangar and set immediately to work, he realized that the reception had been carefully orchestrated to ensure minimal exposure of the arriving items. Billowing black smoke, the pair of heavy capacity forklifts roared into operation, quickly shuttling boxes and crates into the hangar.

The first objects unloaded were two enormous circular crates; by Yost's estimates, they were roughly twenty feet in diameter and five feet tall. He gasped. The size and configuration of the crates lent considerable credence to his theory that Hangar Three was a repository for captured flying saucers. Now he understood how they could be delivered in plain sight without anyone—anyone who didn't know any better, that is—paying any undue attention.

Because he had been woefully unprepared, he had no substantive evidence of the alien spacecraft he had seen a month ago. This time, he was ready. He picked up a Kodak Instamatic 104 camera—pre-loaded with a film cartridge—from the passenger seat. In the ten minutes it took for the three trailers to be unloaded and their contents moved into the hangar, Yost snapped forty-eight exposures—two entire "126" cartridges—of Kodachrome color film. Even as the hangar doors closed as the tractor-trailers pulled away, Yost quickly loaded a third film cartridge, anticipating that there may be yet another delivery.

Wanting to maintain his edge, Yost popped a Dexedrine tablet into his mouth and washed it down with a swig of bourbon. The prescription amphetamines, regularly issued to pilots in combat conditions, had been thoughtfully provided by his friend Carr. The orange pills enabled Yost to endure a typical day with less than four hours of sleep.

Capturing the UFO delivery on film was a coup, but there was yet another reason for Yost to mark this day in red letters. Obviously sensing that he was worthy of their trust, his friends at the warehouse had invited him to an after-hours poker game in Dayton. Although they played cards virtually every night, at least after the lieutenant took his leave and departed for the Officers Club, the stakes rarely exceeded matchsticks or pocket change. Tonight's game was the real deal, according to his cohorts, with some fairly significant pots to be won.

Before putting the lid back on the plastic medicine bottle, he quickly inventoried his supply. He had five tablets left, which should be plenty enough to get him through his shift and still have at least a couple held in reserve. Although Yost considered himself to be a proficient card player and skilled gambler, he wanted to ensure that he was wide awake and alert when there was real money on the line.

18

TRUTH BE TOLD

Aerospace Support Project
10:00 a.m., Tuesday, August 20, 1968

Dreading his impending conversation with Carson, Wolcott gulped down his coffee, grabbed his Stetson, and made his way upstairs to the Blue Gemini's Life Support facility. Manned by three civilian contractors, the facility was the repository for spacesuits, helmets, survival kits and associated equipment. Occupying almost a third of the building's fourth floor, it consisted of a suit maintenance workshop and a climate-controlled storage chamber.

Each pilot was allocated three identical spacesuits: one for training, one for flight, and one as a back-up. Each custom-made suit cost over a hundred thousand dollars. Designed for the harshest of environments, the multi-layered garments would protect the wearer in the absolute vacuum of space, where temperatures could easily range from 250 degrees below zero to 250 degrees above, and could change from one extreme to the other in the blink of an eye.

The suits were of a new design, adapted from a model intended for the MOL program. They were far easier to don and doff than the G3C suits worn by the NASA Gemini astronauts, and were durably constructed to be worn on multiple flights. Lacking NASA's massive budget, the Air Force viewed the ensembles as work clothes, intended to be used over and over, instead of being worn once before being placed on permanent display in an aerospace museum.

Like a gaggle of spinsters at a quilting bee, the technicians doted over the suits, ensuring that every piece was correctly adjusted and in perfect working order. They nagged after the pilots to watch what they ate, since even the slightest weight fluctuation would cause fitting problems. They fussed whenever the pilots returned the cumbersome suits with even the slightest smudge or trace of dirt after wearing them for hours in the Box.

Carson was being fitted for his newly arrived back-up suit. A technician was helping him out of the ensemble as Wolcott arrived. Watching as the lanky pilot wiggled and squirmed to free his shoulders, the scene reminded Wolcott of a butterfly painfully emerging from a snug cocoon.

"New Sunday-go-to-meetin' duds?" asked Wolcott. "You look mighty dapper, pard. Very spiffy. I hope your tailor can do something with those lapels, though."

Carson laughed. "Yeah, and they need to let out the inseams a bit, too. It's a bit snug in the crotch." With his shoulders free, he slowly pulled the suit down until it was bunched around his waist. He then sat down on a reclining chair to allow the technician to finish tugging the suit off.

"Joel, how are you?" asked Wolcott. "Takin' good care of my friend Carson here?"

The suit technician nodded as he fastidiously laid out the suit in a full-length storage box. Wolcott couldn't help but notice that the garment looked vaguely like a headless body in a coffin.

"Can you lend us a minute, Joel? Why don't you grab a smoke or something?"

"Sure thing, Virgil. I need to write up my records on the fitting, so I'll be in the office," replied the technician. He handed Carson a set of fingernail clippers and an emery board. "I know you want to play with those gloves a bit longer, Drew. You should trim your nails so they don't snag on the liners. I'll fetch those mitts in a few minutes and pack them away when you're done."

"Here's the deal, hoss," said Wolcott, sliding into a chair next to Carson. "We need to make some significant adjustments over the next few weeks."

Carson carefully pared his nails as he listened to Wolcott relate the details of his elaborate scheme to meet the January deadline. He set the clippers aside and eased his hands into the pressure suit gloves just as Wolcott described Ourecky's role in the plan.

Stunned, Carson mumbled, "So I'll be flying the Box with *Ourecky* from now on?"

"Just until January. We have to hit this milestone, hombre. Mark Tew and I have hashed this out seven ways from Sunday, and we don't see a way to manage it without Ourecky. You do this thing for us, and we'll fix you up with a new partner in ample time to make the big dance. You can take your pick from the other crews. I *promise* you can have anyone you want."

"*Anyone* I want? Then why not just do that *now*? Sticking Ourecky in the Box is a huge waste of time, Virgil. He's an engineer, not a pilot. He's not going to fly. It sure seems more practical to assign me a right-seater who has some chance of making the trip."

Wolcott was losing his patience. He took a deep breath and then exhaled slowly and deliberately. Carson was an exceptional pilot, but he would argue the hide off a bull if granted an opportunity. "Look, pardner, we're absolutely slap out of options. Yeah, you're right: Ourecky ain't goin' to fly, but right now he's your *last* chance to grab a ticket to orbit. I can assure you that Tew ain't going to budge on this issue. We either hit the mark, or we don't. I'm startin' to believe that Mark would be just as happy if the whole program was scrubbed."

Carson looked at the floor and shook his head. "Then maybe it's just as well," he muttered.

"Look, pard, if you have something on your mind, then just spit it out."

Carson said, "To be honest, Virgil, I'm starting to doubt that we'll ever leave the ground. I used to enjoy the training, but now it's gotten to be so much drudgery. I joined the Air Force to fly, and now it seems like I spend most of my time trapped in a cage like a lab monkey, watching lights flash and pushing buttons. At least the monkeys are rewarded with banana pellets; my reward is being stuffed back into the Box."

"But that's part of the game, buster. That's what you signed on for. We made it very clear that you boys would be in intensive training for up to two years before you made your first flight."

Testing the gloves' dexterity, Carson flexed his fingers. "I grew up in a boxing gym, Virg, so I know exactly how important training is. It's just really difficult to maintain this pace when it just seems like I'll never have a chance to climb into the ring."

Sighing, Wolcott tilted his Stetson back on his head. As invaluable as he was to Blue Gemini, Carson could be painfully frustrating to deal with. He was like a mustang that rode well on the range, but couldn't bear to be pinned up in a corral. But despite all that, Wolcott saw much of himself in the cantankerous test pilot. "Well, Carson, you can write it on the wall that you won't be ridin' a rocket unless we achieve this forty-eight hour mark. I can assure you that we can't do it before January without you and Ourecky workin' together. That's the bottom line, pardner."

Looking pensive, Carson peeled off the bulky gloves and laid them on the table. "Okay. Virgil, I'll put forth my maximum effort between now and January. I'll bring Ourecky up to speed, and we'll make the forty-eight hour mark, if you're willing to grant me something in return."

"Go ahead, pardner. The sky's the limit. Just ask away."

"I want to fly in Vietnam, Virgil. In *combat*."

It took a few seconds for Wolcott to compose himself. "Well, you can ask for danged near anything, pard. But that just ain't goin' to happen. *Ever*."

"But I'm not asking for a full tour over there," countered Carson. "Just a couple of weeks to log some combat time. Then I would fall right back into the training cycle. No one would miss me."

Wolcott shook his head. "Maybe you're just not takin' in the whole panorama here, hoss. This is a seriously classified endeavor. We can't ever risk puttin' you in a situation where there's even the slightest possibility that you could be shot down or captured."

"But…"

"I know you hanker to fly in combat, Carson. That's what all fighter pilots train for and ultimately live for, but you need to accept the fact that it ain't ever going to happen. I feel for you, hoss, I truly do, but we told you this at the outset, and you still signed on."

Wolcott's answer didn't appear to do much to alleviate Carson's disappointment, but the pilot solemnly shook his head, as if resigned to his fate.

"One more thing," said Wolcott, twirling a silver-tipped end of his bolo string tie. "You know we're headed up to Alaska in October to do cold weather drops with the paraglider, correct?"

"I am. I'm looking forward to it," replied Carson, turning to face Wolcott.

"Well, that's good, particularly since I'm puttin' Ourecky in your right seat for the drops. Gunter is working the bugs out of that new landing simulator. As soon as it's ready, I want you to start workin' with Ourecky on it whenever it's available, when you two aren't in the Box."

"You're kidding me, aren't you, Virg?" asked Carson incredulously, shaking his head.

"Nope. I don't know if you've caught on to it yet, pard, but we're just fresh out of pilots here, and they're just not offerin' any more for sale down at the Base Exchange. We have to plant someone in the right seat of that paraglider rig, so why not Ourecky? Are you tellin' me that you can't train him to handle his side?"

"But it's a *live* drop, Virg. It's not a canned ride in a simulator. It's a *hazardous* undertaking."

"So is it any more danged *hazardous* than riding in the back seat of your T-38 when you're flat-hattin' and gallivantin' your way across the country?" asked Wolcott, studying Carson's reaction. "Huh, pardner? Think I didn't know about your dogfighting shenanigans?"

His face reddening, Carson was silent. "So Ourecky finked on me?" he asked quietly.

Wolcott laughed. "He didn't have to, pard. He kept his mouth shut, but you apparently forgot that we pay for your fuel tickets up here. You also must have forgot that I've flown more than a few times, and I know exactly how much gas it takes to fly from Eglin to here. So don't be so durned anxious to blame Ourecky because you got caught with your pants 'round your ankles."

"I'm sorry, sir."

"If you say so, pard. Shucks, you sure weren't doin' anything that the rest of us ain't done in the past. Frankly, I would probably think less of you if I knew you weren't out there mixin' it up every once in a while." Wolcott spit brown tobacco juice in a trash can, grinned, and asked, "So how did our young egghead take to dogfighting? One bag? Two? Three maybe?"

Carson shook his head. "*None*. Not a one."

"None?"

"*None* at all. The guy must have an iron stomach. I've never seen anything like it."

"Interesting," noted Wolcott, scratching his chin. "Very interesting."

8:30 a.m., Wednesday, August 21, 1968

"Hey Mark, the Democrats are kickin' off their convention," observed Wolcott, perusing the headlines of the local newspaper. "Looks like they're expectin' a big showdown in Chicago, especially since LBJ won't run and they can't geehaw on their platform for Vietnam. The police are expectin' a lot of trouble from Yippies and hippies and anyone else obligated to make a nuisance of themselves. What a dadblamed mess! Can't people even act civil anymore?"

Tew didn't look up but continued to read a technical report describing recent problems encountered with the Gemini-I's landing gear. Their use of the paraglider hinged on the proper functioning of the landing gear, which were essentially simple skids that emerged from wells in the underside of the spacecraft after the paraglider was deployed. In recent tests, engineers discovered that the skid well hatches had a tendency to stick when the vehicle was subjected to the blowtorch heat of reentry and then immediately exposed to extremely cold temperatures.

NASA had originally considered using the paraglider to land the Gemini after returning from orbit but had abandoned the scheme in favor of a much simpler parachute splashdown at sea. Tew pondered whether they should have taken the same path, except that the splashdown recoveries required the deployment of a massive armada of ships and aircraft girdling the globe.

As an alternative, the paraglider would allow them to quietly return their Gemini-I spacecraft without fanfare. Moreover, avoiding exposure to the highly corrosive effects of saltwater would permit the spacecraft to be re-used, while NASA's Gemini capsules were effectively ruined at splashdown.

But despite the paraglider's benefits, Tew was immensely concerned that they were staking the lives of these pilots on too much untried and inadequately tested technology. Of course, on the plus side, the long-awaited Paraglider Landing Simulator had finally been delivered by the contractor two days ago and was currently being assembled and tested in the simulator hangar. According to Heydrich, it should be in operation within a month or so.

"Whew. Looks like ol' Ike ain't goin' to make it," proclaimed Wolcott. "Sounds like his ticker is just about ready to give out. Can you believe this? *Twenty* people have offered to donate their heart to Ike. Do you suppose that maybe they just don't understand the consequences?"

Tew was not in the mood for Wolcott's recitation of the morning's news. His stomach was a churning cauldron of bile and acid, and his heart thumped heavily in his chest, even though he wasn't exerting himself in the slightest. His stomach growled audibly. He was hungry, but detested the notion of yet another bland meal of melba toast dunked in buttermilk.

Looming on the desk before him was a four-inch stack of proposed expenditures that required his approval, five thick binders of test results, and a draft plan for the recovery network that had to be in place for the first three missions. Tew was weary of the incessant paperwork, and weary of this project in general. His wife had long since grown tired of his routine, and had returned to their home in California several months ago.

"And if that isn't enough, hombre, just take a gander at this one," exclaimed Wolcott, pointing at an ominous headline. "*Warsaw Pact invades Czechoslovakia*. I just can't believe it. They crossed the border with nearly a million troops. Chisel my words in stone, amigo: this will continue to escalate and we'll be in a shootin' war with the Russians before this decade is out. No more of this danged proxy business. It'll be standin' room only at the O.K. Corral."

Tew sighed and then looked up to see Ourecky standing in the doorway. "You wanted to see me, sirs?" asked Ourecky.

Wolcott greeted him. "Come on in and have a sit-down, brother. Coffee? I think Stan just brewed up a fresh pot. Strong, though. Stout enough to float a colt."

"No thank you, sir," answered Ourecky, taking a seat at the conference table. Attired in black slacks and a short-sleeved white shirt with a dark blue tie, he had apparently adopted the de facto uniform of Heydrich's simulator staff.

Tew set aside the landing gear report and said, "We have a rather awkward situation to contend with." As Ourecky listened intently, Tew described the absolute necessity of completing a forty-eight-hour simulation by January. "And so that's it," he concluded. "We need you to go into the Box with Carson between now and then so we can hit this milestone."

Ourecky said nothing, but just gazed down at the table, as if he was taking a moment to process the new information.

"I know that's a lot to think about," said Tew. "Do you have any questions, Captain?"

"Well, uh, sir, we're way behind on preparing the formulas to integrate into the new computer, and I still have at least two months of work yet on the terminal guidance controller project back at Eglin. General, I'm not questioning your plan, but are you sure that Major Carson couldn't work with one of the other pilots? After all, if you ever have to execute this contingency to intercept a satellite, they'll be the ones to actually fly—"

"We're sure, pardner," interjected Wolcott. "*Absolutely* sure. You're it."

Tew opened a manila folder and then slipped some papers across the table. "Those are amendments to your orders, Ourecky. You won't be going back to Eglin, except to pack up your belongings and move up here. As of this morning, you are permanently assigned to this effort, so you needn't be concerned about juggling your time between projects."

"Son, we're askin' you to make some huge sacrifices for the team, but we ain't offerin' you anything but long hours and misery," confided Wolcott. "The Box might be a novel experience for you right now, but I know you'll dread it in time. But I also know you're aware of the importance of Blue Gemini."

Tew watched for Ourecky to react, but the young engineer seemed to take it all in stride. Although he had set aside his reservations when he deferred to Wolcott's plan, Tew didn't think it was fair to subject Ourecky to the mental and physical agony of the simulator for the next few months, especially since he didn't stand to reap any of the rewards. And if the Box wasn't stressful enough, just coping with Carson could be an ordeal in itself.

Tew suddenly felt as if he had been slammed in the chest with a sledge-hammer. His vision was blurred, his fingers were numb, and he couldn't catch his breath. These spells were becoming more common, but he didn't want to tell his doctor about them because he knew the doctor would advise him to slow down and get more rest. *If only he had those options.*

"Mark, buddy, are you okay?" asked Wolcott. "Should we drive you over to the hospital?"

"I'm fine," croaked Tew, sipping from a glass of water. "Really, Virgil, I'm fine."

"Can this wait, sir?" asked Ourecky. Wide-eyed, he looked like a kid at the movies who had his first glimpse at Superman's vulnerability to green kryptonite. "I can come back later."

"No, I'm *fine*," said Tew weakly. "Captain, you've done excellent work for us so far. We just need you to shift your focus for a few months so we can transition from the theoretical to the practical. Can you do that for us, son?"

"I will, sir."

"And that's just what we wanted to hear, pardner," noted Wolcott.

10:30 a.m.

There was a knock at the door. Tew and Wolcott looked up simultaneously and saw Jimmy Hara standing there.

"Busy, sirs?" he asked. "Virgil asked me to do a follow-up on Ourecky's lady friend, sir."

"And what did you find out?" asked Tew.

"As best as we can tell, they've been dating for about the past month. She's a stew for Delta airlines. She did a year of junior college here in Dayton and then went to work for Delta three years ago. Her usual sched-ule has her flying down to Atlanta late Sunday afternoon, and then she does a cycle of morning flights from Atlanta to Dayton, then evening flights from Dayton to Atlanta, remaining overnight in Atlanta at one of the hotels near the airport. Normally, she gets back into Dayton on Thursday afternoon."

"So does she have a name, Hara?" asked Tew impatiently.

"Yes, sir." He handed Tew a folded scrap of paper. "A friend of mine at Delta's staffing office gave me a peek at her file. She's a dependable worker but looks to be a bit promiscuous. I guess it comes with the territory, doesn't it? Coffee, tea, or me?"

Tew casually unfolded the paper and read the name. The color quickly faded from his face. "Did you happen to glean any other information from her records, Hara?" he asked.

Hara nodded. "General, I did. Are you looking for something specific?"

"Did her records indicate where she was born?"

"Yes, sir," answered Hara, referring to his notebook. "Molesworth, England. July 15, 1944. I'm guessing that her father was an American serviceman of some sort."

"Someone you know, pardner?" asked Wolcott.

"Something like that," replied Tew. He handed the scrap of paper to Wolcott and then fumbled in a desk drawer for his ulcer pills. Over the course of the past few weeks, the drawer had taken on the appearance of a well-stocked pharmacy.

Wolcott read the name aloud: "Beatrice Anne Harper." He closed his eyes, apparently trying to divine why the name was so familiar. "Bea *Harper*? Now ain't this just a grand coincidence?" he exclaimed, laughing. "Surely there ain't too many women out there with that handle."

Tew nodded, choked down a couple of ulcer pills, and chased them with a glass of tepid water. "Very true. And I doubt that you'll find many born at Molesworth in 1944."

"Jimmy, you reckon Ourecky and this girl are serious?" asked Wolcott.

"It appears that way, Virgil."

"This could be a major problem, Mark," Wolcott said, addressing his concern to Tew. "We made danged sure that our pilots weren't hitched and didn't have any serious attachments. Right now, we have too many chips riding on Ourecky to treat him any different. Well, it looks like I'll just have to head this one off at the pass. That shouldn't be too hard to do if Miss Bea is only in town three nights a week. I'll just make some adjustments to Mr. Ourecky's training schedule that interfere with his social calendar, and this fire should die out just as just as quickly as it flared up. Just that simple."

Tew stared at his desk, stomach churning, silent in his thoughts. *No, cowboy, it wasn't quite that simple.*

"You okay, Mark? You ain't sufferin' another spell, are you?"

"I'm *fine*. Look, Virgil, leave Ourecky's schedule alone. I'll handle this situation myself."

"Suit yourself, pardner. But don't say that I didn't warn you."

19

HUMAN NATURE

Apex Minerals Exploration Inc., Dayton, Ohio
8:30 a.m., Thursday, August 22, 1968

Arriving at the Apex building, Henson checked his watch. He and two classmates had rented a modest apartment a few miles away. It wasn't in the best of neighborhoods, and had few amenities, but it was tolerable and far cheaper than living out of a motel.

Henson fastidiously watched his cash flow and still had a sizeable reserve held back. Some of his fellow students were not quite as fortunate. As he strolled down the narrow hallway toward the classroom, he glimpsed the small office that three of them now shared as a bedroom. In a brief spate of living the high life, the three had blown through the majority of their allotments. Now, the destitute trio was reduced to sleeping on the concrete floor, scrounging for meals and bathing out of the sink in the locker room.

Henson was sure that they would soon join the headstrong man who had smirked at Grau's admonitions to avoid women while operational. After squandering away the contents of his briefcase, he begged Grau for more cash. Instead, Grau sent him packing posthaste to Wright-Patterson, where he was sworn back into the Air Force before receiving his back pay, exactly as promised, along with a one-way ticket to a lonely radar station located in a remote corner of the frigid Alaskan wilderness. *Maybe that was Grau's last lesson for the lothario*, thought Henson. *Two years among the polar bears might finally teach him to keep his id in check.*

Henson took his seat in the classroom, and carefully unwrapped a fried egg sandwich. Listening to the persistent buzz of the light fixture above his head, he slowly ate half the sandwich. Even though his stomach still growled with hunger pangs, he handed the remainder to one of the three squatters. The famished man thanked Henson profusely before gobbling down the morsel. Henson closed his eyes and listened to the measured sound of water dripping into the coffee can in the corner. Unlike most of the men under Grau's tutelage, he had once been a college student, so he didn't need too many lessons on stretching a buck.

Well within his budget for the week, he even considered splurging on a Saturday matinee ticket to see *Planet of the Apes* at a theater downtown. Of course, that was contingent on whether he completed his Saturday homework in time; Grau had tasked each man to photograph activities at a specific location. Henson made a mental note to speak to Grau about his particular target for the assignment.

The steady drip was interrupted by a faint but familiar rhythm of scrapes and thumps. Henson recognized it as Grau's cane, and opened his eyes. Patient as an old box turtle, Grau slowly made his way to the front of the classroom. He turned to face the class and started to speak, but launched into yet another protracted coughing spell. He wiped his mouth with a blood-tinged handkerchief and then put it away. His hands fluttered like leaves in a stiff wind.

"Today we're going to talk briefly about human nature," said Grau weakly. His voice grew progressively stronger as he spoke. "Interacting with people has become a lost art, and the man who masters it stands to gain much. In your case, so long as you're with Apex, you must develop the capacity to rapidly adapt to new environments, assess the people around you, and swiftly cultivate the relationships that will enable you to accomplish your mission."

Grau took a sip of water, slowly swirled it in his mouth, and spat it into the coffee can in the corner. "Before we venture too far, there is one critical lesson that I want you to lock into your heads: Learn to value people. Wherever you go, everyone you meet has worth. From the dirtiest beggar groveling in a gutter to a king seated high on a gilded throne, every single life has value, although it may only be to the person living it.

"When you're finally sent afield, you'll be tasked to locate various types of resources, arrange for their use, and to gather information. For the most part, you will be on your own. As you roam the world, it's important that you understand that every single person you meet is a potential key to obtaining the resources and information that you require. The faster that you connect with them, the faster that you will be able to encourage them to provide what you need.

"Moreover," said Grau, scratching his nose, "since this process requires that we make that initial personal connection, we should be aware that all humans share a common need, a common desire. What is this desire?"

No one volunteered an answer.

"Oh, come now," urged Grau, tapping his cane on the floor. "What do all people desire?"

"Are you looking for something out of Maslow's Hierarchy of Needs?" offered Henson. "Oh, so you've had an education, Mister Henson?" Grau said. "Hah, that's a good college answer, but we need something that we can actually apply in the real world. So, here's the answer: every person desires to be *acknowledged*."

"Acknowledged?" asked Henson.

"That's right. Think about it. Regardless of their station in life, everyone wants to be acknowledged for something. They may not consciously ask for it, but they desire it all the same. A crying baby desires to be acknowledged with a bottle or a dry diaper. A brilliant man may be remote in his wisdom, but yearns to be acknowledged for his intelligence. A beautiful woman may act aloof, but yet she also craves to be acknowledged for her beauty. A lowly janitor may strive to fade into the background, but he wants to be acknowledged for the floors he sweeps and the toilets he scrubs clean.

"And when you're hauled down into some murky basement and you're facing a torturer who threatens to yank out your teeth with rusty pliers, you'll unconsciously acknowledge him with your fear," declared Grau. Momentarily grimacing, he closed his eye and rubbed his right cheek. "As I said, gentlemen, everyone desires to be acknowledged.

"A simple gesture of acknowledgement is the first step toward making a connection. Let me tell you, with few exceptions, most people yearn for human contact. They absolutely ache to tell their story. But they've been conditioned to be passive, so they wait for someone to come along and make that connection. Just a simple word of acknowledgement can mean so much. And few things are as painful as the sting of constantly being ignored. And if you understand that, and you make it a consistent habit to speak to people and acknowledge them, if even only to demonstrate that you recognize their presence, then you're well on your way to motivating people to do your bidding."

"So *what*?" asked a man to Henson's right. "What difference does any of this make?" A couple of the others gasped quietly, amazed that anyone would dare clash with Grau. Henson recognized the challenger as the squatter who accepted his leftover sandwich scraps.

"Good question," answered Grau calmly. "Let's imagine that I dispatch you to a developing country. Your first task is to locate a truck to move a heavy object—maybe a ton or so—from a remote location in the bush to a port, so that it can be loaded onto a freighter. Simple enough?"

"Sure," replied the man smugly. "I find a phone book, look in the yellow pages, find a pay phone, and make a few calls. Problem solved."

"Hah. But for most places you'll go, there are no yellow pages, and few phones, if any."

"Okay. So I just start *asking* people where to rent a truck," countered the man.

"And you'll annoy them. And word will spread about the foreigner in town. People will shun you, primarily because you spring right off the blocks presenting yourself as a stranger who is pursuing an object, and who is not concerned or even curious about the people around him."

"Again, I say, so what?" retorted the man. "I'll bet you anything that the word will eventually spread that I need to rent a truck, and someone will come to me with a truck to rent."

"*Hah*! You are *correct*," replied Grau. "But here are two thoughts for you. First, this is a complicated business. You'll rarely be sent anywhere with just one simple task. Most often, you'll have a considerable number of resources to locate before a job is entirely complete.

"Second, you must appreciate the nature of doing business in developing countries. Here, we are spoiled by the luxury of being able to directly ask for the object we seek. Outside our borders, it doesn't work that way. You *never* ask for anything directly. *Everything* is indirect. You have to make the connection with someone and then spend the time, however long it takes, learning about the person's family, where they're from, their aspirations, etc., etc. Only then, and there's really no good way to measure when *'then'* will be, do you casually mention that you might be interested in hiring a truck. And that's how things are accomplished in the Third World."

"I think I understand," ceded the man. Henson sensed that the man really didn't understand but was just abandoning his futile assault on Grau's logic. Henson had noticed that the man, like the others who had been in the military for a few years, were the slowest to grasp what Grau was teaching them. They were apparently too engrained with the regimented military approach to doing things, where everything was simply black and white, and they weren't too inclined to venture out into ambiguous realms of gray.

"Good," said Grau. "Now, most of the connections you'll make will be very superficial. In some instances, though, you'll need to develop a deeper interaction, because what you may require may not be nearly as simple as renting a truck or hiring a few strong backs to clear rocks from an airstrip. Developing this relationship will require you to assess the person to determine what motivates them, what makes them tick. In doing so, you'll learn what you'll need to nudge them in the direction you want them to go.

"What you most need to learn is how and when to apply the nudge," continued Grau. "And that's not something that I can readily show you. It must be learned with experience. In theory, with very few exceptions, virtually *anyone* can be enticed to do *anything*. It just comes down to giving them the appropriate nudge that convinces them to do your bidding."

Grau continued: "Now, if you're trying to motivate someone to do something illegal, immoral, or otherwise contrary to their interests, a significant nudge will likely be required."

"So, Mr. Grau, how do we know how much is enough?" asked the man sitting to Henson's left.

"You listen," replied Grau. "And you pay attention. Much of this comes with experience, but what you're trying to determine is your subject's threshold."

"Threshold?" asked the man. "So you're implying that everyone has a price?"

"I've never met anyone who didn't," asserted Grau. "But don't assume that a nudge will always involve money, because often something far more subtle will suffice. I'll tell you, motivating people with money often creates more problems than it solves, because you immediately and permanently reduce your interaction to a financial transaction."

11:35 a.m.

As the class broke for lunch, Henson waited for Grau. Since most of the men headed to the little diner at the end of the block, which had only five stools at the counter, there was usually a frantic rush for the door when Grau cut them loose. Only a few moments passed before Henson and the old man were alone in the dingy room. "Mr. Grau?" he said quietly.

"Mr. Henson," replied the old man, gazing up. "Something on your mind?"

Henson nodded. "It's about my weekend photography project, sir. You assigned me to photograph kids playing at a playground in Walnut Hills. Are you familiar with that area?"

"I am."

"Then I guess you know that this part of Walnut Hills is a *white* neighborhood." *Surely Grau knows that this is a mistake*, thought Henson, especially since he had given most of the other men assignments at industrial sites, bus stations, and similar locales. "I don't want to question your judgment, Mr. Grau. But I'm more than a little concerned about this assignment."

"You are?" Grau said, raising his eyebrows as if he was surprised.

"Maybe you're missing my point, sir," said Henson, in the calmest voice he could muster. "You're asking me, a black man, to photograph children in a white neighborhood on a Saturday morning. There's sure to be some parents around, and they're likely to make plenty of assumptions. None of those assumptions can be good. On one hand, they might decide I'm some kind of child molester, and on the other hand, they may think I'm an advance scout for the Black Panthers, and that we're cooking up a scheme to snatch their kids. In any event, I would rather not end up facedown in the gutter with my skull smashed open."

Grau studied Henson with his solitary eye. It was so unmoving and emotionless that Henson would have guessed it to be glass, except that he knew that wasn't possible. The corners of Grau's cracked lips turned up ever so slightly, almost like a faint smile. "No, Mister Henson, maybe you're missing *my* point. Your task is to take your pictures without drawing undue attention to yourself. So I suggest that you be careful."

Knowing that his assignment would not be changed, Henson shook his head. "I've learned a lot here, Mister Grau," he observed, starting to turn away. "And I'm sure that I still have much to learn. But there's one thing you'll never understand, and that's what it's like to be a black man in a white world, to always be an outsider who can never blend in."

"Oh, is *that* what this fuss is about?" scoffed Grau, fishing a photograph out of his wallet. It was a faded black-and-white picture of a younger Grau and a strikingly beautiful black woman. The woman wore a distinctive dress, like a kaftan common to West Africa, and looked to be roughly ten years younger than Grau. "That's my wife, Henson, God rest her soul. I met her not long after the War, when I was posted in Dakar. So I know just a bit about being an outsider."

Henson studied the photograph for a moment, noticing that the younger Grau's face held two eyes, and then replied, "You've made your point. But tell me, why do you insist on giving me the most difficult assignments? For once, can't I land something just a little easier, maybe something like you give the other guys?"

Flipping up his eye patch, Grau scratched the pink interior of the vacant socket and chuckled quietly. As much as it bothered him to gaze into Grau's intact eye, it really unnerved Henson to look into the empty hole where once an eye had been. Almost as if on cue, a fly landed on Grau's cheek. Passive as a monk, the old man didn't even flinch as the fly slowly crawled along his face and then disappeared into the dark void.

"Hah! Feeling a bit challenged, are we? Want an easier task?" Grau leaned forward, gave him a gentle nudge, and quietly said, "No. *Prove* yourself, Mr. Henson."

Simulator Facility, Aerospace Support Project
9:10 a.m., Friday, August 23, 1968

"Okay, gents, let's chat," said Wolcott. "We need to sort out a few things. Gunter?"

"*Ja?*" Heydrich apprehensively leaned back in his chair as if anticipating a firing squad's fusillade.

"Pard, I still want you to continue training the other two crews in the Box, but the priority is to support these two gentlemen so they can complete a forty-eight-hour full-up before January." Wolcott gestured to Carson and Ourecky, who were seated at the end of the table. Are we square, Gunter?"

"*Ja*. We are, Virgil," replied Heydrich.

"Carson, your job is to bring Captain Ourecky up to speed on the procedures and the Box. I want you to draw up a tentative schedule and have it on my desk by next Friday. Square, pard?"

Carson nodded obediently. "Will do, Virgil."

"And Ourecky, buster, make no mistake: you're the little engine that's pulling this whole danged railroad. I need you to do *exactly* what Carson says to do, so we can hit this mark by January to move on to the flight phase. If we fail, gents, it's *over*. Savvy? Any issues?"

"Ourecky will need a suit," observed Carson. "The criteria for that forty-eight-hour simulation require that we're suited up in full mission gear."

Like a Fleet Street tailor sizing up a client, Wolcott eyed Ourecky. "He's about Agnew's size, ain't he? We'll wrangle that one, pardner. I'll call Joel today to make the necessary arrangements. Anything else?"

Carson shook his head.

"Then you gentlemen need to get to work."

"Sit down," said Carson, taking a seat on the metal steps leading up to the simulator.

"This is *not* my idea, sir," asserted Ourecky. "I want you to know that."

"I know," replied Carson, wringing his hands. "Look, Ourecky, I don't like this situation, but I promised the boss that I would put forth my absolute best effort to make it happen. Where I come from, a man's word is all he has. I'm not going to lie to you. I don't like you, and I really don't care what you think about me, but I want you to promise me that you'll put forth your best effort between now and January, and that you'll do what I ask you to do. Agreed?"

Ourecky nodded.

"No," growled Carson. "That's not good enough. Promise. Out loud, Ourecky."

"Okay, Major Carson. I promise to put forth my best effort. I will do what you ask me to do."

"Good. First, I expect you to learn every inch of the vehicle, inside and out. I know you've been studying already, but we're going to set aside the books and spend our time with metal. We'll stay in the cockpit until I am absolutely confident that you know it intimately."

Unwrapping a stick of gum, Carson continued. "Then I'll tie a blindfold on you and you're going to show me that you know it with the lights out. When I'm comfortable that you know the vehicle and you know the procedures backwards and forwards, then we'll commence with the simulations, but not until then."

"Uh, sounds like a good plan, sir," replied Ourecky.

"There's more. You *will* meet me at the gym at six every morning, Monday through Friday, rain or shine. Sneakers and sweats. We'll lift some weights, do some running, wind sprints, skip rope, calisthenics, and mix it up in the ring. Whatever it takes to toughen you up, Captain."

"Are you sure that's necessary, Major? Uh, I think I'm in pretty good shape already, and I didn't hear the general say anything about—"

"You didn't hear? You didn't *hear*? Let me jog your memory," interjected Carson. "What *I* heard was Virgil Wolcott putting me in charge of training you. I also heard, *very* recently, a promise coming out of your mouth that you would put forth your best effort and that you would do whatever I ask you to do. Does that sound even vaguely familiar, Captain?"

"Yes, sir."

"Look, Ourecky, you've just dipped your toe in the water. You have absolutely no idea what you're in for. Just wait until you've finally been in the Box for twenty-four hours straight and you're completely wrung out and then you realize that you'll eventually have to go twice as long before this ordeal is finally over."

Carson leaned forward. "Listen to me, Captain. The Box is brutal beyond belief, and if you're not in top physical condition, it will *destroy* you. Agnew was a highly trained test pilot, and the Box reduced him to a blathering idiot, so you can just imagine what it can do to you if you're not adequately prepared."

"The gym. Six o'clock. Monday through Friday," noted Ourecky. "I'll be there."

Wright Arms Apartments, Dayton, Ohio
12:55 p.m., Friday, August 23, 1968

As they had arranged, Jimmy Hara called him from a pay phone in a diner three blocks away from Bea's apartment. The drive from his office had only taken twenty minutes. Tew glimpsed Hara parked on the street, and they acknowledged each other with a quick nod.

He parked his car next to a dilapidated playground and then slowly trudged up the three flights of stairs to her floor. Pausing at the top of the stairs, he caught his breath and chewed on a chalky antacid tablet. A cat screeched from one of the nearby apartments.

As he stood before Bea's door, one of her neighbors—a frumpy middle-aged woman in curlers and a faded house coat—walked by with an over-flowing laundry basket. Seeing Tew, she huffily stuck her nose in the air and cleared her throat, as if pronouncing judgment on him.

With his stomach in knots, Tew verified the apartment number one more time before he knocked on the door. Number Thirty-Four. Finally, he gathered his composure and rapped lightly on the veneer door. He heard faint noises on the other side.

Bea gradually opened the door until it was caught by the security chain, peered out, and then said, "Sorry, but I'm not interested in a vacuum cleaner if that's what you're selling. I don't need encyclopedias or brushes, either."

"Beatrice Harper?" he asked. He immediately recognized her eyes; they were exactly the same shade and intensity as her mother's.

"*Bea*," she declared. "*No* one calls me Beatrice."

"Your mother did. It was your grandmother's name."

With her eyebrows arched quizzically, she looked at him. "Who *are* you? Do I know you?"

"No, but I knew your father. I'm Mark Tew. I think you know Scott Ourecky. He works for me on the base." Tew pulled out his identification card and held it out to her. She peered at it through the narrow gap between the door and the doorframe.

Nodding as she unchained the door, she asked, "Why are you here?"

"I want to talk with you," he said. "Please."

"And Scott works for you?"

"He does."

"Okay. Come in. Excuse the mess. I've been out of town all week and haven't had much of an opportunity to pick up." She waved a hand at the couch. "Have a seat." The television was on; the words "*like sands through the hourglass, so are the days…*" were droning out of the speaker as she reached up to turn down the volume. "Can I bring you anything? Tab? Seven-Up? Coffee? I only have instant. Nescafe, if that's okay. I'll put the kettle on to boil some water."

He shook his head. "No, thank you."

She sat on the opposite end of the small couch. Barefoot, she was wearing bell-bottomed blue jeans and a white cotton peasant blouse with tiny flowers embroidered on the shoulders. A white towel was wrapped around her damp hair, like she had just emerged from the shower. "I'm sorry. I can be so scatterbrained sometimes. What was your name again?"

"Mark Tew," he replied. "General Mark Tew."

"Mark Tew? I'm sorry, but I'm drawing a blank with your name. You knew my father?"

"I did. I knew your mother, also. *And* your stepfather."

"Well, now," she said. "I suppose we could just be a big happy family then."

He pulled a photograph from a manila envelope and held it out to her. It showed several pilots, most attired in flight gear, in front of an F-86 Sabrejet fighter. "I brought this for you."

"Thanks, but I already have that same picture." She pointed at an identical photograph, mounted in a simple black frame, next to the portrait of her parents taken on their wedding day. "It's the only picture I have of my dad and my stepfather together." She studied the picture: all the men were young and vital, smiling with the confidence that they would live forever. "Tell me about my father. Was he a good pilot?"

"He was definitely all that and more," answered Tew. "I guess you know that this picture was taken in Korea, about a month before he was killed. I actually met your father…and your mother…during World War II, back in England."

"Really?"

"Really, Bea," he said. "And that's part of the reason that I came to talk to you today. I need to apologize to you for some things that I've done. I feel like I owe you that."

"You really don't owe me anything."

"Let me explain," he answered. "Here's how I met your father. I had been wounded on a mission over Bremen, and they made me the squadron maintenance officer after I was released from the hospital and still recuperating. Your father was an engine mechanic, so he worked for me, more or less. I remember when he met your mother, and I remember when they were married. And I'm sorry that your father wasn't there when you were born. That was my fault."

Bea shook her head. "It wasn't *your* fault. It was a war. I'm sure that thousands of fathers weren't there for the birth of their children. I'm sure that there are thousands of men who never got to see their babies at all. Anyway, I've gotten over it. My mum did, too."

He nodded. "You're right. But in your case, he could have been there for your mother. It was my fault that he wasn't, and I want to apologize to you."

"If you insist, but how could it have been your fault?" she asked.

"Your father was a mechanic," he explained. "Mechanics didn't fly on the missions because they were too valuable to lose. Your father always begged me to let him go up one time as a door gunner, just so he could know what it was like for the guys."

"Well, after I went back on flying status as a co-pilot, I snuck him aboard for one mission. Just one mission. Your mother was five months pregnant at the time, Bea. It looked like a safe flight, a milk run, so I let him hitch a ride with us."

"So now is the part where you tell me he was some kind of big hero back then?" she asked. "Just to make me feel better about him not being there for my mother? It's really not necessary. It was enough for me that he was my father. You don't have to make him out to be a hero."

Tew laughed quietly. "Hero? No, Bea. None of us were heroes. We all *acted* like brave warriors, but we were just a bunch of scared kids jammed together in that airplane, and we did our jobs because that's what was expected of us. To your father's credit, he stayed at his station until the very end, but I know he must have been frightened out of his wits. If anything, your father was a hero because he had the gumption to go up when he didn't have to."

"What happened that day?" asked Bea.

"Well, it definitely wasn't a milk run after all. The squadron lost three bombers almost as soon as we crossed into Germany. We made it to the target, but we were chewed up by flak during the bomb run. Lost two engines and a big chunk of our right wing.

"Our Fort was so damaged that we couldn't keep pace with the formation. We started falling back, and then we were jumped by two German ME-109's. Three men killed immediately. The plane was falling apart around us, so the pilot ordered us to hit the silk. Four of us made it out. We were all captured by the next morning. The Germans moved us around for a week, and then we landed up near Moosberg, close to Munich, in Stalag Seven-A."

"Was it bad? I don't know much about the war. I've seen some movies and things on TV…"

"Well, it definitely wasn't *Hogan's Heroes*," he replied. "By then, the guards knew that the game was all but over, and they knew the clock was just ticking down to the end."

Tew frowned. "I think we were more a nuisance than anything else, but they were obligated to at least keep us alive. The camp was jammed with Allied prisoners. Everyone was crawling with lice. It was always miserably cold and there was never enough to eat. I didn't see your father very often in the camp because they tried to keep the officers segregated from the enlisted men, but I knew that he was always thinking about your mother and you."

She smiled.

"He made that birth cup for you there in Moosberg," he said, pointing at a crude metal cup on a shelf by the television. "He made it out of a Klim can. Klim was powdered milk that came in the Red Cross ration packages. Empty Klim cans were like gold, because they were so useful. Lord only knows what he traded to land that one, or what he sacrificed to keep it."

Tew picked up the keepsake. Examining it, he saw her name—Beatrice Anne Harper—painstakingly stippled with the point of a nail; on the other side, there was another name inscribed: Charles Jacob Harper. "So, Bea, do you know why there are two names on this cup?"

"I do," she answered. "My mum told me after my dad was killed in Korea. Long before he was shot down, they had already decided to name me after her mother if I was a girl, and name me after his father if I was a boy. When he made that cup, he didn't know whether I came into the world as Bea or Charlie, so he put both names on there to be sure."

He smiled. "That's right. But look, Bea, there's more. I still feel that I owe you."

"And I told you: you owe me nothing," she replied, staring at the picture from Korea, focusing her attention on her stepfather. "My life has been complicated enough by men who showed up at the doorstep thinking that they owed me something."

"Listen to me," he pleaded. "After your father went to college, I convinced him to come back into the Air Force and I pulled the strings that got him into flight school."

She closed her eyes and rocked slowly back and forth.

"And, Bea, I pulled more strings to bring him over to Korea. All my life, I've pulled strings to move people around, like they were marionettes in a show I was staging, and I pulled your father's strings until it finally killed him. If it weren't for me, he would have been there for you and your mother all these years. He loved her so. He loved you both."

She shook her head. "It was his choice. That's how it was meant to be." She handed the picture back to him. "I think we've talked enough. You should go now."

"Bea, I didn't come here to talk about your parents. I came to talk to you about Scott."

"Scott? What about him?"

"I know that the two of you are dating."

"And how you would know *that*?" she demanded.

"I know a lot of things," he replied. "But that doesn't matter. What do you know about Scott? Has he told you very much about what he does?"

"Honestly? Not much. I know he's really smart, and I know he's an engineer."

Tew smiled. "Scott's smart, all right. He's far beyond smart. He's just plain brilliant."

"I know," she said. "I also know he's kind, gentle, and polite. I like him, a lot, especially since he's not like other guys I've been with. Most men are just in a hurry to get what they want, and when they've got it, they're in a hurry to leave."

Bea continued. "Scott's not like that at all. Now, if you're trying to determine if he gabs about his job, then I can tell you that he doesn't. I have no idea what he does on the base and I have no idea what he does in Florida, either. He just doesn't talk about it at all. I've asked him about it, but he just kind of hints that it's too boring to talk about."

"That's good," he said. "But I wouldn't expect anything different from him."

"Yeah, he keeps his mouth shut. Now I have a question for you."

"Yes?"

"So, General Tew, does Captain Ourecky require your approval to date someone?"

"No," he said. "You know, Bea, it's really ironic that we're talking about this, because back during the war, your father needed my official approval to marry your mother."

"You're kidding."

"I'm not. Because he wanted to marry an English citizen, your father had to come to me first to secure my permission, and then the squadron commander's, and then the wing commander's. The three of us had to sign a form so your parents could be hitched."

"That's funny," she said. "Now that you mention it, I have seen that form, but I didn't know what it meant at the time. My mother showed it to me before she died. She kept it in a scrapbook with their marriage license. So, they had to ask your blessings to be wed, oh Great and Powerful Oz. Thank you for my life, as it is. But what does this have to do with Scott and me?"

"Nothing, really. But I'm really curious how serious you are about Scott. I know you two have been seeing each other for several weeks. Do you love him?"

"What a *funny* question," she replied. "Like that's any of your business. Well, if you really must know, I'm not sure yet. I *like* him. I enjoy being around him. I think he's someone I could fall in love with, but I'm not there yet. Is that a good enough answer for you, General?"

He nodded. "Bea, Scott's a good man, and I don't want to see him hurt. I'm not implying that you intend to do that. In a way, I have a selfish motive. What he's working on is extremely important, and I would really prefer that he was able to focus on it without distractions."

"So you would rather just lock him in a closet somewhere?"

"I didn't say that. Bea, trust me, if I wanted to isolate him from any distractions, I could. That would be simple, but I'm not going to stand in the way of you and Scott seeing each other. All I ask is that you remember that Scott is working on something very special and very important, and he won't be able to talk to you about it. That might not seem meaningful now, but I can tell you from my own experience, it can really wear on a relationship."

"So, he's working on something secret? Something classified?"

Tew nodded, and then added, "Just because something is classified doesn't mean that it's particularly interesting or exciting. It's just something that we can't talk about."

"I think I can live with that," she said. "But tell me: Does Scott fly?"

"No, he's an engineer. Most of his work is purely theoretical."

"Oh, really? I assumed that if he was working with Drew Carson, then there must be some flying involved."

"You know Major Carson?"

"I know *of* him," answered Bea. "I know enough to steer clear of him. I only wish that my friend Jill had known to do that."

"Well, yes, Carson is assigned to our project," confirmed Tew. "We have some other pilots as well. They're involved in flight-testing the systems and equipment that we're working on. Look, Bea, if you're going to be involved with Scott, then you should know that we have to test systems in all sorts of conditions, so you shouldn't be surprised if he has to travel a lot. We expect our engineers to be hands-on, to solve problems immediately as they happen."

Bea nodded. "Well, since you've been so kind as to come here to talk, I can hold up my end of the bargain. Whatever Scott's involved with can stay behind the gates at the base; no matter how curious I become, I won't pester him about it. But can you promise me that you won't make a pilot out of him? I've promised myself that I wouldn't allow another pilot to break my heart."

"I don't see that happening, Bea," said Tew. "Scott's a good engineer, and he's content with what he does." He started to check his watch and then felt a momentary stabbing sensation in his chest. "Bea, I had better head back to the base before they miss me. Thanks for taking the time to talk. And listen, Scott doesn't need to know that we've talked. Agreed?"

"Cross my heart," she said, escorting him to the door.

As he drove back to the base, Tew thought of what he could not bear to tell Bea. More so than anyone else in the world, except possibly the MiG pilot who pulled the trigger on him, Tew was responsible for her father's death.

The events of that day stood out starkly in his memory. Harper and Andersen were on strip alert at Suwon on a scorching hot July morning, suited up, sitting in their planes, anxiously waiting a call to action. And Tew brought it to them. There was an urgent request to provide cover for a crippled F-86 struggling to return to base; the Sabrejet was being dogged by three MiGs, and the rest of the squadron was fully engaged further north.

Tew recalled sprinting from the Operations Quonset hut to the gravel alert strip, frantically giving the two pilots the spin-up signal with his finger. Harper's engine started right on cue; Andersen's engine just refused to turn over. Even though it was near homicide to scramble a fighter without a wingman, especially against three MiGs, he ordered Harper to launch anyway. He remembered Harper taxiing for takeoff, grinning at him from the cockpit. A cheerful smile, a quick salute, a playful wave and Harper was gone…forever.

20

MAMA'S BEST CHICKEN

Airport View Apartments, Vandalia, Ohio
5:25 p.m., Friday, August 23, 1968

Henson planted himself on the rickety picnic bench in front of the apartment building. He checked his watch; the guy in 6-B normally got home roughly at 5:30, and by 5:45 he was usually locked into a heated squabble with his wife.

Tugging the tab from a cold can of beer, Henson mentally hashed over the names he had gathered during a brief visit at the Dayton City Hall. Two buildings down, someone was grilling pork chops on a hibachi. The fragrant, greasy smoke hung low in the motionless afternoon air. Salivating, he wished that his dinner menu held more substantial fare than two slices of stale white bread adorned by cold baloney and processed cheese.

Slowly savoring the beer, he watched some kids gathering under a spreading oak tree. Most brandished toy guns, and they took turns at playing US paratroopers or Viet Cong guerrillas. After watching them for several minutes, Henson discerned a recurring pattern. The Viet Cong would disappear into the dank storm drains that emptied into a massive flume adjacent to the apartments. They would lurk patiently, defending their concrete tunnels against the American GI's sent underground to ferret them out. He observed that the VC always seemed destined to lose, but not before reaping their fair share of American lives in the bargain. Surely, the kids knew nothing of tactics and war, but their campaign wasn't much different from the brutal conflict being played out several thousand miles to the west.

A blue city-owned Ford pickup truck cruised into the parking lot just as the kids dispersed to their battle stations. Its motor sputtered and coughed several times before it finally gave up the ghost. A short white man in blue dungarees stepped down from the truck's cab; he clutched a dented black lunch pail in one hand, and a thick set of jangling keys in the other. Grossly overweight, he looked to be in his mid-thirties.

Henson intended to coax the man into a short visit to Walnut Hills tomorrow. He figured that it would cost him roughly ten bucks to close the deal, but he also guessed that the man would likely be eager for any excuse to be unfettered from his shrewish spouse on a Saturday morning.

"You work for the city?" asked Henson as the man drew near. He took a sip from his beer.

"Huh?" answered the man, obviously wondering how Henson could know that. He looked back toward the parking lot. "Oh, the truck. Yeah. Past fifteen years."

"Me, too," declared Henson. "Just hired on this morning. Part-time, anyway. I doubt I'll ever rate a truck, though. Not for a while, anyway."

"You said you were hired part-time?"

"Yeah. Just to scratch out some extra coin. I work another job out by the airport during the week. I came here from New Orleans a month ago, and my uncle arranged it so I could work as an independent contractor on the weekends. I might go full-time this winter, though, handling a shovel on a salt truck. I hear the city benefits are good."

"Can't complain. Hey, brother, that beer's looking mighty good there. Spare one?"

A 707 roared by low overhead, making its final approach to the airport. With the oppressive jet noise reminding him why their rent was so cheap, Henson waited for the plane to pass over before replying, "Help yourself, babe."

"So who's your uncle?" asked the man, sitting down on the opposite bench. The picnic table shuddered as he shifted his considerable girth to face Henson. He pulled the tab from a glistening can of Carling Black Label, discarded it on the ground, and took a long swallow.

"Calvin Washington. Everyone calls him Cal."

"Sure. Cal supervises one of Street Maintenance crews. Hard worker for a n—" The man didn't finish his thought. "So you'll be assigned to Street Maintenance?"

"No. Not yet, anyway. Right now, I'm supposed to work for Bob Hendricks." Henson swatted a mosquito drawing blood from his left forearm.

"Oh, sure. Parks and Recreation. No kidding? What will you be doing for P & R?"

"Mostly I'll be mowing grass at playgrounds and ball fields," responded Henson. "Plus they want me to inspect and maintain the playground equipment. I'm supposed to make sure the hardware is tightened up on the swings, slides, and other stuff. Safety, you know."

"Makes sense," observed the man. He used a pocketknife to peel a curling strip of green paint from the table. "Several kids have been hurt on those old playgrounds. It's a damned wonder that some parents haven't sued the city yet." The man took another deep swig from the beer. "So when do you start?"

"This weekend. I'm supposed to go look at a playground and fill out an inspection form. I think it's really just a test, to see if I'm dependable." Henson reached in his pocket and extracted a scrap of paper, which he handed to the man. "Does this address look familiar? I figure I'll just walk over there in the morning, while it's still cool."

The man studied the address. He guzzled the beer and chucked the empty can into a nearby trashcan. "Oh, man, that's *Walnut Hills*," he muttered. "That's cross town, man, down near the university. That would be an awfully long stroll."

"That's what I was afraid of. I don't have a car and I hate riding the damned bus. Takes forever to go anywhere. Hey, you want another frosty one? Looks like you're a might thirsty."

"If you *insist*," replied the man, popping the top from another can. "So what's your name?"

"Matt. Matt Henson."

"I'm Ted Shouk," said the man, extending his hand to shake Henson's. "Good to meet you, Matt. Hey, I won't need my city truck tomorrow. Since you're doing city business, you want to take it down to Walnut Hills?" Shouk took a swallow from the beer, then removed a key from his ring and handed it to Henson. "Straight down and straight back, no side trips, okay?"

The benevolent gesture wasn't what Henson expected, but it wasn't something he could refuse, either. He nodded and solemnly accepted the key as if it would unlock the gold vaults at Fort Knox. "Straight down and straight back, no side trips. Man, thanks. I'm indebted to you."

"It's nothing. Just drop the key in my mail slot when you're done. Apartment 6-B. There's also a box of tools in the truck bed if you need them." Shouk swiveled around, swung his legs out from under the bench, stood up and stretched. "Hey, thanks much for the brew, Matt. Good to meet you. Take good care of the truck, okay?"

Henson nodded and watched Shouk stroll away and climb up the stairs to his apartment. Shouk had no sooner closed the door behind him before the yelling started. Sipping his beer, Henson listened to the playful shouts echoing from the storm drains; the stalwart Screaming Eagles had yet to finish their assault against the wily VC, but he had achieved his first objective.

Walnut Hills Neighborhood, Dayton, Ohio
10:45 a.m., Saturday, August 24, 1968

Several white kids were cavorting in the playground as Henson parked the truck and climbed down from the cab. He selected a suitable assortment of tools from Shouk's toolbox, jammed them into the pockets of some blue coveralls he had brought from Apex, and set to work. As he cinched the loose bolts on a teeter-totter, he awaited the inevitable, and it was not long in coming.

Six white men came out of nearby houses, armed with baseball bats and other implements. One hefty gentleman in plaid Bermuda shorts wielded a particularly lethal-looking pickaxe. Another—not yet out of his blue pinstriped seersucker pajamas—carried a hockey stick. They shooed the kids to a distant corner of the playground and swaggered toward Henson, brimming with malevolent bravado.

"Boy, are you *lost?*" growled a tall man with curly black hair. He menacingly tapped his left palm with a Little League Louisville Slugger as he fixed Henson with a baleful gaze.

Henson calmly pretended to ignore them as he applied more torque to one of the teeter-totter's bolts. He whistled quietly, wiped sweat from his brow with his forearm, and looked up.

"I…I…I…I think the man asked you a qu-qu-qu-question," stuttered the pajama-clad man with the hockey stick. "Are you l-l-l-lost, b-boy?"

Henson gestured with the wrench, pointing it at the truck. "No, sir. I'm here for the city," he asserted. "I'm here to make sure that all this equipment is up to safety standards."

"Any reason they couldn't have sent someone else?" asked the pajama-clad man with the baseball bat. "Someone with a lighter complexion?"

Henson wiped sweat from his brow as he answered, "I guess not. Look, gentlemen, I would be more than happy to leave if I offend you, but I'm obligated to warn you that if I do, the city's not liable for any injuries that might occur on this property, and it'll be at least another six weeks before someone else makes it out here."

"Six *weeks?*" asked the man with the hockey stick.

"At least." Henson balanced the teeter-totter on its pivot point and examined it before adding, "Well, if you fellows aren't happy with me, I guess I better be pulling off."

"Let's not be too damned hasty about that," retorted the man with the pickaxe. "If the city's paying you good money to be out here, *boy*, then you need to get back to work. But let me tell you something, *boy*, we're going to keep an eye on you."

Henson nodded. "Seems reasonable to me, sir."

The motley group withdrew to a picnic pavilion. Henson could see that the merry-go-round was woefully out of balance. In fact, most of the playground equipment was obviously in need of some simple maintenance. With a few nuts and bolts tightened, sharp edges filed smooth, oil applied here and there, it could be as good as new. *Or at least safe as new,* thought Henson. As the men drew straws to stand guard over him, he resumed his chores.

The tall man with the baseball bat was the first to stand watch. Doing his utmost to appear threatening and vigilant, he lingered in the pavilion's shade. Paying scarce attention to Henson, the children resumed their play. After a while, the tall man set down his bat and joined Henson in his efforts to restore the merry-go-round to its proper kilter.

Within the hour, seeing that the merry-go-round spun as it should, at least a dozen more kids came to play. A stocky red-haired man with a crowbar relieved the first sentry. He was joined by a man carrying a Styrofoam cooler and another bearing a paper sack of charcoal.

Two men loaded a barbecue grill with black briquettes as the third tuned a transistor radio to a popular station and cranked up the volume. The playground was filled with the sound of "Jumping Jack Flash" by the Rolling Stones. Laughing, the three enthusiastically sang along with Mick Jagger. The red-haired man doused the charcoal with lighter fluid, struck a wooden kitchen match, and flicked it into the grill.

Tightening the bolts on a slide, Henson studied the trio out of the corner of his eye. They were no longer making even the slightest pretense of watching him. Observing as the red-haired man started to lay chicken on the grill, he felt a combination of anger and disgust rising in his growling stomach. No longer content to hold his tongue, he summoned his fortitude, jammed his wrench in his pocket, and strolled purposely towards the pavilion.

Walking under the shade of the picnic shelter, Henson pointed at the flaming grill and observed, "Beg your pardon, sir. I hope you don't mind me saying, but you probably want those charcoals to burn down for at least twenty minutes before you put that chicken on."

"So you work for the city and you're also a chef?" smirked the red-haired man.

"No. My mama owns a little café in New Orleans, near the French Quarter, and I used to tend the barbeque pit on weekends." Henson pulled out his wallet and showed them a picture.

"*Man!* This is your mama? She looks just like Aunt Jemima!" commented one of the men.

"But darker," noted another.

"True, but she could flat cook some chicken, baby. And she taught me how to cook, too. If you lay that chicken on right now, before those coals simmer down, it'll be charred crisp on the outside but raw pink on the inside." Henson sniffed the air, smelling the unmistakable scent of yet unburned hydrocarbons. "And it's also going to taste *exactly* like that lighter fluid."

"Well now, that does explain a lot, Harry," noted one of the men, nodding knowingly. He pulled the tab from a beer, and white foam gushed from the opening of the sweating can. "Maybe it wouldn't hurt to listen to the boy."

"Okay, okay," muttered Harry, deftly using a pair of tongs to quickly snatch off the three thigh quarters he had already placed on the grill. "Got any other suggestions?"

Henson pondered the situation. "Do you have any barbeque sauce?" he asked.

"Knew I forgot something," said Harry. "Bob, you want to dash up to the A & P and grab some sauce? Probably want to lay in another case of suds while we're at it."

"Tell you what," offered Henson. "Instead of ruining that yardbird with store-bought sauce, why don't I mix you up some of my mama's best chicken sauce? I'll guarantee you have the ingredients at home, and it's a whole lot tastier than anything you'll find on the shelf at the A & P. We'll need some mayonnaise, vinegar, lemon juice, black pepper, salt and cayenne pepper."

"*Mayonnaise?*" asked Harry. "We're not putting any damned mayonnaise on chicken."

"It's a white sauce, not one of those gooey tomato sauces," said Henson. "It's a Southern thing. I promise you'll like it."

"Well, I suppose we can try it. I'll send the kids to round up the ingredients."

"You want a beer?" asked another man. "I didn't catch your name. I'm Bob McManus."

"Henson. Matt Henson. Good to meet you, Bob." He shook each man's in turn. At this point, with the sun quickly climbing overhead, a cold beer looked inviting, but he declined the offer. "I'm on the clock, but if you'll hold one over for me, I'll tip it with you in a bit."

"So is there anything we can do to help you out with the playground?" asked Bob. "After all, it is our kids using it."

"Well, I could definitely use some extra hands if you're willing to lend them," answered Henson. "By the way, would you all mind me taking some pictures after I'm done? It's kind of a hobby for me, plus I like to keep a record of the work. I develop my own film."

"No kidding," replied Bob. "No, I don't see a problem with you snapping pictures. Not at all."

Apex Minerals Exploration Inc., Dayton, Ohio
6:30 p.m., Monday, August 26, 1968

Changing from his coveralls back to street clothes, Henson was tired, and he wasn't looking forward to the long walk home from Apex. They had spent the entire day in a stifling hangar at the airport, learning how to evaluate aircraft to see if they were properly maintained and airworthy, and would spend the remainder of the week doing much the same thing.

Henson recognized most of the aircraft—including a DC-3 and a De Havilland Otter—from the hangars at Aux One-Oh. Besides showing them how to scrutinize maintenance records, which typically weren't trustworthy documents in Third World countries, the instructors walked them through a detailed pre-flight inspection for each type of aircraft. It was the same inspection that a pilot would perform, so the instructors assured them that they would catch any significant problems if they methodically followed the checklists step for step.

Henson waited patiently for his turn at the sink. By the time it was his, the hot water was long since gone. Diligently scouring his hands with coarse Lava soap, he looked forward to returning to the apartment and spending some time in the shower. Of course, by the time he made it home, his roommates probably would have already exhausted all the hot water there as well. As he turned off the tap and checked his hands for any residual traces of grease, he heard a series of faint bumps and scraping sounds, growing progressively louder, in the hallway.

"Mr. Henson," said Grau, leaning into the doorway of the locker room. Grau wasn't involved in the more technical training, like this week's aircraft evaluation sessions, and had apparently spent the day looking over the results of the weekend photography assignments. "When you're done tidying up, please join me in my office. We need to chat."

Henson nodded, neatly stacked his aircraft inspection checklists in his locker, and then folded his blue coveralls and placed them atop the checklists. On the way to Grau's office, he stopped by the break room in the hall and retrieved a brown paper sack from the refrigerator.

He saw that Grau's desk was partially covered with the black-and-white photographs he had developed yesterday in the makeshift darkroom he had installed in the apartment's bathroom. As Grau finished some paperwork, Henson studied a framed photograph on the wall behind his desk. It was a picture of Grau and his African wife. The Eiffel Tower was in the background. His unsmiling wife held an infant, and Grau's expression was somber. His face, still young, was marked with fresh bruises and scars; a gauze patch was taped over his right eye.

Henson reached into the paper bag and retrieved a large barbecued chicken breast. "Mind if I eat, sir? I'm starving. I didn't know we would be stuck at the hangar all day, and I missed lunch. Left it back here in the fridge."

"Suit yourself," said Grau, looking up from his papers.

Munching on the chicken, Henson held out the greasy bag to Grau. "Would you care for some, Mr. Grau? I have a boxful back at my apartment. I would hate to see it go to waste."

"No, thank you, Mr. Henson. I appreciate the invitation, but I have dinner plans. I did notice that you fed your three wayward brethren this morning, though."

"Every life has value," quipped Henson. "And it seemed like the Christian thing to do. Are you sure you don't want any? This is *really* good chicken. It's my mama's best barbeque recipe."

Grau shook his head. "We need to discuss these photographs, Henson. Did you not understand your assignment?"

"I did, sir. You asked me to photograph activities at a playground. In Walnut Hills. On Saturday morning." Henson pointed his hand at several prints in turn; the photos obviously had been snapped from about ten to twenty feet and depicted gleeful children playing on swings, slides, monkey bars, and other playground apparatus. "Is this not what you wanted?"

"And these, Mr. Henson?" asked Grau, gesturing at several other photos. One depicted a grinning Henson seated at a picnic table brimming with food; Henson's ebony face was in stark contrast to the smiling white faces of the family packed in around him on the bench. Another showed a smiling white man perched on a wooden extension ladder, using an oilcan to lubricate the chain pivots of a swing set.

"It's kind of difficult to explain, sir," answered Henson. "You see, I went out there to shoot pictures of the kids, and after a while the whole thing just evolved into a neighborhood cookout and block party. Strangest thing I've ever seen. Here, this picture is of me and the Smiths. Here's Harry and Bob working on the teeter-totters. That wood there was a little worn and warped, so we replaced it with an eight-by plank Bob had spare in his garage. Here are the guys cleaning the rust off the slide. Steel wool is a natural for that. That one there is me and the Coopers. She has cancer, you know, and won't be with us much longer. I think Jack Cooper already has his eye on that brunette divorcee that lives around the corner on Polk Street."

"Hah. I see your point, Henson," muttered Grau. "You passed your assignment, but there are more than a hundred pictures here, and you were only required to take two rolls of film. I appreciate your enthusiasm, but you're aware that you're not drawing any extra money when yours runs out, right?"

"Certainly. I only had two rolls when I went out there. These people bought the extra film, and they gave me money to buy some more developing chemicals. I told them that shooting the pictures was a hobby, and they just wanted some copies of what I shot. Plus I took several family portraits. That's not a problem is it, Mr. Grau? You don't *mind*, do you?"

Admiring one of the photographs, Grau smiled ever so slightly and shook his head. "So, enlighten me, Mr. Henson. How did you do this?"

Henson grinned and replied, "I guess that some mysteries are better left unspoken, Mr. Grau, but the easiest explanation would be that I listened to you and just did as you said."

Grau sat silently for several seconds, as if he was waiting for something, and finally spoke. "Well, Mr. Henson? Isn't this the opportune moment when you shift into your best shuck and jive Br'er Rabbit impersonation and gloat about putting one over on all those dumb white folks?"

Shaking his head, Henson pondered Grau's comment. "No, sir," he said. "I wouldn't do that. To be honest, Mr. Grau, that was the best time I've had in several months. We had an awkward start, but these people made me feel welcome, and I felt like I belonged there. Yeah, maybe I went out there to deceive them, at least just enough to do my assignment, but events sure took a turn I didn't expect. Thanks for sending me to Walnut Hills."

Grau flipped up his eye patch and scratched his empty socket. "You're more than welcome. Excellent work, Mr. Henson. I must say, you are by far my favorite pupil. You are free to go."

Simulator Facility, Aerospace Support Project
4:25 p.m., Friday, August 30, 1968

It had been another agonizingly long day, and Gunter Heydrich was both weary and famished; he longed to plop down in front of a heaping plate of his wife's sauerbraten, red cabbage and spatzel. With no simulations scheduled for the weekend, he looked forward to the indescribable luxury of a full night's sleep. They probably could have ended the day with the last scenario, but Carson insisted on yet another full-blown launch-to-orbit run. This was their fifth for the afternoon; hopefully, it would be their last.

Heydrich stubbed out his cigarette in his overflowing aluminum ashtray, took a sip of stale coffee, and dropped two Alka-Seltzer tablets in a glass of tap water. As the white tablets dissolved, he studied the six controllers seated at the first two rows of consoles. To a man, uniform in their rumpled white short-sleeve shirts and dark ties, they looked like they were stretching their last frayed nerve.

Like a quarterback addressing an exhausted huddle in a last-ditch goal line stand, Heydrich said, "Okay, launch-to-orbit, one more time. We're hitting them with incident twenty-three, second stage ignition failure. Everyone set?" He waited for each to acknowledge with at least a faint nod or slight gesture of affirmation. Confident that his controllers were ready, he keyed the intercom switch and spoke. "Okay, let's run through the pre-launch from the top, gentlemen."

"Got it, Gunter," replied Carson. By this time, he knew the procedures by rote and had scarce need to refer to a checklist or guide. "Abort Control handle is in Normal position. Maneuver controller is stowed. Altimeter is set to 28.52."

"Confirm altimeter set to 28.52," said the CAPCOM—the Capsule Communicator—a jittery young Air Force captain from Omaha.

Carson continued in a rapid-paced monotone: "IVI is zeroed. Sequence panel telelights are extinguished. Abort lights are extinguished. Att Rate, Guidance, Engine One, Engine Two lights are extinguished. Left panel top three rows of circuit breakers all set to closed position. Boost-Insert and Retro Rocket Squib set to Arm. Retro and Landing switches are set to Safe.

"Initiating gyro run-up," stated Carson, at a relentless rapid-fire pace, like a machine gun knocking down row after row of a human wave attack. "Aligning platform. Conducting computer checkout. Sequence Panel telelights are functioning correctly. Main Batteries switches to On. Retro Rocket Squib batteries switches to On."

Moments passed before the CAPCOM announced: "Launch vehicle is transferring to internal power. Stand by for engine gimballing."

"Standing by for gimballing," replied Carson, obviously primed for the simulated launch.

"T-minus one minute," stated the CAPCOM. He covered his mouth, stifling a yawn. "Stage 2-P valves coming open in five…T-minus forty seconds…T-minus ten, nine, eight, seven, six, five, four…first stage ignition…two, one, zero. Hold-down bolts fired. Lift-off."

"The clock is started," announced Ourecky.

Carson's angry voice boomed over the intercom. "*Stop* the simulation! Switching off VOX!"

The controllers let out a collective groan. Frustrated, Heydrich buried his head in his hands. *Would this day ever end?* The intercom loop went silent as Carson switched it off to have a private conversation with Ourecky.

Inside the simulator, Carson punched Ourecky's left shoulder. "*I* call the clock, Ourecky. *Do you understand?* The Command Pilot reports the clock starting. Always. No exceptions. *Got it?*"

"Ow! I didn't know if you saw it," replied Ourecky sheepishly, rubbing his shoulder. "I just wanted to make sure…"

"*I call the clock,* Ourecky. *Me.* The *Command* Pilot. Do you understand?"

"I understand, sir. You call the clock."

"If you understand that, Ourecky, maybe we can put this train back on the rails. But from now on, stick to the procedures the way they're written. *Period.*" Carson reached up and re-set the communications switches. "We're okay in here. Resume the count at T-minus ten, please."

"Restarting the clock at T-minus ten seconds at my mark," stated Heydrich. His tired voice reflected his exasperation. "*Mark.* Start the clock."

The CAPCOM repeated his countdown: "T-minus ten, nine, eight, seven, six, five, four…first stage ignition…two, one, zero. Hold-down bolts fired. Lift-off."

"Roger liftoff. The clock is running," said Carson.

With the clock re-started, the simulation ran smoothly for the next two minutes. "Engine One indicators are red," reported Carson. That was normal; the change reflected that staging—a literal passing of the flame from the first booster stage to the second stage—was imminent.

"BECO," stated Ourecky, announcing Booster Engine Cutoff.

Just a few seconds later, Carson stated, "We have an Engine Two underpressure indicator. Abort indicator. We have an Abort indicator."

"Pilot confirms Abort indicator," added Ourecky, glancing at the ominous red light.

"Executing Mode Two abort," declared Carson. His voice reflected slightly more stress than before. He reached down with his left hand and shoved the abort handle from the Normal setting to the Shutdown position and then immediately cycled it to the Abort position.

The gesture, much like manually shifting a sports car, would result in a shutdown of the malfunctioning second stage rocket engine, followed by a systematic activation of a series of line-cutting guillotines and pyrotechnic charges to separate the spacecraft from the booster rocket. "*Abort. Abort. Abort,*" he reported. "Retros firing in salvo mode."

Just a few minutes later, the simulation was concluded. Carson and Ourecky climbed out of the Gemini mock-up and headed for the locker room. Sticking to his traditional routine whenever he emerged from the Box, Carson paused briefly at the soft drink machine to grab a cold bottle of Coca-Cola; Ourecky followed suit.

"Major Carson, one thing keeps bothering me. Is there any reason that they haven't thrown a Mode One abort at us yet?" asked Ourecky, opening his locker. Mode One was the first of four possible Gemini abort protocols that required immediate action by the crew; the abort modes available to them varied according to the altitude at which an emergency occurred.

"Mode One abort? Like where the rocket is blowing up on the pad?" asked Carson, pulling off his boots and stripping out of his flight suit. He hung the damp garment in his locker before pulling out a pair of khaki chinos and a powder blue Ban-Lon knit shirt. He sprayed his armpits with Right Guard deodorant, and then quickly sniffed them before donning his shirt.

"Yeah. They've hammered us with just about every conceivable glitch so far, so I figured that a Mode One pad abort must be coming up soon." Ourecky pulled the top of the flight suit down to his waist and examined the fresh knuckle-shaped bruise on his left shoulder.

Carson laughed. "And just what would you do in a pad abort situation, Captain Ourecky?"

"Well, sir, I guess we wouldn't have any alternative other than to fire the ejection seats and blast out of there. The procedures manual is pretty clear on the protocols."

"True," observed Carson, pulling up his chinos. "But you're reading the NASA manual. There won't be any ejection seats on *our* version of the spacecraft. If the booster blows up on the pad, then we'll be vaporized in a reddish-orange cloud of rapidly burning fuel. Correction: *we* won't be vaporized, *I'll* be vaporized. You'll probably be watching my demise from a safely distant blockhouse."

"No ejection seats?" asked Ourecky incredulously. "You're kidding, right?"

"Nope," replied Carson, tugging on his Ban-Lon shirt. "Not in the least. In order to do the intercept mission, we're obligated to haul a lot of extra weight to orbit: new computer, bigger radar, more batteries, extra survival equipment, and other stuff. So some really smart engineers realized that the ejection seats weighed 154 pounds apiece and occupied a considerable amount of cockpit volume."

Tucking his shirt into his trousers, Carson continued. "So, they stripped them out. When the final version of the Gemini-I is delivered, it'll have lightweight seats configured for low-velocity ejection. They'll only be good for post-reentry scenarios, in case the paraglider doesn't deploy. Granted, we won't have the full-blown Weber ejection seats, but on the plus side we'll have a bit more room, plus we'll have a lower equipment bay below the seats."

"It doesn't bother you that you'll be riding a ballistic rocket with no means of escape?"

"Not really." Carson studied his face in a shaving mirror hanging inside his locker door. He pulled a pocket comb through his thick brown hair and then smoothed his moustache with his fingers. "Look, the ejection seats were only good for certain phases of the flight. On the launch sequence, they were viable from the pad to—"

"Seventy-five thousand feet," said Ourecky. "Roughly one hundred seconds into flight."

"Correct," noted Carson. "The ejection seats can be used up to seventy-five thousand feet, but as the command pilot I can also elect to abort by firing the retro rockets once we're above fifteen thousand feet, which is—"

"Fifty seconds into flight."

"That's right. So those seats are really most useful for the first fifty seconds of flight. And I can almost assure you that if the rocket decides to spontaneously detonate on the pad, the ejection seats aren't going to be particularly useful. Besides, I don't think any of the NASA astronauts really trusted the seats anyway."

"Really?" asked Ourecky. "I never knew that."

"Do you know very much about ejection seats?" asked Carson, threading his belt through his belt loops.

"No, other than what they showed me during egress training."

"Well, let me tell you, I've punched out twice, for real, but that doesn't make me any kind of expert," explained Carson. "Look, an ejection seat is a very serious piece of engineering. It has to blast you out of a malfunctioning aircraft but at the same time not kill you in the process. That's tough enough for a supersonic aircraft, but even that doesn't hold a candle to ejecting a guy clear of a malfunctioning rocket on the pad."

"Why?"

"Because rockets *explode,* that's why. I talked with one of the NASA engineers who worked on the Gemini ejection seat design. That seat not only had to be designed to blast an astronaut clear of the spacecraft, but also clear of the explosion fireball as well. And even beyond that, it had to throw him far enough away from the pad so there wasn't any danger of his parachute melting from the heat."

Carson continued. "Needless to say, you're talking about a seat that's exerting an *enormous* amount of G-forces on the human body. The engineers weren't even entirely convinced that the astronauts would survive the ride, and the NASA Gemini astronauts damned sure weren't that confident in the seats.

"Look at what happened on Gemini 6 with Wally Schirra. He and Tom Stafford were sitting on top of a Titan II that fizzled, and by all rights, if they had followed their procedures, Wally should have pulled the eject handle and fired them out of there. Schirra kept cool and elected to stay put.

"As it turned out, the situation wasn't nearly as bad as what they had thought, and they were able to fix the Titan and launch a few days later. Now, had they ejected, it would have been a totally different ball game. Of course, everyone painted Schirra out to be a hero for saving the vehicle, and he certainly was an exceptionally cool customer for a Navy guy, but the chances are good that he knew that they probably wouldn't have survived the ejection seat ride, even in the best of circumstances. Personally, I think he was aware of that, and that's why he didn't pull the D-ring. Of course, he never said that, and there's really no way of knowing what goes through a man's mind in the split-second it takes to make that kind of decision."

Carson took a sip from his Coca-Cola. "Ourecky, you have to look at it from my perspective. Yeah, I don't particularly like it that I won't have any means of escape if my rocket blows up, but I also know that an ejection seat only grants me just the slightest margin of survival anyway. "

He sat crossways on a bench and dusted his toes lightly with foot powder. As he pulled on his socks, he added, "I suppose that the ejection seats were more of a feel-good public relations device than anything else. NASA launched their Gemini spacecraft in broad daylight in full view of the world. From a PR perspective, they couldn't put guys up there without some sort of safety net. Hence, the ejection seats. But our launches will be far from the public eye, and a couple of anonymous Air Force pilots are a lot more expendable than high profile NASA astronauts."

"I just can't ever imagine going up on a rocket without an escape tower or an ejection seat," said Ourecky, shaking his head. "It just seems too much like suicide. After I went through egress training, I was deathly afraid of ever having to eject, but I sure would like to retain the option."

"Yeah, there's an engineer's perspective for you," observed Carson, slipping into his shoes. "You engineers are pragmatic enough to yank out the ejection seats because they weigh too much, provided someone else has to take the ride."

"I suppose you're right, Major. It really doesn't bother you that you won't have one?"

"Honestly, no. If there's even the slightest chance for me to go into space, I'm going. No hesitation, no reservations. If going to space requires that I ride a rocket that has a fifty percent chance of blowing up and incinerating me to a crisp, then I'm willing to accept that risk." Carson slid his wallet in his pocket. "Now, if you'll excuse me, Captain Ourecky, I've got a hot date."

21

WAKING UP

Aerospace Support Project
3:00 a.m., Friday, September 13, 1968

Mark Tew switched on his gooseneck desk lamp, slipped on his reading glasses, selected a black leatherette binder from the shelves behind his desk, and resumed reading. At present, the shelves strained under the weight of eighty-seven identical three-inch black Government Issue binders, painstakingly cataloged and indexed, bulging with reports, memorandums and other documents that described every minute detail and nuance of Blue Gemini.

If they ever went operational, each mission would be a hugely complex undertaking that would require the tightest synchronization to ensure that every single piece was functioning properly at the right place and at the right time. Information in the binders described the production schedule for the Gemini-I spacecraft, as well as the refurbishment plan to refit each spacecraft after it was returned from a mission, so that it would be ready for a successive mission.

Hundreds of pages outlined the intricate scheme for selectively removing Titan II ICBMs from their silos and retrofitting them to launch manned spacecraft instead of nuclear warheads. More paper was dedicated to explaining the elaborate movement plan in which the modified boosters would be transported by rail to the Horizontal Assembly Facility near San Diego, where they would be mated with the Gemini-I vehicles. Each completed assembly would be hermetically sealed in weatherproof transport container, which would then be loaded onto a modified Navy LST landing ship for the 3500-mile voyage to the Pacific Departure Facility launch site on remote Johnston Island.

Besides the flight hardware, it was critical to ensure that all the other essential pieces—communications networks, tracking systems, rescue forces, and the like—were also ready.

Of the roughly three thousand men and women who would be directly or indirectly involved with each mission, only a tiny minority—197, to be exact, including the two men uncomfortably seated in the nosecone—would be aware that a rocket was carrying men into space. Of those 197 insiders, only a third would actually be aware of *why* they were going into orbit.

So to keep the remaining twenty-eight hundred workers in the dark concerning the full extent of the project, several of the binders described the labyrinth of cover stories, partial truths, and outright lies that collectively served as a Churchillian bodyguard of the truth.

So much information! Rockets, spacecraft, launching sites, tracking systems, ships, airplanes, contingency landing strips, people, places and ideas. While a committee approach was effective for managing most of the issues associated with Blue Gemini, Tew felt that it was still crucial that at least *one* man maintain a tight grasp on all aspects of the entire scheme.

Tew imagined that someday there would be massive computers—probably as big as this entire building—that could store all of the information in these binders. A huge thinking machine would eventually be able to house and correlate all of these assorted bits and pieces of information and perhaps make sense of it. But until such a machine became reality, the job fell to Tew. Or, at least, he took the labor upon himself.

To Tew's dismay, Wolcott had long ago abandoned the sleep discipline they had learned at AFSC. While Wolcott did labor long hours—he was usually in the office for at least fourteen hours a day—he had adopted the routine and trappings of a normal life, or at least something roughly akin to one. Sometimes Tew was irritated by his friend's seeming lack of commitment, but the rangy cowboy consistently kept things moving forward. Continuing the relationship forged in decades of past projects, Tew focused on the more mundane aspects of organization and budgets, and Wolcott kept a steady eye on the machines and the men.

In contrast to Wolcott, Tew rarely left the building. He perceived of himself like an ever-vigilant captain of a warship at sea, with a makeshift wardroom adjacent to the bridge, so he was always ready to return to the action at a moment's notice. Of course, since his wife had retreated back to California, there was scarce reason for him to leave his post.

Listening to his stomach grumble, he was also reminded that it had been weeks since he had sat down to a normal meal. His digestive woes had sentenced him to the blandest of diets, so he had little motivation to sup at a restaurant or the Officers' Club. Although Wolcott constantly needled him to join him at the Club to throw back a few and bleed off a little steam, the notion of having a drink was far removed from his mind.

If anything, his recent ascetic lifestyle compelled him to glimpse into the excesses of his past, making him realize just how quickly he could be lured to find solace in a bottle. In any event, there was much work yet to be done. With critical deadlines looming close over the horizon, Tew set aside distractions and stayed the course. Eventually, Blue Gemini might fail or run aground, but it damned sure would not be because he was asleep in the wheelhouse.

Wright Arms Apartments, Dayton, Ohio
5:00 a.m., Friday, September 13, 1968

Like Tew, Ourecky had no need for an alarm clock; without any external stimulation, his mind instantly clicked into gear and he was wide awake. He opened his eyes to check the clock on the nightstand: its dim luminous hands confirmed it was five o'clock. He quietly slid out of the bed—trying not to wake Bea as she softly snored—and slipped into the bathroom.

A few minutes later, after brushing his teeth and shaving, he tiptoed past the bed toward the living room. Sitting on the edge of the couch, he pulled on his gray sweat suit bottoms and put on his socks. He was lacing up his high top Converse sneakers when she appeared at the bedroom door, wearing her pink cotton nightgown. Untied, it was slightly open at the front, revealing a narrow sliver of her naked body in the dim light. He paused to admire her.

"See something you like, sailor?" she said, grinning.

"Well, I don't see anything I *don't* like," he replied. "Just wish I could stay a little longer." Carson would absolutely wear him out if he was just a minute late, and as much as he yearned to go back to bed with Bea, he didn't want to face the wrath of the pugnacious pilot.

"You're up early," she said, stretching and yawning. "Why the hurry, honey?"

"I'm supposed to meet Major Carson at the gym. Just like last week and the week before."

"Why, Scott? Don't they already get enough of your time?"

"Carson's helping me get in shape," answered Ourecky. "He goes to the gym at six, so that's when I have to go, if I'm going to work out with him."

She yawned again. "Well, I can see you've lost some weight and you're obviously getting stronger. I'm sure not complaining, Scott, but are you doing this for some reason? And I'm not too comfortable with you hanging around with Drew Carson so much. Please, please tell me you're not putting in for flight school."

Pulling his sweatshirt over his head, he laughed. "No, nothing like that. We're just assigned to the same project. And he offered to let me work out with him, that's all."

"If you say so," she answered. "I'm headed back to bed. See you tonight?"

"Sure. I'll call when I leave there." He stood up and kissed her. "Bye, Bea."

"Bye," she mumbled sleepily. As he was almost out of the door, she called after him: "Hey, Scott Ourecky. I love you."

Surprised, he stopped in his tracks and turned towards her. She had never said that before, and he wasn't sure how he should respond, only that he had to respond. He paused and then opted for the truth: "I love you too, Bea."

Simulator Facility, Aerospace Support Project
7:30 a.m., Friday, September 13, 1968

After a brisk five-mile run and a strenuous round of weight lifting, Carson and Ourecky showered and went immediately to the Simulator Facility.

"Good news, gentlemen," said Heydrich. "We've hammered out the bugs on the Paraglider Landing Simulator. You can go back into it this morning, if you see fit, Drew."

"It's about time," snapped Carson. "We're doing live drops next month up at Eielson, and Ourecky here hasn't even done a fair weather drop yet. Let's get on with it."

The Paraglider Landing Simulator resembled an oversized child's gyroscope, consisting of two massive rings that supported the reentry portion of a Gemini-I spacecraft. The interlocking rings allowed the mock-up capsule to move in all axes during a simulation, including the roll maneuvers used to achieve lift during reentry, so as to more precisely control the touchdown point, as well as the turns and banking of flying under the paraglider.

Projectors were positioned in front of the mock-up's two tear-drop-shaped windows to provide the pilots with a realistic view of how events were unfolding outside. The projectors were linked to a camera suspended over a highly detailed model that accurately depicted the terrain in the vicinity of a planned recovery site. As the pilots "flew" the paraglider, its actions were replicated by the boom-mounted camera, which then "flew" the same course over the simulated terrain. The simulator allowed for different variables to be inserted into the problem: variations in wind, lighting conditions, and equipment failures, to name a few. The terrain mock-ups could be swapped out to replicate landings at sites around the world.

"Climb up and buckle in," said Carson, nudging Ourecky's shoulder. "I'll be up shortly."

"Drew, what's the plan for the day?" asked Heydrich, adjusting his glasses. "Do you want my crew to run the whole sequence, from orbit to landing?"

Carson shook his head. "Too time-consuming, Gunter. I just want to focus on the post-reentry profiles we'll be flying up in Alaska. Start us at the 60K light and run it to landing. Ourecky's not going to fly to orbit with us, but he'll be sitting in that right seat next month, and I want to make absolutely sure he can put us safely on the ground if I'm incapacitated."

"That's not what Virgil wants," asserted Heydrich. "He wants the full orbit to landing profile."

"As much as I respect Virgil, it's not going to be his ass on the line in Alaska. Gunter, just grant me this one day, and then we can transition to the full profiles tomorrow. Okay?"

"*Ja*, Drew. As you wish."

"Thanks. Do me a favor, and make several runs without any glitches, so both of us can get used to the flow. Ourecky's been studying these para-glider procedures, but he needs to make the leap from the books to reality. He absorbs this stuff quickly, but he also can become a little overconfident and complacent as the day draws on. When he starts getting lackadaisical, I'll let you know, and you can lower the boom on him."

"We'll do it, Drew. Have a good day in there. Break for lunch at twelve?"

"Sounds fine by me," answered Carson, spitting his gum into a waste-basket as he pulled his headset over his ears.

The first series of landings went without incident. Ourecky was happy with himself, since he kept pace with the procedures without any undue difficulty. This paraglider landing business certainly seemed a lot less demanding than their intercept simulations, where he relentlessly worked calculations as Carson flew the requisite corrective maneuvers.

Things changed drastically after lunch. With a large hamburger and French fries weighing heavily in his stomach, Ourecky felt groggy as he resumed his place in the simulator. After the technicians closed and latched the hatches, he heard the whir of motors as the simulated cabin was rotated to a vertical "nose-up" position to simulate reentry.

He heard Heydrich's voice over the intercom: "Clock is starting, gentlemen."

"Clock," replied Carson. "Landing Arm is *On*. Altitude is sixty-five thousand."

The cabin rocked back and forth slightly. "60K telelight. Minor oscillations," stated Carson, observing an instrument light that indicated that they were passing through sixty thousand feet. "Altimeter verifies 60K light. Standing by for drogue."

"60K light. Standing by for drogue," confirmed Ourecky. The drogue parachute would slow the capsule slightly and ensure that it was oriented correctly and stabilized for deployment of the much larger paraglider. Looking out through his small window, Ourecky watched a film clip that showed the drogue chute streaming out and blossoming.

"Drogue is out," observed Carson. The cabin's rocking motions slowed considerably. "Oscillations dampened by drogue. Switching to Rate Command."

"Drogue is deployed. Rate Command."

"Install D-Rings," ordered Carson.

"Installing D-Ring," answered Ourecky. Keeping pace with Carson, he retrieved his ejection D-Ring from its stowage pocket and locked it into place with a pip pin. A swift tug on the D-Ring would activate a low-velocity ejection system; although the escape sequence wouldn't be nearly as traumatic as one with the Weber ejection seats aboard a NASA Gemini capsule, it still wasn't safe to install the D-Rings until the drogue was out.

"Check your restraint straps again. Make sure they're good and tight," stated Carson, as he cinched snug his own shoulder straps.

Ourecky obediently did the same, wondering why Carson always seemed so pre-occupied with this point, since it wasn't explicitly stated on the descent checklist.

"Descending through fifty thousand," noted Carson. "Thirty seconds to main."

Watching the clock and altimeter, Carson manually deployed the paraglider. It took several moments for the bat-shaped paraglider to completely open. To simulate the delayed opening, the cabin rocked back and forth.

"Paraglider is deployed," stated Carson. "Stand by to rotate to two-point landing attitude."

"I confirm paraglider is out," answered Ourecky. "Standing by for landing attitude."

Suddenly the cabin's nose pitched down to simulate the change to the two-point suspension of the paraglider. Ourecky felt himself now sitting upright again.

"Performing paraglider performance checks," noted Carson, gradually pulling back on the hand controller mounted in the center console between the two men. "Stall check."

"Stall check," confirmed Ourecky.

"Stall check is good," observed Carson, resetting the controller. "Executing port turn check."

"Port turn check." Ourecky heard motors whir as the cabin tilted to the left to simulate the banking maneuver.

Carson centered the hand controller, and the cabin followed suit, moving to an upright position. "Port turn and recovery are good," he stated, subtly nudging the hand controller to the right. "Executing starboard turn check."

"Starboard turn check." On cue, the cabin "banked" to the right. Ourecky noticed that Carson seemed to be holding them in the bank much longer than he usually did. He realized that Carson had re-indexed the hand controller to recover to level flight, but the simulated paraglider just wasn't cooperating.

Simulating the ever-increasing centrifugal force, the cabin progressively rotated sharper to the right until Ourecky was awkwardly pressed downwards against the right wall. Disconcerted, he realized that the simulation wasn't exactly true to life; the centrifugal force actually would force him straight downwards in his seat, not against the wall.

"Unable to recover from starboard turn," stated Carson calmly. "Looks like we have a fouled control cable. Executing abort."

Pinned against the right wall, switches jammed into his flank, Ourecky suddenly realized that the next steps were his, since most of the controls and indicators for the paraglider were located on his side of the instrument panel.

Several small items, not adequately secured, fell out of their stowage pockets and cascaded against him. Fighting against the distractions, struggling to remember the paraglider abort checklist, he groped for a toggle switch marked PARA JETT ARM, but Carson's hand found the switch first.

"Paraglider jettison pyro *armed*!" snapped Carson. He immediately threw a second switch, which also should have been Ourecky's responsibility. "Paraglider jettison pyro *fired*! Are you conscious, Ourecky?!"

In their earphones, the men heard a bang replicating the simultaneous detonation of four pyrotechnic charges that would shear the paraglider control cables and bridles from the spacecraft. At that point, free of the malfunctioning paraglider, the spacecraft would resume freefalling through the atmosphere.

Like a carnival ride gone awry, the cabin abruptly snapped upright and then continued pitching over until it was pointed nose down. As simulated aerodynamics took over, the cabin gradually rotated back around to replicate the blunt end falling first to earth.

"I'm awake! I'm okay!" blurted Ourecky. He reached up to toggle a switch labeled STAB DROGUE ARM. "Stabilization drogue armed." He threw a second switch. "Stabilization drogue fired." The small parachute would barely slow their descent; its sole purpose was to ensure that the spacecraft was stably oriented for the next step in the abort procedure.

Ourecky knew what was next in the simulated emergency; he made certain he was correctly centered in his seat, reached down and grabbed the ejection D-ring between his legs.

"Stab drogue deployed," noted Carson. "Stand by to eject. Altimeter indicating twelve thousand feet. *Eject, eject, eject*!" Carson pulled the ring, and the men heard another two loud bangs in their earphones, simulating the pyrotechnic charges that blasted the hatches open and blew their seat pans clear of the spacecraft.

"Simulation concluded," stated Heydrich over the intercom. "A little slow, gentlemen, but the outcome was satisfactory."

Shaking his head, Carson stated, "Switching VOX off," as he toggled the intercom switches. He turned his head to face Ourecky. "Let's clear up something, Captain. Did you hear a knock-out tone?" To simulate a crewman being rendered unconscious, a loud tone would sound in his headset, at which time he was supposed to sit quietly through the remainder of the simulation.

"No. Was I supposed to hear one?"

"I don't think so," snarled Carson. As the cabin slowly rotated to the upright position, he jabbed Ourecky hard in the left shoulder. "You had better *wake* up. What's gotten into you?"

Embarrassed by his failure, Ourecky grimaced from the pain. It would be yet another bruised indentation that he would have to later explain to Bea.

The next simulation proceeded smoothly, right up until the point the para-glider deployed. As the cabin transitioned to the landing attitude, Ourecky noticed that Carson was sitting quietly with his hands folded in his lap. He had apparently heard the knock-out tone in his earphones, so he was sitting out the landing portion of this drill.

Without the slightest hesitation, Ourecky clicked right into gear, smoothly executing the paraglider post-opening procedures. He verified that the paraglider was controllable and made sure that the computer was shut down. Using a "swizzle stick" tool to extend his reach, he threw switches and depressed circuit breakers on Carson's side of the cockpit.

Although Carson remained perfectly silent, true to his part, Ourecky could not help but notice his slight smile. Keeping pace with the land-ing checklist, Ourecky acquired the TACAN beacon for the simulated landing site and flew a course toward it. He opened the skid well doors, lowered the three skids, and then verified that they were locked into place.

It was going better than he could have imagined. He checked the clock and altimeter, scanned the remainder of the instruments to ensure he hadn't forgotten anything, and looked up through his window. In the darkness, he saw the red lights that marked the landing site.

Ourecky confidently initiated the conversation with the notional controller located at the field: "Control, this is Ultra One. Receiving TACAN at Channel Eight. Setting up for right-hand approach on Runway Three-Zero. Five thousand feet indicated altimeter. State field elevation and altimeter, please."

In his headset, a voice replied: "Ultra One, this is Control. Copy TACAN on Eight, right-hand approach. Field elevation is one-one-zero feet, current altimeter is two-seven-eight-seven. Winds out of five-zero at seven knots, gusting to twelve. Landing surface is loose gravel."

"Control, Ultra. Copy field at one-one-zero, altimeter two-seven-eight-seven. Winds out of five-zero at seven, gusting to twelve." Proud of himself for single-handedly flying the approach, Ourecky grinned.

Suddenly, he thought of Bea; the enticing image of her in her nightgown was vivid in his mind, and he remembered her words from this morning. *Was he right in saying what he did?* He remembered listening to Carson and the other pilots joking about the pitfalls of stumbling into a woman's "I love you, too" trap; in their words, a man might as well slap handcuffs on himself when he positively responded to a woman's declaration of love.

It was just a momentary distraction, but even a moment was too much. A quick scan of the instruments jolted him to reality. He was a few seconds late executing his turn to the base leg of his approach. Immediately recognizing his blunder, he yanked the hand controller hard over to the right. "Ultra One turning right to base," he blurted.

"Ultra One, Control copies right turn to base leg," came the dry reply over the intercom.

Ourecky felt flushed. Almost immediately he pulled the stick again to execute the ninety-degree turn to his final approach. "Turning right to final. Gear down and locked," he said.

He glanced up through his window; the simulated view showed a series of bright lights—portraying railroad flares marking the touchdown area—bobbing in the darkness. Short of a miracle, after making the late turn, there was *no* chance that he would make it to the runway.

The controller's monotone voice answered his call: "Ultra One, I copy you on final with gear down and locked. You are cleared for Runway Three Zero. Good luck."

Ourecky quickly glanced to his left. Still feigning unconsciousness, Carson's scowling face was beet red, and a vein throbbed in his forehead; he was obviously seething with anger, but remained absolutely silent. His eyes were open, but his faint smile was long, long gone.

Instead of a violent impact accompanied by the traumatic sounds of a crash, the end of the botched simulation was marked by Heydrich's heavily accented voice booming over the intercom: "You're in the trees six hundred meters short of the runway, Captain Ourecky. You and Major Carson may have survived, but it's unlikely."

There was a long moment of awkward silence as Ourecky waited for Carson to detonate. He instinctively shielded his left shoulder with his right hand, hoping that he could ward off or at least soften the blow of Carson's fist. "Sorry," he mumbled softly. "Uh, I…"

Interrupting him, Carson finally spoke. "Are we done in here, Gunter?"

"*Ja.* I should say so," replied Heydrich's voice over the intercom.

"Fine. Switching VOX off." Carson reached up and disabled the intercom.

"I'm really sorry, sir," yammered Ourecky. "I just…"

"*Quiet,*" ordered Carson. His voice was flat calm, without emotion or inflection. "I'm not going to bash you, Ourecky, but you need to listen to me, and you had better make absolutely sure that it sinks in. Do you understand?"

"I do, sir."

"Ourecky, next month we're going to Alaska, and we'll be flying this thing for real. They're going to haul us up to thirty thousand feet and eject us off the ramp of a C-130. I've logged forty-two drops on this rig, and I can assure you that it doesn't always work exactly as advertised."

Carson continued. "You free-fall for a few seconds while you're begging for the drogue chute to pop open, and then you sweat for a few more seconds while the paraglider deploys. Up there, it's not like riding this simulator. You're getting slammed back and forth, and if your shoulder restraints aren't lashed down absolutely tight, your head is going to get banged against the instrument panel. Can you picture that?"

"Yes, sir," replied Ourecky meekly.

"Good. I'll tell you, I would gladly shoot a hundred dead stick landings in a T-38 rather than make a single paraglider drop. And while I think that ninety percent of the stuff they throw at us in the Box is unrealistic annoying crap, I can assure you that it's not unrealistic for one or both of us to be knocked unconscious during a paraglider deployment. For your information, of the forty-two drops I've made, I was knocked out on three when the paraglider came out."

"Ouch."

Carson nodded. "Yeah. *Ouch.* Anyway, on one of those three, I didn't regain consciousness until Agnew had already flown the full approach and had landed us. You *cannot* afford not to take this training seriously. You can't allow yourself to be distracted, not for a split second. I know that most of this is just a big game to you because you're not going to fly, but I can assure you, Ourecky, this is *not* a game. If we're not on our game in Alaska, we can *die.*"

"Major Carson, I'll pay attention," said Ourecky. "We won't end up in the trees again."

"That's good. Ready for a mulligan?"

Ourecky nodded.

Switching on the intercom, Carson spoke: "Gunter, we're ready. Same run, please."

22

THE FROZEN NORTH

Aerospace Support Project
1:30 p.m., Tuesday, September 24, 1968

Ourecky had been in the Life Support Facility all morning, being sized up for the space suit he was to wear during extended simulations due to start in November. His physical stature was almost identical to Major Agnew's, so the suit technicians were confident that they could alter Agnew's training suit to fit him. His fingers were slightly longer than Agnew's, so a new set of gloves had to be custom-made at the plant in Connecticut.

To produce the gloves, he had to sit still for two entire hours with his hands encased in a foul-smelling latex rubber compound, which the technicians used to mold exact plaster models. While it was a welcome break from the grind of flying the Box with Carson, he despised the notion of idle time, so he tacked a set of Gemini-I electronics schematics to the wall and memorized them as the gummy latex cured.

With that ordeal behind him, he and Carson reported to the supply room to be fitted for the cold weather gear they would wear up north.

Divided into an intricate warren of plywood bins and shelves, the store-room seemed to hold enough gear to outfit an army for any potential environmental extreme, from the desert to the jungle and all forsaken points in between.

"Okay, gents," growled the supply sergeant, a hefty man with a short crew cut and a Brooklyn accent. "Make damned sure that this stuff fits properly before you leave my shop. When you're comfortable with it, then we'll do the paperwork and it's all yours."

Puffing on a thick cigar, the sergeant wielded a yellow cloth tape to gather Ourecky's key statistics. He went into a back room, and then emerged a few minutes later to dump two large piles of winter clothes on the counter.

"So we'll wear this gear when we go out for the survival training?" asked Ourecky, trying on an Arctic parka. The heavily insulated coat had a "snorkel" hood rimmed with soft white wolf fur.

Carson laughed; he had been to Alaska and was obviously familiar with the equipment. "We'll wear it for survival training? Ourecky, we'll wear this stuff *all* the time. It's cold up there, like you can hardly imagine. You don't dare take a step outdoors, not even to stroll down to the chow hall or the O Club, without suiting up first. We'll even wear this gear when we fly."

"We'll wear this while we're flying? As tight as the cabin is, it can't be too comfortable."

"Yep. We'll wear it even when we fly. And if you're worried about being comfortable, here's a crucial flying tip to file away: Don't dress for the *flight*; dress for the *crash*."

Nodding, Ourecky realized that he was right; he just hadn't considered it like that before. Oddly, Carson's observation reminded him of something Bea had related one Sunday afternoon as she dressed to go to the airport. She said that she hated their new stewardess uniforms, because they were made mostly of polyester synthetics, which would melt in a fire. He had just never considered that Bea and Carson would share anything in common, least of all having to contemplate the potential occupational hazards of airplane crashes.

"Look, Ourecky," said Carson, slipping into his own parka to check the fit. "I've graduated from the survival courses at Stead and Eielson and just about every other crappy place you could name. Survival training is miserable, but nothing to be afraid of. Just stick close, do what I tell you, and you might even come back with most of your fingers and toes."

Ourecky nodded as he jammed the bulky parka into his kit bag. "I appreciate that, sir."

"So, have you had any kind of survival training at all?" asked Carson, flipping up the hood of his parka. "I can't imagine that you engineers spend any time doing field training."

"Well, uh, I was a Boy Scout back home in Nebraska. I actually made Eagle Scout."

"*Really*? You were a Boy Scout back in Nebraska? And an *Eagle* Scout to boot? Well, all that wilderness training should come in mighty handy if we're ever forced to bail out over a cornfield or if we're taken prisoner by a herd of angry cows."

Ourecky examined a set of elbow-length mittens. They were connected by a cloth lanyard, like the ones he used to wear sledding as a kid, and had a large square of soft brown fur sewn on the back of each hand. "Why do they put this fur here? That doesn't make a lot of sense."

"Well, *theoretically*, you use those fur pads to warm your face," explained Carson. Demonstrating, he donned his mittens and then crossed his arms in front of his chest, placing the backs of the mittens against his cheeks. "But In reality, these fur pads serve an entirely different purpose. When it's fifty degrees below and you have a runny snot locker, you sure can't pull a handkerchief out and blow it, so you just wipe your snout on the fur pad."

"You're kidding."

"Nope. It freezes immediately, and you just brush off the ice crystals. *Voila*." Satisfied with the fit, Carson took off the mittens and stowed them in his kit bag. "This stuff was apparently designed by someone with common sense."

Wolcott walked into the supply room and greeted the supply sergeant. "You have time for me to draw my gear, Bob? I'm accompanying these boys to the frozen north."

"Sure, Virg," replied the supply sergeant. "Do you need the full issue?"

"Lay it on me, pard." Wolcott tilted his Stetson back on his balding head, then lit a cigarette and inhaled deeply from it.

"So you're travelling up with us, Virgil?" asked Carson.

"Yup. I am indeed, pardner. I need to keep an eye on all you young bucks." Wolcott glanced at Ourecky, who had taken a seat on a bench behind the counter while trying on a pair of insulated "Mickey Mouse" vapor barrier boots. Lowering his voice, he asked Carson, "So, how's our young captain doing? You got any qualms about makin' the paraglider drops with him up there?"

"Well, Virg, he had a rough start in the landing simulator, but he's been doing much better lately." Carson crammed the last of his gear into his kit bag before zipping the canvas carry-all closed. "I've been watching the other crews, and he's performing at least as well as they are. I'm just really concerned about him making the transition from the simulator to real life."

"Understood, pardner," said Wolcott. "But I guess you know that we would have an empty seat otherwise, and right now I can't yank anyone else out of my hat."

"I understand, Virg. Look, I'm confident I can handle the drops myself, but I still have to think about what will happen if I'm cold-cocked again on the opening shock."

Wolcott took a long draw on the cigarette. "We have that covered," he said, exhaling a cloud of gray smoke. "The contractors have wired a remote activation setup to the emergency egress system, hoss. I'll be riding the chase plane for all of your drops. If I don't hear from you and I ain't confident with Ourecky flying it to the ground, then I'll throw the switch and fire the seats. We'll be watching out for you. Just consider it your fail-safe system."

"I really appreciate that, Virgil."

Eielson Air Force Base, Alaska
8:45 a.m., Tuesday, October 8, 1968

After a week of Arctic survival training at Eielson, Carson and Ourecky climbed onto an HH-3E "Jolly Green Giant" rescue helicopter to be dropped off at a remote wilderness site. Ourecky had enjoyed the training and anxiously looked forward to putting the theoretical lessons into practice. He had been somewhat surprised with the attitudes of Carson and the other Blue Gemini pilots, who smugly considered themselves to be authorities on wilderness survival.

Even though the instructors did their utmost to maintain their interest, the pilots appeared bored. While the jaded pilots considered the training a nuisance, Ourecky paid rapt attention, taking detailed notes and applying himself diligently to every opportunity for practical exercise. They all but snickered at him as he stacked kindling to start a fire, tied knots, whittled snare mechanisms to trap small animals, and manufactured a simple fish-net from parachute cord.

And now it was time to see if it all actually worked. Following the trace of a valley, the HH-3 slowed as it approached the site that would be their home for the next few days. In the rear of the aircraft, crouching on the open ramp, the crew chief pointed at a dark shape in the stark white landscape below. Shouting over the painful whine of the twin turbine engines, he explained, "There's your pod out there by the creek. Like we briefed you, just step off the ramp and keep your feet together. You'll sink right into the snow. Make sure you're okay, no broken bones, and then give me a wave with your right hand. Ready?"

Kneeling by the ramp, awkward in his heavy gear, Ourecky nodded briskly. He was apprehensive; as anxious as he was to start the wilderness exercise, he wasn't too keen on the notion of jumping out of the helicopter. He watched as the crew chief flashed a thumbs up and pointed at the ramp. Carson nonchalantly pushed himself to his feet, took three steps to the edge, and casually walked off into the void, vanishing into the swirling white blur outside.

"Okay," shouted the crew chief, placing his hand on Ourecky's shoulder. "Go!"

He hesitantly waddled to the sheer edge of the ramp and then stepped out vigorously. Clamping his feet together, he plummeted through the churning cold of the rotor wash. After a seemingly endless fall, he plunged deep into the snow below.

Momentarily breathless, he wiggled his arms and legs to make sure they were functional, and then wagged his right hand over his head. The helicopter's downdraft felt like a sandblaster, scouring his face with millions of minute frozen crystals. Squinting through the icy gale, he glimpsed the helicopter creep forward, gathering speed and ascending into the dismal gray morning sky. Minutes later, the overwhelming cacophony was replaced with utter silence.

Ourecky wriggled to the surface. Crouching on his hands and knees, he ensured that his "bail-out" kit was still stashed in the left thigh pocket of his field pants. Packed in a cloth bag, the kit contained essentials—a hank of parachute cord, a signal mirror, a reflective survival blanket, a waterproof match case, a pocket knife—just in case. His right thigh pocket held a URC-64 survival radio—roughly the size of two packs of cigarettes—and a Mark 13 "day-night" flare.

About fifty feet away, Carson's head and shoulders popped out of the snow. Sputtering, he blew snow out of his nose and then asked, "Ourecky, you okay?"

"I'm all right," answered Ourecky, shivering. "*Freezing*, but okay."

"Good. Start working your way to the pod. I'll meet you there."

Half walking and half crawling, Ourecky gradually made his way to an old ejection pod from an F-111 fighter-bomber. For everyone else participating in the exercise, including the rescue forces, the isolated pair was an "Aardvark" crew. Despite that, he knew that the two-man pod contained two survival rucksacks that would eventually be packed in the Gemini-I spacecraft.

It took Ourecky almost thirty minutes to traverse the three hundred feet to the pod. He was amazed at how much energy he expended; the frigid cold seemed to drain him, like it insidiously sapped energy out of batteries. He tugged off his mittens, letting them dangle from the lanyard around his neck. Underneath the mittens, he wore "anti-contact" gloves made of cotton fabric. The thin gloves did little to keep his hands warm; in this sub-zero environment, their sole purpose was to prevent his skin from inadvertently freezing to any metal that he might touch.

He unlatched the canopy and tried to swing it open. Creaking loudly, the Plexiglas hatch budged open slightly more than an inch, but no more. *Not good*, he thought; if they couldn't open the canopy to gain access to the survival kits, they were in for a miserable experience.

Grasping for options, Ourecky mentally reviewed the rules of the exercise. With their URC-64 radios, they could summon rescue forces in the event of an emergency, but if their dilemma wasn't significant enough to actually warrant a prompt extraction, they could be cast back out for another five-day iteration. Likewise, in addition to the contents of the two survival rucksacks, there was an emergency bundle containing high-calorie cold weather rations, a tent, multi-fuel stove and fuel. But if they so much as cracked the seal on the bundle, they earned another plunge off the HH-3's tail ramp, followed by a five-day refresher in frozen purgatory.

On the night prior to coming out here, over beers in the little Officers Club at Eielson, Carson vowed that he had no intention of spending any more than the requisite five days in the wilderness. If they had to suffer, he declared, they would, but they would not crack open the emergency bundle or switch on the URC-64s under any circumstances.

Ourecky considered their alternatives. If they couldn't pop the canopy open, they could probably smash out the Plexiglas with a sizeable rock. Shivering uncontrollably, he decided to try the stubborn canopy again. Stomping his feet, he tamped down the snow so he would have a solid platform for more leverage. It took him several attempts, but he finally forced the canopy open; the severe cold had caused the hinges to stick considerably.

Peering inside the cockpit, he saw that the two tandem seats were still in place, but the control sticks and most of the instruments had been stripped out. An old cargo parachute—simulating the Gemini-I's paraglider—was crammed behind the seats; they had been briefed that they could salvage the parachute for their shelter or other purposes. The nylon survival rucksacks were jammed into the foot wells of the cockpit.

"How's it look?" asked Carson, teeth chattering, standing up behind him.

"Good," commented Ourecky. He was overheated from the exertion and flipped back his parka hood to bleed off some excess heat. "Nice of you to come by."

"Climb in," ordered Carson quietly.

"Huh?"

"Climb into the cockpit, Ourecky. *Now.* There's no sense standing out here in the cold."

Clambering through the open hatch, Ourecky did as he was instructed. He slid headfirst into the cockpit and crawled into the left-hand seat.

Carson climbed into the right-hand seat, where the F-111's WSO—Weapon System Operator—would normally sit.

"Home, sweet home," commented Carson, closing the canopy. "Of course, it'll be a bit awkward to sleep sitting up, but I can live with that."

"You're planning to *stay* in here?" asked Ourecky skeptically. "The survival instructors said that staying in the pod was a bad idea."

"Yeah, those Boy Scouts want us to exert ourselves building a log cabin, but the smart play is to conserve our energy. I've done this drill a few times, so I know the rituals. We're out here by our lonesome for five days, so we make our own rules. We'll make ourselves a cozy little nest. When the weather's decent, we'll climb out and bash around a little bit to make it look like we made a sincere attempt to earn our merit badges, but otherwise we hunker down in here."

"But what if they come out to check us?" Ourecky noticed that the moisture from their breaths had already formed a thin layer of frost on the interior of the Plexiglas.

"Paranoid, are we? They're not coming to check on us, Ourecky. Like I said, I've already been here and have the cute little patch to show for it. Midway through Day Five, they'll roll the SAR guys. Then it's a chopper ride back to a hot meal and cold beers at the O Club."

"It just seems like we're cheating," replied Ourecky, unzipping his survival rucksack.

"And *that*, Mister Ourecky, forms the basis for the most valuable lesson about all forms of survival training: if you're not cheating, you're not trying, and if you get *caught* cheating, you're not trying hard enough."

"If you say so, Major." Ourecky pulled an odd-looking gun from his rucksack. It was a collapsible M6 Scout Aircrew Survival Weapon, an "over-under" combination of a .410 gauge shotgun and .22 caliber Hornet rifle. It would be the only firearm between the two men.

"Hand it over, Ourecky. I've been through the training, so I'll handle the shooting iron." Carson dug through his rucksack and located a stiff plastic object about two feet square and two inches thick. "Just what I was looking for," he said, twisting a metal latch in the center of the square. The object made a loud hissing noise; a vacuum-compressed sleeping bag materialized from the unlikely package. "Find yours and make yourself comfortable."

Ourecky found his sleeping bag and stuffed it behind his seat. As he inventoried the contents of his rucksack, he said, "I guess we can stay in here tonight. If you don't mind, after a while, I'll go outside and scout the immediate area. I won't stray too far. Okay with you?"

"Sounds good. I might even go with you, but don't be too offended if I stay put."

7:45 a.m., Wednesday, October 9, 1968

It wasn't yet light when Ourecky woke up. Despite the sleeping bag and several layers of parachute material, he was colder than he had ever been in his life. He had never imagined that it was possible to be *this* cold and still be alive. His extremities were heavy and numb, and to cap it all off, he had a splitting headache. Shivering uncontrollably, he remained in his seat for almost an hour, waiting for the sun to come up, as he contemplated what he should do.

Finally, mustering his courage, he climbed out of his nylon cocoon. Maybe Carson was content to be cooped up in this frigid cockpit, but he wasn't. He unlatched the portside hatch canopy, jammed his shoulder against it, and then stood up. The canopy reluctantly creaked open. He heaved his survival rucksack out into the snow and then rolled headlong out after it.

Shaking violently, he stood up, closed the canopy, and then stomped his feet to coax some circulation into his toes. He scanned the desolate landscape and spied a slightly wooded area adjacent to the creek, about a hundred yards east of the pod. The trees weren't much to speak of, mostly new pines and some stunted hardwoods, but they would offer at least some modicum of protection from the wind, as well as the raw materials he would need to establish a campsite.

Instead of crawling on his hands and knees atop the snow, as he had done the day before, he intentionally stayed upright, methodically packing down a path from the pod to his planned abode. Like yesterday, his transit was painfully slow and arduous. Several times along the way, the bitter cold and constant wind almost forced him to retreat to the pod in surrender.

Making it to the sparse tree line, he rested. His head still ached and his swollen tongue filled his mouth. He was badly dehydrated but resisted the urge to munch on snow to slake his thirst. His rucksack contained four metal containers of emergency drinking water, each about the size of a soda can, but they were surely frozen solid by now. If he had remembered to put one under his parka last night, he would now have some water to quench his thirst.

To start a fire, he foraged for twigs and sticks, snapping dry "squaw wood" from the trunks of nearby trees. Recalling the Jack London short stories he had savored in junior high school, he took care to set his fire away from any overhanging tree branches, lest it be snuffed out by falling snow. With shaking hands, he carefully arranged his tinder and kindling to fashion a small teepee and stockpiled extra wood to stoke the fire as it grew.

He successfully ignited the tinder with a single wooden match, just as he had been taught in the Boy Scouts, and gently blew into it to entice the flames. Crouching by the fire, soaking in its meager warmth, he opened the jam-packed rucksack and dug out a "Woodsman's Pal"—a multi-function tool that looked like a wide-bladed machete with a pruning hook at the end—and a length of parachute cord. He placed the gray water cans near the fire to thaw them.

Over the course of the next few hours, he built a lean-to sufficiently large to accommodate two men, complete with a thick mattress of springy evergreen boughs. On the open side of the lean-to, he gouged out a burrow for a fire and fabricated a small wall to reflect the fire's warmth into the low shelter. Then he surveyed his handiwork. It wasn't much, but he was confident that even the stoic Carson would realize the merits of being able to stretch out in front of a fire instead of stubbornly remaining in the cramped pod.

After applying some finishing touches, he backtracked to the pod to grab additional gear. He knocked on the canopy and waited for Carson's response, but there was nothing; the pilot was either sleeping very soundly—which was highly unlikely—or something was *very* wrong.

Clumsy in his heavy Arctic gear, he popped open the canopy and swung it up. He anxiously pulled the parachute blanket and orange sleeping bag away from Carson's face. The pilot was pale and unresponsive. Ourecky checked his pulse; his heart was beating, but just barely so. Carson's body was cold to the touch, so Ourecky knew he had to act promptly, or the least of his worries would be spending five extra days in the wilderness.

Climbing into the cockpit, he managed to wedge himself under Carson. Pushing mightily, he shoved the pilot out into the snow. In what seemed like an eternity, he gradually dragged Carson's inert body to the lean-to shelter. Placing him inside, he covered him with his sleeping bag before stoking the fire with more wood. Then he tramped back to the pod to grab the second rucksack, the parachute, and some scrap pieces of aluminum from the remnants of the instrument panel. Periodically checking on Carson, who was now conscious and showing signs of improvement, he used a rock to bash some salvaged aluminum into a semblance of a pan, which he then used to melt snow to produce drinking water.

Before the sun went down, Ourecky had succeeded in fully reviving Carson. "Better now?" he asked, fanning the coals of the fire.

"I'll say," mumbled Carson. Sitting by the fire, he sipped from a water can, which was filled with a tea Ourecky had made by steeping pine needles in hot water. "I guess I owe you one. I probably would have frozen to death if you hadn't looked in on me when you did." He took another sip and looked at their snug sanctuary. "So where did you learn how to do all this?"

"Boy Scouts. Plus I paid attention in class. You just never know when you might actually have to put some of this stuff into practice." Ourecky leaned over the fire and looked into his makeshift pot. "Hungry?"

"Famished."

"Well, it's not much, but I'm cooking up a couple of bouillon cubes. I pulled them out of a survival ration I found in the kit. I also set a few snares on a rabbit trail down by the creek, so maybe we'll have a more substantial meal at some point."

"Babe, I'm just grateful for what we have," Carson said, leaning over the crude pot and taking a quick whiff. "By the way, Ourecky, what happened today…would it be a problem if we just kept that to ourselves?"

Ourecky smiled. "My lips are sealed. I guess it can be our little secret. But just one little thing, though…"

"What's that?"

Grinning, Ourecky thumped Carson on the shoulder and exclaimed, "Pay attention!"

Embarrassed, Carson laughed. "Point taken. I deserved that. I've definitely learned my lesson on this trip."

Sipping the tea, Carson was silent for several minutes, contemplating the flickering coals. Finally he spoke. "Why do you do this, Scott?"

"Do what?"

"You know. The Box. This survival training. I endure all this crap because there's a remote chance that I might go into space someday, but why do *you* do it?"

"Well, I guess I've not really thought about it."

Darkness was falling quickly. As he stirred the fire with a stick, Carson said, "You *should* think about it. You must know that ol' Virg is exploiting you to make this last hurdle. After that, you'll probably be exiled back to your closet. That's not fair."

Ourecky watched red embers rise and flicker out. He was thankful that the winds had died down. "Fair? Life's not about what's fair, Major Carson."

"*Drew*. Call me Drew, Scott. I think you've more than earned that right by saving my butt."

"Okay. Drew, it is. Anyway, I've reconciled myself to the fact that I'm not meant to be a hero. I'm just cut out to be someone who makes it possible for others to be heroes. I figured that out after they rejected my last appeal to attend flight school. It just wasn't meant to be."

"Yeah. Not everyone is destined to be a pilot," noted Carson.

"That's true, but I'll tell you, the past few weeks have been the best time in my life, and if this is as close as I ever come to orbit, then it's all been worth it. Just to be a part of this makes it all worthwhile to me. This is history in the making."

"So you're willing to do all the heavy lifting that makes this thing possible, and then just fade anonymously into the background?" asked Carson.

Ourecky nodded. "Look, no one knows the name of Lindbergh's mechanic, but he was still a part of history all the same. And that's the way it is for me. If any of this ever ends up in the history books, I can point to it and tell my kids and grandkids that I was a part of it."

"But it's never going to be in any history book."

"I can accept that."

"Well, Scott, I sure admire you for that. You're definitely enduring a lot of pain and agony just to remain on the ground and watch others go into space."

Ourecky didn't reply. Anticipating the long night ahead, he banked glowing coals so that the warmth would remain for the cold hours between now and dawn.

"Wow, look at that," exclaimed Carson, gesturing toward the north. The ghostly red and green plumes of the aurora borealis danced and shimmered in the sky. "Isn't that something?"

"It surely is," replied Ourecky, awestruck by the sublime beauty in the heavens. "It most surely is."

10:20 a.m., Friday, October 11, 1968

The ride from Eielson to the field training site took almost an hour by Sno-Cat. Accompanied by a PJ and a security man from the 116th, Wolcott quietly unloaded his gear for the two-mile trek to the observation post. The security man bore a 30.06 hunting rifle; the bolt-action weapon was intended for the grizzly bears that sometimes frequented this environs.

Wolcott was a little wobbly on his snowshoes, so he used an old pair of bamboo ski poles to help maintain his balance. Like his companions, he carried a sleeping bag and other survival gear in an Army mountaineering rucksack. The pack's narrow straps weighed heavily on his shoulders, even through the padding of his parka, and its metal frame grated at his hips.

The outpost was set just below the ridgeline, overlooking the expansive valley where Carson and Ourecky had their survival camp. Keeping vigil in a blind manufactured of rough timbers and white cloth, a man intently peered through a tripod-mounted spotting scope, periodically jotting notes in a journal. He wore camouflage "overwhites" atop his olive drab parka and field pants. The interior of the blind was spare and tidy; a scoped bolt-action rifle leaned against a timber, next to a PRC-25 field radio. A small multi-fuel mountain stove sat in a corner, heating a pot of liquid. The hard-packed snow floor was covered with a sheet of canvas.

Wolcott approached the man and said, "So you must be the one keepin' an eye on this big spread." He stuck out his mittened hand. "I'm Virgil Wolcott."

The man turned, slowly put a gloved finger to his lips, and quietly shushed Wolcott. "Sorry, sir, but sound carries far over this snow," he explained quietly, in a voice just barely above a whisper. "I'm Tech Sergeant Halvorsen. I'm a survival instructor from Fairchild. We're observing your aircrew for the duration of the exercise. We keep an eye on them, but we keep our distance. They won't ever know we're here unless they have a significant problem."

"Sounds good, pard," replied Wolcott. He heard a slight rustling noise and turned to see a small lean-to, constructed of freshly hewn pine saplings and boughs, next to the blind. Inside the lean-to, another instructor was bundled in a heavy Arctic sleeping bag, sleeping soundly.

"That's Tech Sergeant Jackson, sir. Do you want me to wake him, sir?"

"No, pard, don't disturb him," said Wolcott. He stooped down, took off his mittens, and cupped his hands around the blue flame emanating from the mountain stove's burner. "That ain't coffee, is it?" he asked, gazing into the small aluminum pot perched atop the stove.

"Yes, sir. Help yourself. Use my cup there. You have to drink it down pretty fast, though, because it will start freezing a couple of minutes after it comes off the stove."

Wolcott reached into his parka and extracted a pack of cigarettes and a pack of matches. Teeth chattering, holding a cigarette between his lips, he flicked open the Zippo.

"Sorry, sir. No smoking up here. They'll be able to smell it down in the valley."

"Well, pard, you sure know how to make someone feel welcome." Wolcott gingerly poured the plastic cup full of coffee. He stood up slowly, chose an aiming point outside and behind the blind, and let fly a string of tobacco juice. Within seconds of hitting the snow, the spit froze into brown crystals. "So how are my boys doin' out there?"

Kneeling on the canvas floor, Halvorsen consulted his log. "They've been out three days. As you look at their shelter down there, the big bump in the snow about a hundred yards to the west is their F-111 pod. About six inches of snow fell last night, and it covered it up fairly well."

As Wolcott looked through the spotting scope, Halvorsen continued. "Your crew scouted around on the afternoon of the first day, but didn't venture more than a couple of hundred yards from where we dropped them off. They don't have snowshoes, so it's mighty hard slogging around out there, but they did make it as far as that little creek you see off to the east."

"About mid-afternoon of the first day, we observed both of them climb back into the ejection pod. Initially, we figured that they were just sorting out their survival kits, but later we realized that they were seeking shelter in the pod, which really isn't a good idea."

"Why's that, pard?" asked Wolcott. He sipped from the coffee.

"That pod's aluminum, so it soaks up the cold like an ice cube tray. No insulation at all. Of course, a good layer of snow will insulate it somewhat, but you run the risk of suffocating if you can't open the canopy. Plus, if you remain in there, you can't build a fire. You're not going to last very long out here without at least a small fire to stay warm. Normally, when we see an aircrew take refuge in the pod, we know we'll eventually be heading down there to evacuate them."

"But they did build a shelter, though. Looks like a good one," noted Wolcott. Through the scope, he could clearly see details of the lean-to, built of saplings and boughs much like the instructors', as well as gray wisps of smoke. "And they have a nice campfire goin' as well."

"Right, sir. One of your men came out of the pod on the morning of Day Two. He built that shelter and firewall by himself, and then gathered wood for a fire. That took him most of the morning. After he started a fire, he went back into the pod and all but carried your other man to the shelter. We were pretty relieved, because really we don't like hauling people out of here."

Carson, thought Wolcott. After this ordeal was over, he was sure to get an earful from Carson, bitching about how useless Ourecky was during the exercise. In retrospect, maybe he had made a mistake putting Ourecky out there in the snow; there really wasn't any need to, and now it looked like Carson was devoting most of his time and energy to babysitting the engineer instead of perfecting his survival skills.

"Okay, pardner, let me make sure I have this straight," said Wolcott. "You're tellin' me that one of my troops did all the work while the other one just sat on his haunches and loafed?"

"That's pretty much it, sir," replied Halvorsen.

Wolcott thought for a minute. *How could he be sure that it was Carson who did all the work? It certainly made perfect sense, since Carson was a grad-uate of several survival courses before he ever set foot up here, but how could he be sure?* "Son, I see you take detailed notes. Can you answer a question for me?"

"Fire away, sir."

"You said you saw a man climb out of the pod on Day Two, and that he did the lion's share of the work. Did you happen to see where he was sittin' in the pod? Left side or right side?"

Halvorsen consulted his notes. "Left side, sir. Definitely left side."

That clinches it, thought Wolcott. The left seat was the hallowed bastion of the command pilot, and Carson would not surrender it to anyone, under any circumstances. "Thanks," he said, looking back through the spotting scope. "Hey…is that a rabbit cooking on a spit?"

"It is, sir. Your guy set some snares by the creek and caught a good-sized snowshoe hare this morning. They'll be eating good in a little while, much better than us."

Wolcott shook his head, wishing that he had some way to mass-produce copies of Carson. *With a few more guys like him, maybe they would make it to orbit after all.*

23

SAVING THE SHIP

Officers Club, Eielson Air Force Base, Alaska
7:30 p.m., Saturday, October 12, 1968

Carson took a sip from his Olympia beer and looked around. Cramped and dreary, the Eielson "O" Club would be a perfect broom closet, if only there was ample space for a broom. In the far corner, an ancient jukebox seemed to only spin records by Merle Haggard and Buck Owens. If he had to listen to "Mama Tried" just one more time, Carson would be plenty happy to climb back onto that HH-3 eggbeater for another five-day sojourn in the wilderness.

He leaned over to read a newspaper clipping preserved in a black frame hanging beside the bar. The article recounted a horrific tragedy in1955, in which an F-84 Thunderjet had smashed into a base housing complex during lunch hour, killing the pilot and several people on the ground. The dead included a set of eleven-month-old triplets, an Army master sergeant and his three daughters aged two, three and four; and an Army sergeant, his wife, and mother-in-law. In all, fifteen people, including seven toddlers, died as the result of the inferno at Eielson. Most of the dead were Army personnel or their dependents.

Although Carson had thawed out from the survival trip, he kept his parka on, simply because not more than ten minutes elapsed without someone coming in from the cold. Sure enough, as the clock's minute hand swept the bottom of the hour, the door swung open.

Shivering, he tugged his parka around him as a frigid gust whisked napkins from the bar, setting them aflutter like giant snowflakes. Looking up, he didn't immediately identify Wolcott, because the old man wore a thick wool cap in place of his trademark Stetson. Wolcott was accompanied by a man that Carson didn't recognize. The two stomped their boots on the doormat and then underwent the laborious drill of shedding their outer garments.

"Carson!" exclaimed Wolcott, unzipping his parka. "Good to see you're in one piece, pard! You mind if me and my friend bend an elbow with you?"

"Be my guest, Virg," replied Carson, waving toward a pair of empty stools. He extended a hand to the newcomer. "Drew Carson. I guess you can surmise that I work for Virgil here."

"Major Ed Russo," replied the newcomer. Russo stood an inch or two shorter than Carson, but had a similar athletic physique. "I've heard a lot of great things about you, Carson. Sounds like you're a real wizard at surviving in the wilderness up here."

Taken aback, Carson raised his eyebrows. "If you say so, I suppose."

"Ed here is our liaison officer from El Segundo," explained Wolcott, unwinding an Army green scarf from around his neck. El Segundo was a coded reference to the MOL project, headquartered in the Los Angeles suburb of the same name. "He came up here to watch how we do business. He's a test pilot like you, Carson, so you two should have plenty to chew the fat over. Now if you two gents will excuse me, I need to drag my old kidneys to the head."

Watching Wolcott amble away, Carson asked, "So you're a graduate of ARPS?" ARPS was the prestigious Aerospace Research Pilot School at Edwards Air Force Base in California; initially under the direction of legendary test pilot Chuck Yeager, it was the Air Force's preeminent grooming program to prepare pilots to become NASA and military astronauts.

Russo nodded. Looking around the bar, he answered quietly, "Affirmative. I graduated from Edwards and then went straight to the Navy's school at Pax River. I've already been accepted for the next astronaut group for the MOL. That's at least a year away, so they sent me to Wright-Patt to work as a temporary liaison. But from the looks of things, I might end up flying with you folks before I go up in the MOL."

"*Really*? Is that so?" asked Carson, raising an eyebrow. "That's news to me."

"That's the impression I gleaned from General Wolcott," observed Russo. "He said that when we all head back to Wright-Patt, I'll probably be working with you in the simulator facility for the next few weeks, to get up to speed on procedures."

"That's odd. I've been working with Ourecky in the Box, and we're slated to continue working together until January."

"So I hear," said Russo. "Ourecky? He's the engineer, right? The egghead? Math whiz kid? Kind of an idiot savant? He's not actually slated to fly, right?"

"Well, the two of us are supposed to fly the paraglider trainer next week."

"Really? Virgil gives me the impression that he's none too happy with Ourecky," Russo confided. "He didn't lend me any specifics, but he did tell me that since you lost your last right-seater, you'll need a replacement. Who knows? You might land up as *my* right-seater. I've had a peek at your records, and I am two years senior to you, so it makes sense that…"

"It doesn't work that way on the Project. Rank is irrelevant to seat and mission assignments."

"Sure, Drew. If you say so, but I seriously doubt that there's anyone here who's logged as many flight testing hours as I have, or flying hours in general. It's just a matter of proving that I've got the goods, and that shouldn't take too long. I hope you don't take it personally."

"So noted. So where were you before you went to Edwards?" asked Carson, quickly changing the subject. "SAC? TAC? Bombers? Fighters?"

"Fighters after flight school. First F-100 Super Sabres, then F-4's. I was in the right place at just the right time, and flew in Vietnam right from the start. I've logged eighteen months over there. Splashed two MiG-17's and was shot down once by Triple-A. You? Been over?"

"Nope," answered Carson, certain that Russo was already aware of his lack of combat experience. "Not that lucky. I got tagged with interceptors initially, then transitioned to F-4's just before I was picked up for ARPS. I'm hoping to jump into the fight as soon as I can, though."

"So, hoss, where's your partner in crime?" asked Wolcott, returning to the bar.

"Ourecky? Back at the VOQ. He hit the hay early. The trip beat him up pretty bad. He was just absolutely tuckered out. He could barely stay awake during the debriefs."

Wolcott nodded. "What are you drinkin', Ed?" he asked. "Buy you a snoot?"

"I'm abstaining, sir," answered Russo. "It was a long flight and a long day."

"You ain't partakin', Ed? Suit yourself. Another Oly for my friend Carson here, and fetch a scotch and soda for me," called out Wolcott to the bartender. "And a pack of smokes."

The bartender, an enlisted man earning some extra cash, dipped into the cooler to retrieve a beer. He poured Wolcott's drink as he asked, "What kind of cigarettes do you prefer, sir?"

"Whatever you have that's cheap and unfiltered, pard. Surprise me."

"Thanks for the beer, Virg," said Carson. "But it will have to be my last one. I'm pretty tired myself. It's been a long week."

"I know, son." Wolcott lit a cigarette, puffed on it, and then took a sip from his drink. "Man, *Merle Haggard*! I could hunker here on this stool and listen to Merle *all* night."

Trying hard not to roll his eyes and groan, Carson imagined that Wolcott would have that opportunity, provided he could endure the intermittent blasts of ice-cold air.

"Carson, pardner, I need to make you an apology."

"For what, Virgil?" Carson looked at the second Olympia, now unsure that he could even finish it without slumping off the barstool, fast asleep.

"Well, son, I'm really sorry that I saddled you with Ourecky. Looking back on it, I think it was just a bit too much to ask of you. It's one thing for him to be flying the Box with you back at Wright-Patt, but it's obvious he ain't cut out for the rest of this business. I apologize for stickin' you out in the field with him. Sincerely, I mean it."

Carson looked at him quizzically. "He did okay, Virg. He really wasn't any hindrance at all."

Displaying his crooked array of chipped teeth, Wolcott flashed a knowing smile. "If you say so, hombre, but I'll let you in on a little secret. We had folks watching you while you were out there in the snow, mostly for safety reasons, so there ain't any sense trying to convince me that Ourecky pulled his weight out there. I know you built the shelter by yourself, and I know you got a fire started. And that was a fine trick with the bunny. One of the Fairchild instructors told me he's been up here six times and has never caught a rabbit. Good on you, Carson."

Taken completely by surprise, Carson was befuddled; how could he tell Wolcott the truth without telling him the *whole* truth? He decided at this point that it was probably just as well to let the whole matter drop. *After all, what good could come of letting Wolcott in on the secret? Nothing, except possibly to erode Virgil's confidence in him, and that meant falling to the back of the line for any hope of a flight.* "Uh, thanks, Virg. I appreciate it."

Wolcott cupped his ear and craned his head to hear the first strains of the next song coming from the tired guts of the jukebox. "Merle *again!*" he exclaimed. "*Dadblame!* It's my lucky night." He sucked in the last of the cigarette and dropped the almost non-existent butt in an ashtray. "Let me tell you, Carson, I've got a good mind to stick Ourecky back out in the cooler for another five days, but we have too many irons in the fire. You start your drops tomorrow night, right?"

"Yes, sir." Carson felt really tired; his head swam with the notion of flying the paraglider all next week. He looked over at Russo, who was silent but intently listening to the conversation, and wondered what schemes were in the making.

"Well, pardner, let me chuck out something to you," said Wolcott. "I'm more than willin' to delay the drops to change the line-up. You can pick any right-seater you want out of the other crews, and I'll put that varmint Ourecky on the first thing smokin' back to Ohio. I only regret that we didn't have sufficient time to bring Ed here up to speed, or he could drop with you."

Now, Carson felt both tired *and* nauseous. Unlike him and the other four pilots who dreaded making even one more plunge in the paraglider trainer, Ourecky was eager for the experience. Sure, he was apprehensive, but he really *wanted* to make the drop. Now Wolcott was leaning on him to deprive him of that chance.

Carson chose his words carefully. "Virgil, if it's all the same to you, I'll still drop with Ourecky next week. I appreciate your offer, but I don't want to take a chance on having to learn someone else's moves this late in the game, especially in mid-flight. Does that make sense, sir?"

Stirring his drink, Wolcott nodded. "I understand, son. Your funeral. Just don't forget that I offered you an alternative."

Carson nodded solemnly. "I need to hit the trail, Virg. I need to grab some shuteye tonight."

Wolcott clapped him lightly on the back. "Understood. Have a good evening, son."

"Nice meeting you, Drew," said Russo. "Good luck on the drops."

"Thanks. See you in the morning." Carson zipped up his parka and then donned his wool watch cap and mittens. Leaving Wolcott, Russo, the unopened beer, and Merle Haggard behind him, he jammed his shoulder against the door and shoved his way out into the cold. Then he walked a few steps, collapsed to his knees, and threw up in the snow.

Eielson Air Force Base, Alaska
1:45 a.m., Thursday, October 17, 1968

After submitting to an exhaustive medical exam, Carson and Ourecky had a day off to relax and recuperate before beginning their series of five paraglider drops. The drops went well, exactly as planned, right up until the very last one.

In the dimly red-lit cabin, Carson watched the UMB light blink off as the power and communications umbilical was separated from the C-130. Now, they were on their own. He and Ourecky made a final scan of the instruments, ensuring that everything was set for the big drop.

Satisfied that they were ready, Carson spoke to the C-130 pilot over the radio: "Big Box, this is Ultra Three, on internal power. We're ready for push-off back here."

"Ultra Three, this is Big Box," answered the transport's pilot. "Control has cleared us for drop operations. Release in thirty seconds. Good luck."

"Big Box, Ultra Three. Ready for release. Thank you."

"All stations, this is Big Box. Five, Four, Three, Two, One, *Mark.* Releasing package." With a brisk shove by the loadmasters in the tail of the C-130, the little craft slid tail-first off the end of the ramp and lurched into the darkness. Bracing themselves as best they could, the two men were thrashed and slammed around in the cockpit. By this time, it was no longer a novel experience for Ourecky; after four drops, he feared it just as much as the other pilots.

"Drogue is out," grunted Carson, peering through his window to watch the small canopy blossom in the darkness. With the parachute pulling them upright, the vehicle fell blunt end to earth and slightly more stable as they awaited the deployment of the paraglider.

Moments later, they watched the reefed paraglider stream out into the slipstream, tugged out of its cylindrical container by the drogue chute. "Paraglider is out and de-reefing," said Carson. "Longitudinal struts are inflating."

Carson watched the last strut inflate and then declared: "Cross strut's inflated. Ready to go to two-point suspension. Scott, brace yourself and let me know when you're ready."

"I'm braced," answered Ourecky, jamming himself down into his seat. "Ready."

"Switching to two-point." With a jolt, the cockpit pitched forward. Carson went through his post-opening procedures, steering the paraglider through a series of checks to assess controllability. "Stall check… stall check is *good.* Port turn… port turn is *good.* Starboard turn…starboard turn is…" The vehicle whirled to the right in an ever faster and tighter spiral.

"What's wrong?" asked Ourecky anxiously.

"Not sure," said Carson. "Might be a stuck cable." He spoke on the radio: "Chase One, we're having a problem recovering from the starboard turn controllability check."

Wolcott's voice came back on the radio: "Ultra Three, this is Chase One, are you declaring an emergency?"

"Chase One, Ultra Three. Not yet."

Trying diligently to coax the paraglider into normal flight, Carson cursed under his breath as he wiggled the hand controller and assessed the situation. The paraglider was steered by altering its shape; control cables, connected to turn-control motors, tugged down or let up on the rear corners of the delta-shaped fabric wing. Their problem was obviously either a kinked control cable or a malfunctioning turn-control motor; either way, the craft was now doomed since their procedures called for immediate ejection after they jettisoned the paraglider. Except for the faint popping sound from the paraglider, there was only eerie silence.

"Chase One, Ultra Three," said Carson. "I think we have a jammed turn motor. I'm working with it, but you should encourage the rescue forces to move in our direction." Still manipulating the hand controller, he swiveled his head toward Ourecky and said, "Make sure you're cinched up tight. Stand by to eject. It's going to be mighty cold out there."

Ourecky checked his harness and made sure he was poised square in the seat pan. Then he fell silent for a few seconds before speaking. "Drew, I don't think the turn-control motor is jammed. I think it's a sequencer glitch. I think we can fix this."

In his mind, Carson quickly analyzed the dilemma. As long as they were caught in a spin, their rate of descent was twice as fast as normal. At that rate, in two minutes they would be beyond recovery and wouldn't even be able to punch out. If the control lines fouled over the hatches or the hatch pyro didn't fire, they were dead anyway.

"Scott, this isn't a *simulation*. This is *real*," said Carson. "We don't have time to play. We have to punch out." A veteran of two previous ejections, he wasn't afraid of bailing out. On the other hand, he wasn't overly thrilled about ejecting into the frigid cold and the prospects of possibly spending a night out in the frozen wilderness.

"Drew, we can fix this. Really. I'm sure."

"Okay. You have *thirty* seconds, and then we're heading out into the cold." Carson looked at the attitude indicator and added, "We're just shy of seventy degrees angle of bank right now, and eighty is the limit. If we brush close to eighty, I'm calling it. Right now, make sure your oxygen mask is tight. You don't want to lose it outside, unless you want frostbite."

"Checking mask," said Ourecky quickly. "Okay, Drew, when I tell you, chop the circuit breakers for the landing sequencer. They're on your side."

"I *know* where the damned landing sequence breakers are, Scott," snapped Carson.

"Okay. I'm resetting pyro control breakers one, two and three. Re-set. Now, on your side, set the landing sequence breakers control one and control two to off."

"Resetting landing sequence control one and two to off. What's next?"

"On *my* side, resetting paraglider turn-control motor breakers one and two. Now, re-set the PG Reef tele-light. When I tell you, trip the Landing Attitude."

"PG Reef is *green*. Resetting PG Reef tele-light. PG Reef tele-light is now *amber*."

"PG Reef tele-light is amber," confirmed Ourecky. "Now, trip the Landing Attitude switch."

"Landing Attitude tripped. Scott, *ten* more seconds and we punch out. Nine, Eight, Seven…"

"Try the hand controller," said Ourecky confidently.

Carson wrapped his fingers around the hand controller and tugged it to the left. The paraglider immediately responded, and they gradually recovered from the spin. He breathed a sigh of relief, then completed the controllability checks for the paraglider.

"The turn-control motors were fine," observed Ourecky. "The problem was that the sequencer energized a redundant relay that cut off the power to the starboard side motor let-up. We just tricked the sequencer into believing the relay was not activated, and that restored the power to the motor. Not a big problem to fix."

"And just *when* did you know this?" asked Carson, gently manipulating the hand controller to align them on the high approach heading for Eielson.

"It just came to me. I guess I should have figured it out before, but the thought of ejecting into sub-zero temperatures in the dark probably spurred me to come up with it a little faster."

Carson laughed. "You're really quite some piece of work, Scott. Hey, do me a favor. Grab the controller. I need to adjust this shoulder strap. It's really killing me."

Ourecky took the hand controller in his left hand and said, "I *have* the controls."

"*You* have the controls," answered Carson, relinquishing the controller and calmly placing his hands in his lap.

Watching the compass, Ourecky steadily held the unpowered craft on heading as he patiently waited for Carson to resume control. He glanced down and noticed that Carson's hands were still in his lap. "I thought you had to fix your shoulder restraint," he said.

"Oh, it's *much* better now," answered Carson. He reached out and playfully tapped Ourecky on the left shoulder, then put his hands back in his lap. "You saved the ship, so you land it."

"I fly the approach?"

"Yeah. You earned this one. Home, Jeeves."

Officers Club, Eielson Air Force Base, Alaska
8:30 p.m., Thursday, October 17, 1968

As he entered the club with Carson, Ourecky spotted the other four Blue Gemini pilots—Parch Jackson, Mike Sigler, Tom "Big Head" Howard, and Pete "Squeaky" Riddle—seated at a table in the corner. Rounding out the table was the MOL project liaison officer, Major Ed Russo. With seven people seated at the bar and the two available tables full, the place was as busy as he had ever seen it. Seeing that there were no empty seats at the table with the pilots, he spotted two vacant chairs by the jukebox; he and Carson carried them to the table and sat down.

"All hail the conquering heroes," announced Howard, standing up and raising his Olympia bottle. Of the five pilots, he was the tallest, standing an inch shy of six feet. He had broad shoulders, a lantern jaw, and a head that seemed far too big for his body. Besides his massive cranium, his hands were also outsized, so much so that the suit contractor initially claimed that it would be impractical to make gloves suitably large enough to fit him.

The other pilots—excluding Russo—stood and lifted their drinks as well. Embarrassed, Ourecky grinned sheepishly, waved, and sat down. Looking around the table, he couldn't help but notice that the two-man crews instinctively sat together, with the command pilot of each pair seated to the left. Carson sat down to Ourecky's left, although in the awkward shuffle of arranging the extra chairs at the table, Russo managed to insert himself between the two.

"Your first Oly's on me, Ourecky. You too, Drew," said Riddle, Howard's right-seater, waving to the bartender. As a crew, they were a study in contrast; at five foot five, Riddle was the smallest of the five pilots. He had dark red hair and soft features, almost effeminate. Despite his small stature, he was renowned for his flying skills; with over five years working at Edwards, he was the most experienced test pilot amongst them.

"Thanks, Squeak, but just one for me. I need to make a short night of it," said Carson. "Scott and I are zooming back to Wright-Patt early in the morning, and I'm cutting close to my bottle-to-throttle window as it is."

"Did you guys listen to the news today?" asked Sigler. "Nixon said that if he's elected, he wants to have us out of Vietnam by the end of January."

"I sure hope not," said Carson. "I'm still working angles to latch onto some of that action. Of course, that's if Virgil will cut me loose for a few weeks to log some trigger time."

"Never, *ever* happen," said Howard. "We're stuck here for the duration. You might as well accept that, Drew, and quit pining for dreams that will never be." Of the five pilots, only Howard had flown in Vietnam, but only for eleven missions in 1966, before he abruptly received orders to attend ARPS at Edwards. The fact that Chuck Yeager—formerly the commandant of ARPS—was Howard's wing commander at the time probably had much to do with the abrupt transition to California.

"I flew overseas," gloated Russo. "One hundred and twelve missions. Two MiG kills."

Ignoring Russo, Riddle said "I guess we're pulling out of Vietnam then, because it's a sure bet that Nixon's going to win."

"Oh, hush with that malarkey. It's not a sure thing by any means," said Howard.

"Well, Big Head, maybe *you* don't think so," replied Riddle, tearing open a bag of potato chips. "But I sure do. George Wallace can't win, but he's grabbing up Democrat votes that will skew the election results in Nixon's favor. So I declare Nixon the winner by a nose."

"And what a nose it is," noted Carson. "Anybody heard anything new on Apollo Seven? Those guys are upstairs right now."

"One of my old squadron buddies works in Houston at Mission Control," confided Russo. "He said that the Apollo guys are sick as dogs right now, just puking their guts up, and that the NASA brass is on their last nerve with Schirra. Wally is apparently dictating to them which experiments he's willing to do and which ones he's going to shitcan. He's being pretty damned obstinate, and Houston's fed up with it."

"He *is* the commander," observed Carson. Stroking his moustache, he turned to smile and wink at a full-figured Inuit girl sitting at the bar.

"But it's an orbital mission," countered Russo. "It's the first flight with a brand new ship, with a full flight plan and a whole battery of experiments."

"Yeah, that's right," noted Howard. "But he's still the *commander*. Schirra's the guy up there in the cockpit. *No* one should be second-guessing him. If his guys are under the weather and can't keep pace with an overly ambitious flight plan, then they should make adjustments, and the *commander* is always the guy who should call the ball. Not Mission Control."

"Well, I disagree," retorted Russo. "Normally, that might be true. Now, if I was up there…"

Listening to the conversation as he sipped his beer, Ourecky felt a tap on his shoulder. He looked behind him to see one of the pilots from the recon squadron assigned to the base. Five other recon guys occupied a table on the opposite end of the club. Not more than five minutes ago, they had been raucously singing along with a sappy country song on the jukebox. Now they were silent, intently watching their counterpart, apparently in anticipation of a scuffle.

"You're sitting in a red stripe chair," growled the recon pilot.

"Excuse me?" asked Ourecky.

"I said, you're sitting in a red stripe chair," reiterated the recon pilot. He tapped his finger on a two-inch stripe painted in glossy red across the chair's back. "Those chairs are strictly reserved for recon guys who've flown over the line."

"Sorry, man," mumbled Ourecky, beginning to stand up. "My mistake."

Howard stood up, gently nudged Ourecky back into the chair, and then turned to confront the belligerent recon pilot. "Well, my friend here hasn't flown over the line. Since he's not from around these parts, he obviously doesn't know anything about your stupid traditions. But let me ask you, chief, since you're making such an issue of it, have *you* flown over the line yet?"

"I haven't, but that's not the point..."

Interrupting him, Howard said, "Point? Here's the *point*, buddy. The *point* is that you're about half a second away from flying over the line yourself. You need to execute a swift about face and move out smartly, just like they taught you at Colorado Springs. And if you or one of your compadres waltzes back over here to dislodge my friend from his chair, red stripe or not, you're going to be wearing your precious seat for the rest of your days on earth, and it's going to be very awkward when you go to the latrine to take a crap. Got me? Now, scram."

Obviously noticing the other Project pilots glowering at him in a show of unity, the recon pilot abandoned his quest for the chair and slinked back to the far side of the bar.

Howard extended his hand to Ourecky. "Tom," he said. "But everyone calls me Big Head."

"Scott," answered Ourecky, smiling to himself. He already knew the first names of the pilots, but also was aware of an unspoken protocol among them, something that transcended the formalities of rank and regulations. An outsider had to earn the privilege to address a member of the inner circle by his first name, and it was a privilege that could only be personally bestowed, much in the same manner that Howard had just granted it to him.

"Look, Scott, I just wanted you to know that was a righteous save on the paraglider trainer," said Howard quietly. "As much as I despise flying that damned thing, I would sure hate to think about doing it for real without the opportunity to practice."

"Thanks. So you and Major Riddle are supposed to fly it next week?"

"It's doubtful," replied Howard, slowly peeling the label from his beer bottle. "The contractor that built it has a boatload of engineers working on the problem, trying to rig a work-around, but it sounds like we're looking at a week's delay at a minimum. I'll tell you, Scott, it's damned lucky that you figured it out when you did. That snake could have bit any of us on a live mission, and we would have been immensely screwed."

Riddle devoured the last of his chips, crumpled the bag, slurped down the dregs of his beer, belched, and then chimed into the conversation with, "Hey, I was the test lead on that contraption back at Edwards when they were still developing it, even before I got sucked into this project. I've logged more time under that batwing that any man alive, and I don't have the slightest clue about what you did or how you figured it out, but we're indebted to you. It's Pete, by the way, but I also answer to Squeaky."

"Scott."

"So, Scott, how did you break the code?" asked Riddle.

"Well, Pete, we were trapped in this hard right bank," described Ourecky, gesturing with his hands like a fighter ace describing a momentous combat engagement. "And I'm flashing back to the simulator, trying to recall what kind of clues I should be seeing or hearing to help me resolve the problem. But I was drawing a blank. I knew that there were two drives for each control line, a take-in drive and a let-out drive."

Ourecky continued: "So I assumed the take-in drive had reeled in the control line until it completely stopped, but the let-out drive wasn't spooling it back out. Of course, the cable could have been snarled or fouled, but when that happens, you should hear a slight grinding noise on that side of the cockpit. So it dawned on me that *if* the right let-out drive was malfunctioning, I should be hearing *something*, but I wasn't hearing *anything*."

"You know," observed Howard. "I don't remember that from the Box, either."

"It's a discrepancy in the simulator," noted Ourecky. "They've done a good job with piping in all the simulated sounds from the pyrotechnics and retros and the other noisy stuff, but there's no noises from the take-in and let-out drives for the paraglider, even though there should be. Since you're not getting those noises as cues during training, it would be easy to miss in real life, because you're just not conditioned to hear it. Or hear the lack of it."

"Okay," said Riddle. "Then what?"

"Well, once I determined that the let-out drive wasn't running, I decided it must be malfunctioning because it wasn't drawing power, and then I back-traced the schematics in my head to determine what might have caused the problem. I guessed that the sequencer prematurely chopped the power to the let-out drive, and as it turns out, I was right."

"Whew. Very shrewd. I would have never figured out that quirk in a *million* years," Riddle said. "I might have given it a few seconds, but I would have punched out long before I started running schematics in my head."

"I don't think so. I think you would have resolved the problem," said Russo, injecting himself into the conversation. "I think that any of you would have. I know that I surely would have."

Riddle laughed. "Russo, that's mighty high talk from someone who hasn't even flown the damned thing. All I know is that I have the utmost respect for Drew Carson's flying abilities, and I've *never* seen him back down from a problem, and even *he* was primed to punch out of there."

"But I was just saying…"

"And *I* was just saying that you don't have a damned clue what you're talking about," countered Riddle, scowling. He turned toward Ourecky and said, "I don't know if you've caught it, but Carson has gone on and on about you how you saved the ship. That's quite a feat."

"Saving the ship?" asked Ourecky.

"No. Making a favorable impression on Carson is quite a feat."

Remembering the not too distant past, Ourecky kneaded his left shoulder, laughed and said, "Well, Drew's made quite an impression on me a few times, also."

Suddenly, the jukebox's incessant drone of country music was interrupted by the opening strains of "Bad Moon Rising" by Credence Clearwater Revival. In unison, the recon pilots groaned, and two of them stood up to examine the jukebox. Ourecky couldn't help but notice Riddle grin and wink at the enlisted bartender, who smiled and winked back.

"Something going on?" asked Ourecky. "You sure have an evil look on your face, Pete."

Tapping his fingers on the table in time with the music, Riddle grinned. "I dropped by this afternoon when the bartender was loading the cooler. I slipped him a sawbuck to look the other way while I made some adjustments to the jukebox. I restacked the records and installed a hidden switch in the back so I can turn off all those damned twangy country songs whenever they start to grate on my ears, which usually doesn't take very long. Now, whenever someone punches in a number for Merle, Buck Owens or Johnny Cash, we'll be hearing the musical stylings of the Rolling Stones, Three Dog Night, CCR, the Beatles and some others instead. Just my little contribution to the dismal cultural scene up here. I can also switch it back to normal anytime I want."

The bartender walked over and announced, "Hey gents, the cook just came in and he's about to throw your burgers on the grill. How do you want them?"

"Rare," replied Jackson. "The redder the better. Want a burger, Drew? Scott? It's on me."

"Medium," declared Ourecky. "Extra onions cooked on top, if he can."

"I'll pass," said Carson. "I ate chili con carne in the chow hall earlier. It might be a week before my stomach's back to normal. Besides, I'm watching my figure."

"As are we," quipped Riddle. "Two for me, both medium rare, smothered in onions, with cheese on top."

"Do you have a menu?" asked Russo, looking over his shoulder at the bartender. "Do you have anything else besides hamburgers?

"Nope. Just burgers," said Sigler. "Well done for me. Cook it good."

"Same for me. Well done," added Howard. "Charred black."

"Hey, stick buddy," said Carson a few minutes later, standing up as he nudged Ourecky's shoulder. "Let's slip into the corner over there where it's a little quieter. We need to review a few things before we zoom back to Ohio in the morning."

"Sure," replied Ourecky, pushing up out of his chair and following Carson to the corner. "But didn't we cover the flight plan in detail this afternoon? Haven't you already filed it?" He paused, smiled, and added, "Oh, you're tacking on an extra stop, right? Stopping en route to see another one of your hot chicks? Got a dogfight lined up somewhere? Maybe both?"

"That's not it, Scott. We're sticking with the plan as submitted. No deviations on this trip."

"Then what, Drew?" asked Ourecky. He took a sip from his Oly. "Are you okay? You don't look so hot. You look sort of peaked, like my mama used to say."

Looking at the floor, Carson quietly said, "Look, Scott, I need to talk to you, and it has nothing to do with flying back to Ohio. Something's been weighing on me, a big misunderstanding, and I need to come clean with you. I need to apologize."

"Apologize? What on earth for?"

"Hear me out. After we came back from our little camping trip in the snow, Virgil sat me down for a chat. He's under the impression that I did all the heavy lifting out there on the survival exercise, and that I saved your ass, instead of the other way around."

"So you told him what actually happened?"

"No," mumbled Carson self-consciously. "I didn't, and I feel like a real heel for that. He still doesn't know any different. He was already hell-bent on yanking you out of here, even before we made the paraglider drops, and sticking Russo in your place. At least I talked him out of that swap, thank God."

Ourecky chuckled. "Yeah, I don't think you would be content with just simply slugging Russo. You probably would have transitioned to homicide by the second drop."

Solemnly nodding, Carson wrapped his green wool scarf around his neck before pulling his Arctic parka from a wooden peg.

"Where are you going?" asked Ourecky. "I thought we were going to hang out with the guys."

"I'm going to Wolcott's hootch," explained Carson, donning the sage green parka and zipping it up. "I'm going to set the record straight with Virg. He should know what really happened out there in the snow."

"Really?" asked Ourecky. "You're going to confess to Virgil? And just what would that accomplish, Major Carson? What difference could that *possibly* make?"

"Well, maybe if I took the beating that I deserve, then I wouldn't feel like such a damned jerk. Maybe if I finally cleared my conscience, I could log a full night's sleep for a change."

Grinning broadly, Howard walked up to interrupt the conversation. A rather large hamburger looked tiny in his massive grasp. He took a huge bite, chewed, swallowed, and then said, "Scott, your chow is getting cold over there."

"Give me a minute, Tom," answered Ourecky. "I'll be over there shortly."

"Sure, Scott. Headed somewhere, Drew? It's a bit early to be hitting the hay, isn't it?" With his thumb, Howard subtly motioned back towards the bar. One of the recon pilots, the very same one who had confronted Ourecky earlier, was chatting up the husky Eskimo girl, the only woman in the club. Obviously sensing that someone else was paying her some attention, she looked towards the three men and smiled; at least half of her teeth were black with decay, no doubt the result of a recent intake of sugary foods supplanting her traditionally bland diet of seal meat and fish. "I've never seen you abandon such a svelte and fair-haired maiden to fall into the hands of the enemy. What's up with you, Drew? Is this a sign of the Apocalypse?"

"No, just the end of a very long day, Big Head. I have an upset stomach and I want to grab some sleep before I fly home tomorrow."

"Suit yourself," replied Howard. "I'll see you guys back at Wright-Patt in a week, maybe sooner if they don't clear the trainer to fly again. Not that that would bother me too immensely." Howard shook hands with Carson, then turned and walked back to the table to join the other pilots. A Johnny Cash ballad—"Ring of Fire"—blared from the jukebox, and the recon pilots stood in their red-striped chairs, hoisting their drinks as they boisterously sang along. Obviously, Squeaky was showing them a little bit of mercy.

Carson flipped up his parka hood and started to tug on his fur-backed mittens. "See you in the morning, Scott," he declared, starting for the door. "Chow hall, zero six. Sleep tight, brother."

"Wait," said Ourecky. "Answer my question, Drew: What possible good could come out of admitting everything to Virgil? If nothing else, you'll undermine his confidence in you, and that will just set you that much further back on the flight assignment roster. I'm not going to fly, but you certainly deserve to, so what does it matter if Virgil thinks lesser of me?"

"But…"

"And have you considered the possibility that you might anger ol' Virg to the point that he sends you out for another five days in the wilderness? Are you sure that you want that, or do you want to get back to the Box and try to make some headway? Besides," said Ourecky, slowly shaking his head, "sitting out there in the frigid cold was an interesting ordeal, but I can't tell you that I enjoyed it, because I didn't. I don't intend to go back out there of my own volition, so please excuse me if I don't volunteer to share your penance."

Carson heaved a tired sigh. "I really don't think Virgil would send me back out. We're too far behind the power curve as it is. Look, Scott, I really want to clear my conscience of this."

Ourecky stepped forward and subtly grasped Carson's upper arm. "Are you that *sure*? Maybe you should think about it a little more. If you're so damned intent on compelling Virgil to believe that you're a weak sister, don't be too surprised if you find yourself flying the Box with Russo sooner than later. Moreover, you may find yourself over on the right side of the cockpit. Is that really something that you want? Flying right-seat to Russo?"

Carson flipped his hood down and swiveled his head to look towards Russo. "I hadn't considered that scenario," he said softly. He closed his eyes and whistled quietly.

Ourecky nodded. "Drew, you and I know what happened, and that's enough, isn't it? I know you've learned your lesson, and I would trust my life with you in any circumstances. Why don't you let this one go? Like you said out there in the cold, let's just let it be our little secret. Okay?"

"I suppose you're right," answered Carson, unzipping his parka. "Our little secret."

Headquarters of the General Staff of the Soviet High Command
Arbat Military District, Moscow, USSR
10:15 a.m., Monday, October 21, 1968

Lieutenant General Rustam Abdirov cursed under his breath. He had been waiting in this vestibule, outside the meeting chambers of the General Staff, for over an hour. He had requested an audience with them over a week ago, and now that he was finally on the agenda, they kept him cooling his heels, obviously to further assert their authority.

A comfortable chair was provided for him, but he didn't want to sit down. As painful as it was to stand for long periods of time, it was even more agonizing to rise from a seat. He also hadn't taken his morning dose of codeine, which exacerbated his misery, since he wanted to be completely clear-headed for this critical meeting. He had much to accomplish in a short period of time, and his objectives probably wouldn't be attained if he were suspended in an analgesic fog.

The door creaked open, and an aide gestured for him to enter.

"Grab those materials, please," said Abdirov, motioning to stack of documents and a cardboard tube that occupied the chair in the hall.

Glancing up from a report, the Chief of the General Staff said brusquely, "State your business."

Standing at attention, Abdirov reported: "Sir, I am here to respectfully request significant changes to the *Skorpion* mission."

"Are you not content with the task entrusted to you?" asked the Chief, removing his reading glasses to scratch the bridge of his nose. "Are you not receiving the resources that we pledged to accomplish it?"

Conscious that his time was limited, Abdirov struck right to the heart of the matter. "Respectfully, General, I fear that the *Skorpion* design, in the form that I received it, will not be operational in a timely manner."

"So you feel that you've been dealt a rotten onion?" asked the Chief. "Am I to assume that you have a plan to rescue *Skorpion*?"

"In a sense, yes," answered Abdirov. He motioned for the aide to disperse the document and charts among the members of the General Staff. "I propose that we scrap the *Skorpion* design, and substitute an alternative design that I call the *Krepost*." As Abdirov waited for the senior officers to scan the documents and examine his drawings, he contemplated the name he had chosen for his project. A *krepost* was a fortified outpost, like a citadel or fortress. An individual *krepost* was typically part of a larger system of unified fortifications, like the Krepost Sveaborg that the Russians built to protect Helsinki in the First World War.

"This appears to be a manned space station," noted the Chief.

"That's correct, Comrade General. It will be manned by three military cosmonauts, who will be responsible for deploying the nuclear warheads. Although the crew will receive their orders from the ground, the *Krepost* will also be capable of autonomous operations, in a severe emergency, such as if the command and control network was neutralized by an American first strike."

"Interesting," mumbled the Chief, frowning. "But not practical. We appreciate your diligent work, Abdirov, but your task was not to develop and send up a space station, and we're not going to entertain these notions. You need to improve upon the *Skorpion* design and have it operational as soon as possible. You are dismissed."

But Abdirov remained. "I thought I was granted absolute autonomy over this effort, Comrade General," he stated. As he calmly spoke, he stared past the Chief at a patriotic illustration—"On Approaches to Moscow," painted by Vladamir Bogatkin—on the far wall.

"You *were* granted absolute autonomy, but your actions are still subject to our approval, and you can rest assured that we will not indulge you with your own private manned space program." The Chief paused and added, "Did you not hear me say that you are *dismissed*?"

Abdirov stood fast; he and his staff had spent countless hours working on this proposal, so he was not inclined to turn and meekly walk away. "But if the ultimate intent was to build a space-based nuclear bombardment system, and I have determined that the *Skorpion* design is flawed, then wouldn't it make sense to substitute a superior design to accomplish the intent?"

"Be *very* wary where you tread, Abdirov," warned the Chief.

But Abdirov did not relent. "If the General Staff doesn't have the authority to approve these changes, then shouldn't this matter be elevated to a higher authority?" he asked.

Looking toward the window on the east side of the meeting room, the Chief frowned and swallowed. It was obvious that he was cognizant that Abdirov was making a veiled threat to involve his patrons at the Kremlin, if need be. But Abdirov knew that short of killing him outright, there was little that they could do to punish him anymore than he had already been punished. His every day brimmed with agony, so even death would be a respite rather than a punishment.

The Chief sighed and said, "Strictly for the sake of theoretical discussion, let's hear the rest of your plan."

Abdirov continued: "The *Krepost* will be outfitted with a state-of-the-art control system that will allow nuclear warheads to be delivered to any location on earth, under its orbital path, with a high degree of precision. This targeting system will facilitate constant updating of potential target locations. Because we intend to equip the *Krepost* with this new capability, I recommend that the station be armed with multiple smaller nuclear warheads, rather than the single nuclear warhead of the *Skorpion* design."

"*Nyet*," stated the Chief. "That's not something we'll approve, at least for the time being. You'll stick with the warhead you have."

Seeing that he might be at least slightly overcoming the Chief's resistance, Abdirov concluded, "Along with some specialized components that will be produced by my bureau, most of this concept draws from equipment being designed by Korolev and Chelomei bureaus. The life support components will be built around Chelomei's *Almaz* space station. We also intend to employ the Chelomei bureau's TKS unmanned resupply craft for replenishment. The *Krepost* will remain operational indefinitely. It will be crewed on a rotating basis, with the cosmonauts arriving and departing on the *Soyuz* spacecraft, produced by the Korolev bureau. My bureau will focus on the control module for deploying the warheads."

"I think you meant to say *warhead*," interjected the Chief.

"That's correct, Comrade General."

"You're expecting the Korolev and Chelomei bureaus to cooperate on this *Krepost* of yours?" scoffed an Air Force general. "That's not likely."

"If this plan is approved, I will send liaison officers to their facilities, and they will produce their components independently. My personnel will take care of integrating their contributions."

"Still, it's very ambitious," said the Air Force general.

The Chief loudly cleared his throat and then said, "Since our time is limited, let's be clear: I understand that you're proposing this *Krepost* scheme because you've arrived at the conclusion that the *Skorpion* design is flawed. Is that accurate?"

"*Da,* Comrade General," replied Abdirov. "Frankly, given our current technological capabilities, I don't think that the *Skorpion* design can be realized. Perhaps within the decade, maybe, but now."

The Chief nodded and asked, "Then your *Krepost* is offered as a stop-gap, correct?"

"Affirmative, sir."

Signifying his assent, the Chief scrawled his signature on the front page of Abdirov's report and then declared, "Your *Krepost* concept is tentatively approved as an interim measure. You are hereby granted all authority necessary to see it to fruition. If you encounter *any* resistance from *any* quarter, you will direct the offending parties to report to us for clarification."

The Chief continued: "You will pursue this action with three caveats. First, even as your *Krepost* is made operational, you will continue to refine the *Skorpion* design, so that the manned *Krepost* is eventually replaced with the unmanned *Skorpion.*"

"As you wish, Comrade General," replied Abdirov.

"Before we get to the second caveat, I must ask: Are you aware of the *Perimetr* system?"

"*Da,* Comrade General," answered Abdirov. He had once received a technical briefing on *Perimetr.* Obsessed with the specter of their command and control systems being annihilated by an American first strike launched from submarines, some Soviet strategic planners contemplated a vast network of detection and control systems, *Perimetr*—also nicknamed the Dead Hand—which could automatically trigger a retaliatory strike with minimal human input. Ironically, such an outlandish notion had already been ridiculed by popular culture in the West; the 1964 film *Dr. Strangelove*, a dark comedy, portrayed a theoretical Doomsday device that inadvertently initiated global annihilation.

"Good. As you work on *Skorpion*, you will ensure that it can be integrated into the *Perimetr* system."

"And the third caveat, Comrade General?" asked Abdirov.

"I like your sketches," commented the Chief, examining Abdirov's drawings. "But there's something missing, and I would like it added. We believe that the Americans have not abandoned their satellite interceptor program. If that's the case, this *Krepost* would be vulnerable to a robotic or remotely piloted satellite interceptor. To mitigate this contingency, you will ensure that the *Krepost* is equipped with adequate means of self-protection."

"*Da*," answered Abdirov. "Comrade General, I will pursue the development of the Krepost and continue development of the Skorpion, and will abide by your caveats. May I ask for your assistance in a personnel matter?"

The Chief nodded.

Abdirov handed a folder containing personnel dossiers to an aide, who presented them to the Chief. "I request that these men be assigned to my bureau," he explained.

The Chief flipped open the folder and riffled through the dossiers. Pausing at one, he shook his head and slid it to the senior representative of the RVSN, who read it and then passed it to the representative of the GRU. Both men looked up and shook their heads.

"Gregor Mikhailovich Yohzin?" asked the Chief.

"*Da*," replied Abdriov.

"Yohzin's a very competent officer and a superb engineer," noted the RVSN general. "But we've always had questions about his allegiances. That's why we keep him at Kapustin Yar, so that he can be kept under close scrutiny. I know that he might be invaluable to your efforts, Rustam, but we would prefer to keep him working in his current capacity."

"Besides," added the GRU general. "We have some chores for him in the near future."

"So Yohzin is off limits to me?"

"At least for the time being," answered the Chief. "But we will take this matter under advisement, and perhaps he can transferred to you later."

"As you wish, Comrade General," replied Abdirov.

"So, do you have anything else for us?" asked the Chief.

"*Da,*" replied Abdirov. "The Americans have a leg up on us in many aspects of aerospace technology. While I am confident that our glorious Socialist system will quickly catch up with them, it would certainly be useful if we were able to learn from their experience and not repeat their errors. If I had access to some of their technology, our program would be greatly accelerated."

"What is it that you require, Rustam?" asked the GRU general.

Abdirov handed a large envelope to an aide, who brought it to the GRU general. "Ideally, we would like you to procure the actual objects, if at all possible," he explained. "But in lieu of that, anything would be appreciated: documentation, blueprints, technical specifications, and the like."

"We will do our utmost to obtain these objects you request, but I don't see this as very feasible." The GRU general smirked, rolling his eyes. "After all, you're asking for some components from their Gemini spacecraft. That program just ended, and I can't imagine that we could gain physical access to those materials. Certainly this technology is still of great value, especially"—he coughed and continued—"*especially* if it is superior to ours. I have to imagine that these spacecraft and everything associated with this program must be warehoused for contingency operations."

Abdirov sighed. The GRU officer was probably correct, particularly since he was likely looking at the issue from a Soviet perspective. In the Soviet system, *nothing* was obsolete. Even when new tanks and aircraft were widely fielded, the older models were kept in reserve, ready for immediate action if required. If the Soviet soldier of the future was equipped with a laser-firing assault weapon, it could be safely assumed that there would still be several hundred thousand Mosin–Nagant bolt action rifles coated in cosmoline and tucked in crates, patiently awaiting their return to the modern battlefield.

"No promises, but my men will do their best," offered the GRU general.

"*Spasiba,*" replied Abdirov. "I am confident that your men will do their utmost, and I can ask no more than that."

24

COPING WITH ADVERSITY

Simulator Facility, Aerospace Support Project
4:02 p.m., Thursday, October 24, 1968

Confined in the Box since yesterday afternoon, Ourecky and Carson were in the final hour of twenty-four hour full-up mission simulation. In what had become almost a ritual for the last hour, they woke up from a twenty-minute nap, which replicated a "loiter" state in which the Gemini-I was powered down and adrift until an appropriate reentry window was available.

In this scenario, now that they had successfully restored the spacecraft to full operational power, they would receive final instructions for reentry, simulate the descent in the procedures trainer, and then immediately transition from the Box over to the Paraglider Landing Simulator to simulate an approach and touchdown at a remote field.

"Okay, Scott, we're coming into our next comms window," said Carson. "Can you handle it, or do you want me to cover it?"

Ourecky was so exhausted that he could barely hold his eyes open. He glanced up at the clock; in just a few minutes, they would talk with an AMC—Airborne Mission Controller—aboard a notional EC-135E ARIA aircraft. The ARIA—Advanced Range Instrumentation Aircraft—was a strange-looking jet transport commonly known as a "Droop Snoot" or "Snoopy Nose" for the bulbous antenna housing that protruded from its nose.

During actual missions, a small fleet of the ARIA aircraft would shadow the spacecraft as it orbited the earth, flying 2000-mile long tracks that would parallel their orbital path at critical intervals. The AMCs would act as an extension of the Mission Control Facility at Wright-Patterson, relaying instructions to the crew in orbit. In this simulation, the "AMC," played by Ed Russo, was sitting at a desk less than fifty feet away from the simulator.

"No. I'll do it, Drew. It's *my* job." Ourecky twisted to his left, reached over his shoulder, and found a book-sized control panel that was cabled to a larger cryptographic device mounted in the equipment bay directly behind his seat. The equipment was similar to the voice scrambling equipment currently carried aboard US combat aircraft flying in Southeast Asia.

Fighting to remain conscious, he yawned and squirmed in his seat. His back ached like he had been manhandling a jackhammer for the past twenty-three hours. He flipped the power switch on the control panel. "Crypto power is on. Power light is *green*."

"Confirm crypto power is on," verified Carson, looking over to make sure that the switch was thrown and the power light was green. "Scott, give me a read back on the crypto variable."

"Crypto variable is Seven-Two-Six-Three-Six…uh, wait, Drew…I think that last number is a Zero," said Ourecky, squinting to read the small display in the poor light between the seats.

Carson consulted an index card that referenced the cryptographic variables for the different contact periods. The five-digit variable controlled the scrambling of their voices as they transmitted on the radio; in order for their voices to be properly de-scrambled at the receiving end, the numbers had to match on both machines. "No, Scott, that's not it. Check it again."

"Okay, Drew, let me make sure." He switched on his penlight and verified the number. His fingers throbbed with pain, and it was difficult to hold the small flashlight steady.

In the weeks of working in the spacecraft mock-up, he had become painfully aware of a design flaw that probably had not bothered NASA's astronauts, but was absolutely agonizing for him. The immobile cockpit was mounted vertically, so that he and Carson lay on their backs for the entire extent of the lengthy simulations. The inflexible seat constantly chafed at pressure points on his back, thighs, and buttocks, yielding painful abrasions that took days to heal. After just a few hours, his entire body felt like one enormous charley horse.

The pain in his back was severe, but the worst indignity was reserved for his hands. He would have never imagined the excruciating pain that could result from having his hands above the level of his heart for hours at a time, but the resultant cramps were almost unbearable. Sometimes it took every modicum of strength he possessed just to grasp a pencil.

Early on, Carson suggested that he let his hands dangle at his sides, below the level of his heart, to improve the intermittent circulation to his fingers. He had since realized that Carson had the luxury to often do just that, but as the right-seater, he was constantly working calculations or manipulating the computer, so he had little respite for his aching hands. By now, he clearly comprehended the physical and mental suffering that drove Agnew over the edge.

Ourecky read the numbers and stated, "Yeah, Drew, you're right. Sorry. That was a Nine, not a Zero. Crypto variable is Seven-Two-Six-Three-Six-*Nine*."

"I confirm crypto variable," noted Carson, tucking the reference card into a stowage pouch. "ARIA callsign is Pacific Sentry. Our callsign is Scepter One. Commence comms."

"Pacific Sentry, this is Scepter One, over," said Ourecky, thumbing the transmitter button for the VHF radio. There was no reply, just intermittent static. He persisted. "Pacific Sentry, this is Scepter One, over." Still nothing. "Pacific Sentry, this is Scepter One, over."

"Scepter One, this is Pacific Sentry. I have retro and reentry guidance when you are ready to copy," came the fast-paced reply. Ourecky cursed quietly as he realized that Russo was on shift as the CAPCOM who would also be playing the role of the AMC. Filtered through the voice-scrambling circuitry, Russo's voice sounded cartoon-like, as if their simulated mission had suddenly been hijacked by the likes of Donald Duck.

Unfazed, Ourecky positioned a fresh index card on his kneeboard, checked his pencil, yawned, and answered, "Go ahead, Pacific Sentry. I am ready to copy, over."

"Scepter One, you are cleared for PRZ One-One on this rev. Current weather is four thousand scattered with ten miles of visibility. Winds out of One Eight Zero at eight, gusting to twelve. Current altimeter is Two Nine Seven Nine. TACAN is Channel Four. How copy, over?"

Although the comms windows were tight, and they constantly trained to pass information as quickly and accurately as possible, Russo had developed an annoying habit of speaking much faster than he actually had to, probably in an effort to throw Ourecky off his game.

Ourecky struggled to capture the relentless torrent of data, then keyed the transmit switch to verify the information. "Pacific Sentry, I read back that we are cleared for PRZ One-One. Current conditions are scattered at four thousand, ten miles visibility. Winds from the south at eight, gusting to twelve. Altimeter is Two Nine Seven Nine. TACAN is Four. Over."

Russo immediately replied, "*Good* copy, Scepter One. Next available reentry is Contingency Recovery Zone Two-Four. I have your Pre-Retro data, if you are ready to copy."

Clinching his pencil in his teeth, Ourecky flipped over his index card. He gripped the pencil and prepared to take notes. "CRZ is Two-Four. Go ahead with Pre-Retro, over."

"Roger, Specter One," quacked Russo in his rapid-fire cartoon voice. "GET RC is 70 plus 41 plus 36; RET 400K, 20 plus 12: RET RB, 26 plus 39. Bank left 50; bank right 60. Your begin blackout, end blackout, drogue and main times remain the same. Normal IVI's 225 aft, 115 down. Read back, please. Over."

Ourecky wrote furiously. His hands ached beyond belief. "I read back GET RC 70 plus 41 plus 36. RET 400K, 20 plus 12. RET RB is 26 plus 39. Bank left 50; bank right 60. Begin blackout, end blackout, drogue and main times constant. Normal IVI's 225 aft, 115 down, over."

"Scepter One, good readback," blurted Russo, not slowing in the least. "Initial deflection bank angle at 0, 225 up. At 55 degrees, 72 up. At 90 degrees, 70 down. Your 400K pitch angle remains the same and your pitch angle at Retrofire, minus 20 degrees."

His hand throbbing, Ourecky tenaciously jotted down the deluge of information. He finished the last line, and keyed the transmitter to recite: "Pacific Sentry, I read back initial deflection bank angle at 0, 225 up. 55 degrees, 72 up. 90 degrees, 70 down. 400K pitch angle constant. Pitch angle at Retrofire, minus 20 degrees. Over."

"Scepter One, *good* readback. Be aware that you'll retrofire in the dark. Nunki in Sagittarius will be 20 degrees above the horizon at your Retrofire point. How copy? Over."

Sucking in a deep breath as he flexed his fingers, Ourecky remembered that he had to unstow the sextant for the star shot before they were too deep into the retrofire procedure. One lesson he had learned was that there was no catching up if he slipped behind, even in the slightest. With a few pencil strokes, he transcribed the data for the star sighting and then read back, "Dark retrofire. At Retrofire point, we'll see Nunki at twenty degrees above the horizon."

"Good copy, Scepter One. Stand by for data upload, over."

Ourecky breathed a sigh of relief, before reaching out to set the computer to receive the reentry data. He verified the settings and then said, "Pacific Sentry, Scepter, upload when you are ready. Over." Then he groaned quietly and placed his hands at his sides in the hope of coaxing some circulation into his aching fingers.

Smiling, Carson tapped him lightly on the shoulder and commented, "Good job, brother."

Less than an hour later, the simulation was complete. After a short debriefing, Carson and Ourecky retreated to the locker room. Wracked with pain, Ourecky could barely stand as he showered. Leaning his shoulders against the tiled wall, he toweled off, and then sat naked on a bench while a medic daubed antiseptic ointment on the open sores on his back and thighs.

As the medic taped gauze bandages over Ourecky's wounds, Carson dried his hair with a towel. "Are you *sure* you're okay to drive?" he asked, looking into a partially fogged mirror to check the symmetry of his moustache. "I can give you a lift to your room. It's on my way."

"Thanks, Drew, but I'm headed to Bea's place from here," said Ourecky, grimacing as he pulled a white T-shirt over his head. "I haven't seen her since we got back from Eielson, so…"

Carson smiled. "I understand. Hey, let's take a breather until Monday. I think you've earned your pay this month. You can sleep in tomorrow, spend some extra time with Bea, and we'll hit the ground running on Monday. Fair enough?"

Barely conscious, Ourecky nodded and replied, "*More* than fair. Thanks."

Wright Arms Apartments, Dayton, Ohio
7:35 p.m., Thursday, October 24, 1968

Pulling into the parking lot, Bea saw Ourecky's car and grinned. He had bought the white 1962 Ford Fairlane strictly for transportation, and although it was mechanically sound, its body was mostly salt-rusted sheet metal and tan splotches of haphazardly applied Bondo filler compound.

She hadn't seen or talked to him since he had left for Alaska almost a month ago. Switching on her overhead light, she checked her makeup in a compact mirror before applying fresh lipstick and then blotted her lips with a tissue.

She climbed out and retrieved her small suitcase from the passenger side. Years of flying had taught her to pack only what was essential. As she started to swing the door closed, she remembered the bottle of wine she had bought on the way from the airport. She scooped it up from the passenger seat, locked the door, and strolled toward the apartment building. A brisk wind rustled leaves in the dark parking lot; it wouldn't be long before she would need her heavy coat and scarf to venture outside. In the distance, a pair of tomcats screeched and fought.

Singing quietly to herself, she climbed the stairs with anticipation. She tried her door; it was unlocked. As she entered the apartment, her flowery visions of a romantic reunion were quickly dispelled. Snoring loudly, with a string of drool trickling down his chin, Scott was sound asleep on the couch. Sighing, she put her suitcase by the door and the unopened wine in the refrigerator.

The television was on; Fess Parker, playing the stalwart woodsman Daniel Boone, filled the black-and-white screen. Wearing an outfit of fringed buckskin, with his trademark coonskin cap planted squarely on his head, he solemnly negotiated with some semi-hostile Indians to secure safe passage for a group of settlers traversing the Appalachian Mountains.

She clicked the television off and then quietly sat down to watch Scott as he slept. In the past few weeks, he had apparently developed the knack of falling asleep in the most awkward of positions. Right now, he was seated bolt upright on the couch, with his hands resting at his sides. In a deep stupor, he wore boxer shorts and a T-shirt. There were three empty Carling's Black Label cans on the coffee table in front of him. She wrinkled her nose; she wasn't fond of the smell of beer. And it was also highly irregular for him to drink like that; he rarely drank more than one or two beers at a time, and he never—to her knowledge—drank by himself.

At least he had been courteous enough to cover the couch's upholstery with an old sheet. For whatever reason, it seemed as if he always had sores on his back. They were always in the same places—on his shoulder blades, the small of his back, the backs of his thighs—like some sort of odd stigmata that never completely healed. Leaning forward, she peered at the top of his back. She could see the faint outlines of gauze bandages through the thin cotton fabric, and the small dark blotches where blood had seeped through the dressings and had dried on his T-shirt.

She slipped out of her shoes and then gently daubed the saliva from his chin with a tissue. He had changed immensely in the past few weeks; the transformation was obvious even before he had left for Alaska. Even though he'd been in relatively good shape when they had met, he had shed a considerable amount of weight as a result of his daily workouts with Drew Carson.

He had packed on muscle and was now toned and strong. As much as that pleased her, it bothered her as well. It was as if he were a horror movie werewolf undergoing a painfully slow metamorphosis, gradually becoming a physical copy of Carson.

There were other things, far more subtle, that concerned her. When they had first met, Scott had been painfully shy. He frequently got flustered and often had a hard time finding his words, almost to the point of stammering. But those days were long gone. Now he spoke more deliberately and confidently, as if he had attended some supercharged Dale Carnegie course.

Listening to him breathing, she pondered on how he was becoming so much like Carson. Her friend Jill had gravitated toward guys like Carson. Bea had come to realize that Carson's appeal had little to do with his looks—and granted, he was an exceptionally handsome man—but more from the sheer confidence that he exuded, a casual self-assurance that bordered on outright arrogance. There sure wasn't anything else attractive about his personality; in what little contact Bea had had with him, she found Carson abhorrent.

But despite Scott's transformation, he still retained the qualities that had drawn her to him in the beginning: he was intelligent, kind, polite, and gentle. And so long as those things didn't change, she was more than willing to leave things as they were. That included the informal bargain that she had struck with Mark Tew, where she had agreed not to ask Scott about his work. But as time passed, and as he seemed to always come home progressively more beat up, she found it ever more difficult to abide by that agreement.

Suddenly, his breathing sped up slightly, as if he was on the verge of waking up. A frown crossed his face and his eyelids twitched rapidly, like he was having a vivid dream, perhaps a nightmare. She leaned close and whispered in his ear, "Scott? Are you all right, baby?"

His eyes flew wide open as he awoke with a start. He seemed to be in a panic, and his hands immediately flew from the couch to his lap, like he had lost something. His distress lasted only a second or two, and then he looked at her and smiled weakly. "Bea," he muttered.

"It's me, Scott. I'm home." She slid over to sit next to him and put her arm around him; his shoulders were knotted with tension. "Are you ready to go to bed?"

"No. I need to sit here for a little while. Is that okay?" His voice was tired and hoarse, and his eyes were bloodshot like he had been on a week-long bender. His hands trembled in his lap.

"Sure. You sound like you're thirsty. I'll bring you some water." She stood up, went to the kitchen, and filled a glass from the tap. She brought it to him, and he sipped from it. Sitting down next to him, she asked, "Are you sure you're not ready for bed?"

"Not yet," he replied. His posture had not changed. "I'll just stay here right now. Please?"

She nodded and clicked off the lamp. She went back to the bedroom, got a blanket and covered him. Then she put on her nightgown and came to sit with him until he fell asleep. "Night, Scott. See you in the morning," she whispered, cuddling her head against his shoulder.

"Bea?"

"What, dear?"

"Would you marry me, Bea?" he asked, in a raspy voice just slightly above a whisper.

Surprised, she sat upright and looked at his shadowy figure in the darkness. She started to speak, but then realized he was already fast asleep. Anyway, she wasn't exactly sure how she should answer, since she was a little uncertain where the question was coming from. *I guess it will wait until the morning,* she thought, leaning back against him.

8:35 a.m., Friday, October 25, 1968

Ourecky sat at the kitchen table, watching Bea fix breakfast. She wore her pink cotton nightgown, tied snugly at the waist, and an old pair of thick woolen socks with frayed tops rolled down to her ankles. Humming quietly, she fried bacon and eggs for him and poached an egg for herself. As the bacon sizzled on the stove, she filled two mismatched cups with steaming water from the kettle; she dropped a tea bag in one and spooned Nescafe into the other.

From the living room, a television news anchor commented on the marriage of Jacqueline Kennedy and Aristotle Onassis earlier in the week. Two slices of toast popped up in the toaster; she jabbed them with a fork and dropped them onto a plate, then used a spatula to slide the eggs out of the cast iron frying pan. "Did you sleep okay?" she asked, putting the plate before him. "I'm glad you eventually made it to bed last night. Was something wrong?"

"Just a tough week," he answered, chopping up the eggs with his fork before dusting them with salt and pepper. His temples throbbed with the onset of a headache, and his hands still ached from the prolonged stint in the Box. "The tests are a lot more involved than what I thought. They go on around the clock, and I have to stay with them, kind of like a babysitter."

She stirred his coffee and handed it to him. Then she used the spoon to dip the tea bag from her cup. "Did you fly back from Alaska in Drew's trainer? A T-38, right? It looks like that ejection seat was pretty rough on you again."

He rubbed his left shoulder and nodded. "I'm also going up a lot with the flight tests. It can be pretty brutal. Drew said I would eventually get used to the seat. He said that over time, you just grow calluses in all the right places."

She leaned across the table and took his hand. "You just ride in the airplanes, Scott, but you don't *fly* them, right? That hasn't changed, baby, has it?"

"Well, sometimes Drew lets me take the controls when he's eating or adjusting stuff, but all I have to do is hold it on course." He desperately wanted to share with Bea how his quick thinking had saved the ship in Alaska, and how Drew had trusted him enough to land the paraglider himself. He wished that he could cast away this whole façade and just tell her everything.

"So you'll be flying more often?" she asked, spooning honey into her tea.

Taking a bite from the toast, he nodded. "Looks like it. Between the flight tests and the ground evaluations, it's really becoming an ordeal. Bea, you really shouldn't be surprised when you come home and find me a wreck like last night, at least for the next few weeks."

He really hated having to mislead her, but in a sense, he really wasn't. Most of what he told her was true, at least to some extent, but he still wasn't overly comfortable with having to constantly spew half-truths and white lies to obscure the true nature of his work.

He glanced at the empty beer cans in the trash can next to the refrigerator; recalling how drained he had been last night, he wondered just how much he might have already told her and what else he might have said.

But if he had disclosed anything truly significant last night, she wasn't showing it. If anything, she seemed a little antsy, like she was anxious about something.

Bea sipped her tea as she watched him finish his eggs, and then she said, "Scott, is there something you wanted to ask me?"

Setting down his fork, Ourecky looked at her; his mind spun as he contemplated her question. *Was there something I wanted to ask? What does she mean by that? What is it that she wants? Did something happen last night that I can't remember?* Finally he asked, "So do you want to catch a movie this afternoon?"

"*What?*"

"A movie," answered Ourecky, immediately aware he was following the wrong course. "Before I left for Alaska you said that we never seem to make it to the movies anymore. I thought maybe we could hit a matinee. I think *Funny Girl* is still at the Strand. Didn't you want to see that? Isn't that the one with Barbra Streisand?"

He watched her closely, trying to gauge her reaction. For just an instant, a strangely quizzical look crossed her face, and then she laughed. "Oh, Scott, you're a *funny* one. Are you sure there wasn't something else you wanted to ask?"

Now he was really perplexed. *What in the world was she after? Some sort of clue would be very welcome at this point.* "Uh, dinner?" he asked, reaching for his orange juice. "You want to go to a movie and then dinner afterwards? Maybe that Italian place again?" He gulped down the orange juice but was still terribly thirsty. He stood up and filled the glass with water from the faucet. He drank it quickly, but his mouth felt like it was stuffed with cotton and his headache was getting worse. He refilled the glass and quaffed it, and then filled it again.

Turning back to face her, he sipped from the glass. *What was it that she was looking for?* Something obviously had happened last night, but try as he might, he couldn't snatch that memory from the folds of his brain.

As much as he didn't want to rely on Drew's advice on dealing with the opposite sex, he recalled one of their late night campfire conversations in Alaska. Drew asserted that women always desired some form of affirmation, so Ourecky decided on another tack to steer out of the unfamiliar waters. "Bea, do you love me?" he asked.

Her face lit up as she hugged him. "Yes, Scott, I love you. I love you, I do!"

It was a little more of a reaction than he expected. He placed the glass on the counter, wrapped his arms around her and kissed her. Her lips tasted of her breakfast tea, sweetened with honey, and she smelled like soap scented with flowers. "I love you, too," he said.

Simulator Facility, Aerospace Support Project
9:17 a.m., Thursday, October 31, 1968

Carson studied the clock, desperately wishing that he could somehow will it to move faster. They were almost twenty hours into a twenty-six hour simulation, and to say that he and Ourecky were absolutely wrung out would be a gross understatement.

To make matters much worse, they should have climbed into the Box early yesterday morning, after a full night's rest, but technical problems had postponed the start until later in the afternoon. Gunter Heydrich had wanted to scrub the simulation altogether, but Carson insisted that they press on, so as to achieve their objective of logging two additional hours every week until January. Now, watching the hands tediously creeping across the clock's face, Carson wondered if he had made a mistake.

As he scanned his instrument panel, he listened to his stomach growl. As they worked toward the forty-eight hour goal, one of their chores was to develop efficient procedures for eating. It was really more of a time management exercise than anything else; in the cramped, chaotic environment of the capsule, it was virtually impossible to find time to eat.

Carson reached behind his seat, unlatched his food storage container, and yanked out a glassine bag. Lacking NASA's seemingly unlimited budget, Blue Gemini's dietary plan relied on far more pedestrian fare than the freeze-dried meals eaten by NASA astronauts in orbit. The only criteria was that the food be simple to prepare, that it produce the least amount of crumbs and mess, and that it result in the least amount of residue in the lower GI tract. The last issue could not be taken lightly; with the relatively short duration of the planned missions and the incredibly kinetic workload, there would be little time for a potty break. Consequently, any bodily waste would remain in the adult-sized diapers they wore until they returned to earth and emerged from their spacesuits.

And unlike the NASA missions, where every rehydrated shrimp and strawberry was painstakingly accounted for, and every calorie inventoried, there was no structured nutrition plan. Whenever they were hungry, they were free to eat whatever they grabbed out of their individual pantry box.

Carson looked at the transparent bag to determine the meal du jour; as best as he could determine, it was a collection of sandwich squares—about an inch on a side—made of peanut butter smeared on tortillas. Famished, he jammed two of the squares into his mouth and chewed. He was pleasantly surprised; the stuff actually tasted reasonably good. He held a square toward Ourecky. "Sandwich?" he asked. "Sorry, but we're all out of the filet mignon."

Ourecky was busy plugging entries into the computer keyboard. Although still waiting on his hand-me-down spacesuit, he had received a pair of custom-made gloves to train in. He wore them as much as he could bear, so he could become used to them, particularly for the tasks that required the greatest dexterity. Hindered by the clumsy gloves, each computer input sequence took roughly twice as long as normal. "Drew, I thought the rules were that we were only supposed to eat meals from our own food stash. Won't that…"

"Eat, Scott. If you're too busy to unwrap your chow, the least I can do is feed you." Ourecky nodded and then opened his mouth; Carson popped the sandwich square between the right-seater's teeth. "Someone else can balance the books on peanut butter later."

Still inputting data into the computer, Ourecky chewed quickly and swallowed with some obvious difficulty. "Give me a quick shot of water?" he asked.

Carson reached between the seats, found the water dispenser nozzle, and squirted some into Ourecky's mouth. "More sandwich?"

"Yeah. One more ought to hit the spot for a while."

Carson obliged and then ate one himself. "Okay. Just yell if you want more."

"Thanks," answered Ourecky. "Hey, I'm done punching in this sequence. We have seven minutes until our next star shot. Would you mind if I check out for five?"

"Go ahead, babe. I'll stand watch until you're back." He watched as Ourecky immediately fell asleep. Perhaps even odder than his ability to suddenly lapse into deep unconsciousness, Ourecky would blink out of his slumber right at the time he stated—five minutes, almost to the second—and go right back to work as if he had been awake the whole time.

Both men had developed this ability, primarily as a survival mechanism to cope with the excruciatingly long simulations. Their flight surgeon was intrigued with it and wanted to wire them up with equipment to monitor their brain waves, but Wolcott had intervened to squelch the idea at least until they had successfully completed the first forty-eight hour simulation.

Exactly five minutes later, true to form, Ourecky jolted awake. "Okay, Drew, time for our astronomy lesson," he said, taking off his gloves and jamming them between his legs. He dialed a setting into his sextant. The eight-pound navigational instrument was bulky and awkward in the cramped space of the cabin. They were still using the NASA version, although a smaller and easier-to-use version was being developed.

As Ourecky peered through the sextant at simulated stars in his small window, Carson made small adjustments to the spacecraft's attitude to ensure that the horizon was in view and that everything lined up correctly.

Next came a communications window, in which they received instructions and minor updates from a simulated tracking ship at sea. Carson forced himself to remain awake as Ourecky collected the data. As he watched, he noticed that Ourecky was not writing down the information, but as he repeated it back to the mission controller, it was obvious that he knew the details verbatim. With the communications session concluded, Ourecky punched in corrections into the computer and made adjustments on the instruments.

"You're not writing down the instructions," observed Carson, stifling a yawn. "Are you sure you got them right?"

"I am," answered Ourecky impatiently. "Look, Drew, my fingers are about to give out on me. It's all I can do to hold a pencil. Just trust me: I have the information in my head, and I'll give it to you when you need it. Can you trust me, Drew?"

"Guess I had better. Hey, I'm frazzled over here. Can you give me ten minutes?"

"I have it," replied Ourecky, scanning his panel for any aberrations. "Take a nap."

7:15 p.m.

Ourecky awoke to a loud tapping noise punctuated by blaring horns and shouted curses. Gradually regaining consciousness, he saw a police officer tapping a silver flashlight on his side window. Grimacing at the stabbing pains in his hands, he slowly rolled down the window. The horn noises were much louder now, and he realized that they were coming from several cars behind him. Realizing that he was stopped at a traffic light, he vaguely recalled finishing the simulation over an hour ago before leaving the base to head to Bea's apartment.

As the police officer motioned cars around Ourecky's Ford, he said, "I've been banging on your window for the past five minutes, pal. You drunk?"

Ourecky shook his head. "I haven't had anything to drink, sir. I'm just very tired. I'm sorry for falling asleep here." In his peripheral vision, he saw strange figures strolling by on the sidewalk. Turning his head, he saw Batman and Robin, the Green Hornet, ghosts, and a variety of strange figures that he didn't recognize. He thought he was hallucinating, and then he realized it was Halloween and that kids were just now venturing out for an evening of trick or treat.

"You best not be fibbing, buddy, or you'll be eating asphalt in about half a second." The police officer used his flashlight to gesture at an Esso gas station. "When the light changes, pull over there. Don't do anything stupid. I've already called your plates in."

Striving to remain conscious, Ourecky nodded. The light changed, he pulled through the intersection, parked his car next to the building, and stepped out. An icy gust of wind immediately sliced through his windbreaker; he wished that he had worn something more substantial. The police officer pulled in next to him and climbed out of his cruiser.

"Driver's license," ordered the police officer curtly. He positioned himself directly in front of Ourecky, obviously close enough to smell his breath.

Struggling with uncooperative hands, Ourecky eventually found his license. He held it out with shaking fingers. Woozy, he just wanted to climb back into his car and fall asleep.

The officer snatched it from him. "Florida? Care to explain why your driver's license doesn't match your car tag? What are you doing with a Florida license, pal?"

Florida? thought Ourecky. *What about Florida?* Then it clicked. "I got the license when I was stationed in Florida. Air Force. I was at Eglin, on the Florida Panhandle."

"Air Force?" asked the officer. "So when did you get out?"

"I didn't," replied Ourecky. "I'm still in. I'm on the base."

"You got your ID on you? Let me see it."

Looking for his military identification, Ourecky fumbled with his billfold. His hands were knotted in pain. Giving up, he just handed the entire wallet to the police officer.

The officer examined his identification. "Are you *positive* you weren't drinking, Captain?"

Ourecky nodded. A chill came over him, and he shivered uncontrollably.

"Well, I don't have a lot of options here," said the officer. "You were asleep at that intersection for at least five minutes. If there's something wrong with you, like your heart, we need to drive you to a hospital. Otherwise, I can call the SPs on Wright-Patt, and they can come pick you up, or I can take you to jail and have your car impounded. If there's someone you can call, they can pick you up, but you're in no state to drive. So is there anyone you want to call?"

He pulled a scrap of paper out of Ourecky's wallet and held it out. "How about this number?"

Ourecky looked at it and nodded. As anxious as he was to see Bea, he was beginning to think that he should have followed Carson's suggestion and taken a nap before hitting the road.

"Okay. Look, Captain, I'm going to go inside and use their pay phone. Let me have your keys. Climb in your car, and we'll just wait right here until someone comes to pick you up."

About twenty minutes later, Bea arrived at the gas station. She spotted Scott's weather-beaten car and pulled alongside. Just as the police officer had described over the phone, Scott was sound asleep in the driver's seat of the Ford. The officer was inside the gas station, drinking coffee and chatting with the owner, when he saw Bea. He stepped outside just as Bea was getting out of her car.

"So this is your friend?" asked the police officer.

"Afraid so," answered Bea, buttoning her coat to ward off the autumn chill. "Is he all right?"

"I think he's fine," replied the officer, shining his flashlight into the Ford to check on Ourecky. He switched off the light and looked at Bea in the flickering illumination of a streetlight. "Hey…you look mighty familiar. Aren't you on TV? You look just like that girl on…you know the one. The girl who plays the genie. You look *just* like her, except that her hair's longer."

"I get that a lot. I'm very flattered," Bea said, smiling. She looked at Scott through the side window of his car. "Are you sure he's okay?"

The police officer nodded. "He appears to be healthy, just really tired. I can't let him drive away, though. I've never seen anyone so out of it. He fell asleep at a red light, so there wasn't anyone in immediate danger, but he could have passed out while he was driving, and that could have been an awful calamity. On top of everything else, it's Halloween, and there are a million kids out tonight."

"I'm really sorry, officer. I'll take him home and put him straight in bed."

"Lucky him. By the way, is there any chance that he has some medical condition that might be to blame? Something that he might have to go to the hospital for? Narcolepsy, maybe?

She shook her head. "Not that I know of. He just works very long hours."

"So what does he do on the base?"

Bea laughed. "I have no earthly idea. He's an engineer, but I'm not quite sure if that's what he does now."

"Are you positive that he's not taking any kind of drugs? LSD? Marijuana? Barbiturates?"

She giggled. "Drugs? Oh no, officer, Scott doesn't take any drugs. I'll tell you: wild Indians don't make arrows as straight as him."

"Okay. Well, I'm in a generous mood tonight, and he is in the Air Force, so let's just settle this episode with a warning. Just drive him home, and we'll forget any of this ever happened."

"That's very kind of you," she replied. "Would you mind helping me put him in my car? Also, I would like to use a phone, if you don't mind."

"There's a pay phone inside, miss. Do you need some change?"

She shook her head as she reached inside her purse for some coins and a business card. She walked into the store, smiled at the store clerk, deposited a dime into the phone, and dialed a number scrawled on the back of the card. Next to the phone, a color poster showed a grinning tiger over the words "Put a Tiger in Your Tank!" Next to the cash register, a transistor radio played "Midnight Confessions" by the Grass Roots. Clutching the handset to her ear, she listened to the phone ring several times. Finally, a familiar but groggy voice answered.

"Hello?" she said, speaking quietly into the mouthpiece. "This is Bea Harper. Yes, that Bea. I'm calling about Scott Ourecky. You said I could call anytime. We really need to chat." Listening to the reply, she looked at a tidy display of oil and brake fluid. "Yes, I know that place. Tomorrow's Friday. He usually leaves by seven, so I'll meet you there at eight."

She listened again for a few moments. "Okay. I'll see you then. Good night." She hung up the phone, walked back outside, thanked the policeman, and then drove Ourecky home.

25

BREAKFAST WITH BEA

Waffle n' Egg Diner, Dayton, Ohio
8:20 a.m., Friday, November 1, 1968

ooking around, Bea unbuttoned her Navy pea jacket. The diner
was a noisy place, with harried waitresses shouting orders at
short order cooks, accompanied by the constant din and clatter of steel spatulas stirring eggs in skillets. The air was thick with
the smells of coffee, frying bacon and overdone toast. A jukebox,
barely audible above the noise, played "Little Green Apples" by O.C.
Smith. Listening to the buzz of conversations, it was obvious that the
prevailing topic was the potential outcome of next week's presidential
election.

She spotted Tew in a booth near the back of the diner. He wore
a staid brown suit, much like something an insurance salesman might
select to call on prospective clients, and was reading the front section
of the local newspaper. The paper's headline told of President Johnson's
surprise announcement—made just last night on national television—of
an abrupt halt to all bombing in North Vietnam. Tew's florid nose was

adorned by a pair of black-framed reading glasses. He stood up when he saw Bea, smiled broadly, and beckoned her over.

"Bea," he said, politely waiting for her to take her seat before he resumed his. "So nice to see you again. You look lovely." He folded his paper and placed it on the seat next to him, then took off his reading glasses and put them in his jacket pocket.

"Nice to see you again, General Tew."

"Call me Mark," he said. "Please."

Smacking on a cud of pink bubble gum, a raven-haired waitress strolled up. She wore a light blue dress and a frazzled expression. "And what can I bring for you two?" she asked hurriedly, extracting a grease-stained order pad and pencil from the pocket of a white apron.

Tew nodded at Bea and gestured for her to order.

"Oh," she said. "Breakfast would be nice. I'll have hot tea, one poached egg, and toast. No butter on the toast, please." Bea used her thumb to wipe a small circle in the condensation on the window beside her and peered out at her red Kharmann Ghia in the parking lot.

Scrawling down the order, the waitress asked, "And you, sir?"

"Cream of wheat," said Tew. "Farina would be fine, if you have it. A bowl of plain oatmeal if you don't. And a refill on the milk. Drop a couple of ice cubes in the glass, if you don't mind."

"Ice cubes? Sure. Coffee?"

Tew paused to contemplate the question and then answered, "No. I had better not."

"Jelly or jam for your toast, darling?" asked the waitress, looking at Bea.

"Orange marmalade if you have it," replied Bea. "Otherwise, just bring the toast plain."

The waitress nodded and then delivered their order to the cook. Bea pulled two napkins from a dispenser and spread them across her lap. "It's funny," she observed. "But my mum would never dream of coming into a place like this, just because they don't have cloth napkins."

"Some people are just set in their ways," noted Tew. "My wife won't eat any food with her fingers. She says only savages eat that way. Needless to say, we don't attend many picnics."

"You're not having coffee? On a cold morning like this?"

"I would love a cup of coffee, but I can't." Patting his stomach, Tew lamented, "Chronic ulcers. I haven't had coffee or real food in over two months. The docs tell me if my stomach doesn't get better soon, I'll need to have surgery."

"Sorry. I had no idea that you were ailing so," she said, reaching out to touch his hand. It was cold, like he had just come from outside.

"So how have you been, Bea? How is the airline treating you?"

"Fine, but I'm not looking forward to the holidays. It's so hectic, and there are a lot of families travelling with children. I dearly love children, I do, but they don't fare well on airplanes. It's always something: if they don't get airsick, then their ears hurt from the pressure changes. The parents believe that we stewardesses can work miracles to stop their kids from crying, but the truth is that there is only so much we can do, and it's really not that much."

"I think I know the feeling," said Tew, glancing upwards at the ceiling. "Look, Bea, I don't want to be short with you, but my time is limited. What is it that you wanted to talk about? Last night you said something about Scott Ourecky. Is there something I need to know?"

"No, but you said that I could call you anytime if I ever had a question."

"Fair enough. I *did* say that. What's on your mind?"

"I'm really concerned about Scott," she said. "He always seems to be physically exhausted at the end of the week. It's usually all he can do just to stay awake, but he normally snaps out of it by Saturday. He always seems to have fresh injuries, like scrapes or bruises. I think that he also suffers some memory loss, at least temporarily, and I'm really concerned that he may be having hallucinations. I'm just really scared, Mark. Can you understand that?"

Balancing a serving tray, the waitress walked up to the table. Her saddle shoes squeaked on the dirty linoleum. "Poached and toast?" she asked, sliding a white china plate in front of Bea. She placed a steaming bowl of Farina before Tew. "And hot tea for you, darling, and fresh milk for the gentleman. I hope you two enjoy."

Tew poured some milk into the Farina, added a spoonful of sugar, and stirred the white mush with his spoon. He furtively tried a spoonful and then frowned. "This is all news to me, Bea. Really. You might not believe it, but I have very little personal contact with Scott."

Bea sliced into the egg with her fork, dipped her toast into the oozing yolk, and then daintily bit off a corner. She glanced at her reflection in the napkin dispenser to check her mascara. Chewing slowly, she cooed, "*Mark*? That's what you want me to call you, right?"

He nodded, spooning up Farina.

Gently fluttering her eyelashes, she quietly said, "Mark, I might come across as some naive bimbo stewardess, but I've spent enough time around the military to know how things work. Scott and I are close, but we're not married, so I don't feel obligated to act like some junior officer's wife who's terrified to question a general. I'm *genuinely* concerned about Scott, and that's why I asked you to talk. Now, are you going to be straight with me or not?"

He didn't speak, as if stunned that she would dare challenge his authority. Then he answered, "Sincerely, Bea, I'm *not* trying to deceive you. I rarely have routine contact with him, but if you're worried about his well-being, then I should be concerned also." He drank some of the milk. "I'm curious, though. What makes you believe he suffers from memory loss?"

"Memory loss?" she asked, sprinkling pepper on her egg. For a moment, she considered describing what had transpired last night, but thought better of it, because she wasn't sure whether she might inadvertently land Scott in trouble. "Well, Mark, sometimes when he's really worn out, he says things, and the next day, after he has had a good night's sleep, it's very obvious that he doesn't remember saying them." She bit off a piece of toast as she watched a wave of panic come over his face and saw that his hands were trembling.

"What *exactly* does he say?" he implored, self-consciously placing his hands in his lap. He cringed as his face lost color, as if he were witnessing a ghost hovering over the table between them. "I mean, what is it that he says at night that he can't remember saying the next morning?"

She smiled coyly. "Just between us, Mark? Our little secret?"

"Just between us, Bea," he insisted. "Please."

"Well, he asked me to marry him, but he obviously didn't remember it the next day."

"That's *all*?" croaked Tew. The color gradually returned to his face.

"That's *all*? You don't think that's significant? I certainly do. Trust me, Mark, I've received more than a few marriage proposals in my time, but it just really bothers me that a man like Scott could ask me that one night and then completely forget it by the next day." She put down her fork and pointed at her heart. "That kind of stings right here. You know what I mean?"

"I can see how that would be bothersome. But he didn't say anything about…"

"What he does at work?" she asked curtly.

Swallowing, he nodded.

"No. He never talks about it, except in the vaguest of terms. Any secret that you have with Scott is safe with him." She took a sip of her tea. "Mark, I read the papers and I hear the rumors, so I'm aware that a lot of strange stuff goes on in the government and the military. Just the other day, I read an article about an Army sergeant who claims that the CIA slipped him LSD in some far-out mind control experiment. I dearly love Scott, and he's still as sweet as when we first met, but sometimes he is so out of it that it just gives me pause for thought."

"Bea, when we first talked, I urged you not to get too curious about what Scott did," Tew said quietly, leaning over the table toward her. "I can't talk about any specifics, but I will tell you what I can. Right now Scott is involved in a crash project to get a vital system operational as soon as possible. I'm sorry that it's so demanding for him, but we're at a crucial juncture and it's hard on everyone involved. Now, if you promise me that you can keep a secret between us, I'll tell you some things that should dispel some of your fears. I'm still working on some of them, so you can't share them with Scott. Fair enough? Can you keep a secret with me?"

"I can," answered Bea. "I will, if you're being sincere with me."

"I *am* being sincere, Bea. Truly. First, regardless of whether we're successful or not, Scott's involvement will be over in January. Second, we asked Scott to come here because he had some very special skills that we needed to tap into. To motivate him to participate, I made him some promises. He fulfilled his end of the bargain, and I'm going to make good on mine."

"And what did you promise him, Mark?"

"He said that he wanted to go back to school for an advanced degree, so I'm making arrangements for him to study for his PhD at MIT. MIT is a prestigious school. It's near Boston." Grimacing, he pulled an aspirin tin out of his pocket, opened it, and selected two pills. "He'll go there as an Air Force officer, so we'll pay for everything."

"But I suppose you're not doing this strictly out of the kindness of your heart," she observed. "I expect that the Air Force expects something in return."

Nodding, Tew slipped the pills into his mouth and washed them down with milk. "That's correct, Bea, but it's still a tremendous opportunity for him. Now, mind you, he won't go directly to MIT from here. Sometime after January, we'll send him to California for a few months, to work with an Air Force agency out there, mostly to get a jump on his thesis work. Then sometime around mid-summer, he'll move to Cambridge to start classes in the fall."

Bea laughed. "And the reason you're telling me this is because…"

"Bea, I sense that you two are getting a lot more serious, so I feel like I owe you at least some inkling of what your life may be like in the future, so you can plan accordingly."

"So you're assuming that I'm going to pull up stakes and follow him? What if I want to stay right here in Dayton? My job and friends are here, so what would motivate me to leave?"

"I'm not telling you what to do," Tew said. "I'm just asking you to think about Scott's future. He's a brilliant man with incredible potential. I'm expediting the process to put him at MIT, but I strongly suspect that he's so smitten with you that he would readily pass up a great opportunity if you asked him to. All I'm asking is that you help me to help Scott make the right decisions, even if it entails some sacrifices on your part."

"I'll think about it. I won't make you any promises, but I'll at least be open to the idea."

"Thanks," he replied. "Remember, he doesn't know about MIT yet. I need him to concentrate on his work until January, so I urge you not to share any of this with him."

She nodded. "I'll keep my end of the bargain, Mark, if you keep yours."

"Done. Now, Bea, as much as I've enjoyed chatting with you, I have to head back to the office before they decide to dispatch a search party."

Simulator Facility, Aerospace Support Project
8:00 a.m., Tuesday, November 5, 1968

Outfitted in his training suit, Carson lounged on a Barcalounger in the suit-up area. As he studied his mission profile notes, a suit technician checked pressure fittings and made minor adjustments. Stowing a pack of Juicy Fruit gum in a leg pocket, he glanced to the other side of the room. Wearing cotton long underwear, Ourecky was cradled in an identical recliner.

The alterations on Ourecky's training suit had just been completed. Once they suited up, they were scheduled to climb into the Box for a twenty-eight hour run. Since this was his first time to wear the hand-me-down suit during an extended simulation, the plan was for him to wear it for as long as he could tolerate it and then take it off for the remainder of the test.

Russo walked into the suit-up room and watched as Carson underwent the final stages of the suiting process. He nodded toward Ourecky. Holding a thermometer in his mouth as a medical technician took his blood pressure, Ourecky smiled and waved back.

"I don't understand why they're wasting so much time on him in the Box if he's not going to fly," stated Russo quietly, leaning over Carson and shaking his head.

Carson looked to Russo and said, "As far as I'm concerned, buddy, he's *already* flown. Maybe he won't fly into orbit, but he's certainly paid his dues here."

A suit technician interrupted the conversation. "Ready for your helmet, Major?" Wearing pristine white cotton gloves, the technician gingerly held the helmet before him, like a loyal page tending to a valiant knight before an epic battle.

Carson nodded. He tugged on the recliner's handle to bring the seatback upright and then checked the hold-down cable assembly on the suit's neck ring, making sure that it was positioned correctly. If the hold-down was set too loose, his helmet would bob around awkwardly; if it was set too tight, the helmet would be painfully jammed down on his head.

Clenching his fists, he was silent as the suit technician lowered his helmet onto his head, fitted the helmet's ring to the matching neck ring, and then clicked the ring's fastener to seal the helmet to the suit. He checked the placement of the microphones in front of his lips and swiveled the clear faceplace down to check that it moved freely before pushing it back up.

"So why are they fitting Ourecky for a suit?" asked Russo skeptically.

"We need Ourecky to complete the forty-eight hour sim. It has to be done in suits, and it has to be completed by the end of January. And I would appreciate it if you refrain from saying anything negative about Scott. He's gone above and beyond to contribute to this project."

"If you say so, but it's a lot of effort for someone who isn't going up."

Heydrich appeared at the doorway. "Drew, I need your assistance with a pre-flight systems check. Captain Ourecky will join you as soon as he's kitted up."

"*Jawohl*, Herr Gunter," answered Carson. "I'll be out on the floor in a minute."

With covetous eyes, Russo studied the details of Carson's cumbersome suit. "Need a hand up?" he asked.

"Please." Carson scooted forward in the seat of the recliner and stuck out his hands.

Russo grabbed them and pulled Carson to his feet. "Man, I envy you. I would kill to take your place. You're definitely in the catbird's seat."

Carson smiled. "Be patient, Ed, and you'll have your turn. But let me warn you: after you've spent a full day locked in the Box, you would kill to be *anywhere* else." He picked up his gloves, turned to Ourecky and waved. "See you in a bit, Scott. Don't forget to hit the head one last time before you suit up. It's going to be a long, long day."

Aerospace Support Project
12:05 p.m., Wednesday, November 6, 1968

Puffing on a torpedo-sized Macanudo, Wolcott dropped his helmet bag on his desk. "Marcus! *El jefe!* Your prodigal children have returned from the Alaskan wilderness!"

Tew slowly stood up and extended his hand. "Good to see you again, Virgil. Celebrating?"

"This?" asked Wolcott, pulling the cigar out of his mouth. "Oh, yay and verily, I'm celebratin' all right. I'm holding another stogie, if you feel like joinin' in."

Tew cringed at the thought of smoking a cigar. "So, I guess you heard the news then."

"Yup. We overnighted in South Dakota, at Ellsworth. The networks didn't make the final call until this morning, though, after I stayed up all danged night watchin' the returns. So Dick Nixon's in? Shucks, maybe there's a glimmer of hope for us after all."

"A Nixon win is definitely excellent news for us," confirmed Tew, sitting down. "Glad you're back, Virgil. Have you thawed out yet?"

Wolcott placed his Stetson atop his desk, sat down, and leaned back in his chair. "Barely. Pardner, I've been cold, but I ain't *never* been that cold."

"Well, I'm just glad you're back in one piece. And you managed to qualify the other two crews on the paraglider?"

"Yup. So now we have a grand total of two and a half flight crews." Wolcott sorted a stack of mail from his overflowing inbox. "If this kid Russo works out, we'll fold him into the mix and we'll be back to fielding three full crews. It ain't optimal, but it's certainly better than nothing."

"Virgil, I want to talk to you about Ourecky," said Tew abruptly.

"Ourecky? Yeah, Mark, we *do* need to talk about him. If he hadn't figured out that glitch on the paraglider, we probably would have lost the drop trainer. And that ain't all. The contractor figured out that the same post-retro sequencing error could have shown up on the flight hardware as well. So, I might have had my doubts about Ourecky after the survival field trip fiasco, but he sure saved our bacon."

Tew nodded in agreement.

"And pardner, that's why I want to shift him into Mission Control after January." Wolcott puffed on his oversized cigar. "I s'pose you're going to tell me that you've come to the same conclusion, right? After all, great minds do think alike, don't they, pardner?"

Frowning, Tew shook his head. "No, Virgil, that's not what I have in mind for Ourecky. For the past week, I've been making arrangements for him to start work on his doctorate next fall. In the meantime, he'll go out to El Segundo to work on the MOL."

"That's good, pard, but there's still a mess of work yet to be done here, and I would like to hang onto Ourecky for a tad bit longer—"

Tew interrupted him. "*No*. Come January, he will have served his purpose here, and we need to fulfill our commitments to him." Tew closed his eyes and then said, "Virgil, have you ever given much thought to the people we've killed and the lives that we've destroyed?"

Tapping his cigar on the edge of his ashtray, Wolcott whistled. "Yeah, Mark, betwixt us, we've killed a heap of folks, but do you think any of them would have hesitated an instant if they had gotten the whip hand on us? That's the nature of war, pardner."

Tew shook his head. "No, Virgil, I'm not talking about the Germans or North Koreans or Chinese or all the innocent people that just happened to be under the bombs we dropped. I mean the guys that we've killed or the lives we've destroyed because we made bad decisions or we just weren't paying attention when we should have."

"Shucks, so now you've gone and grown a dadblamed conscience, pardner? Granted, we've tromped on some backs along the way, but just how does that pertain to this situation? We ain't harmed Ourecky, and I'm sure that we'll take very good care of him when it's time for him to hit the trail, but in the meantime—"

"As I've said, Virgil, we've squeezed everything we can squeeze out of that young man, and I'm not going to permit you to wring one drop more. He's yours between now and the end of January, and then we make good on our promises to him."

"We didn't make any promises to him, pardner."

"*You* didn't make him any promises, but *I* did, and I expect you to honor my wishes."

"As you wish, pard. As you wish."

26

INTERLOPER

Parking Lot 20, Wright-Patterson Air Force Base, Ohio
3:25 p.m., Tuesday, December 10, 1968

Shivering from the cold, Eric Yost crouched in the back of his Chevrolet panel van, balancing on a metal milk crate while peering through a small hole. Recently, he'd configured the van for extended sessions of clandestine surveillance. He had obscured the back windows with black spray paint and scratched out several peepholes, each roughly an inch in diameter, that were virtually impossible to detect from outside the van.

Beside him, a second milk crate was jammed with essential supplies: cans of chili and beef stew, batteries, film, bread, peanut butter, two cartons of Lucky Strikes, a flashlight, and a set of Army surplus binoculars. It also contained his collection of espionage novels; an avid fan of the genre, he intently studied the works of Ian Fleming, Graham Greene, and John le Carré.

Covered with clutter—food wrappers, dog-eared Playboy magazines, and empty liquor bottles—a moth-eaten Army blanket spread over the bare metal floor of the van. A rancid stench reminded him that he was long overdue in emptying the dented milk can that he had stolen from a dairy and pressed into service as an improvised urinal.

He took a slug from a half-empty fifth of Old Crow, belched, and then mentally assayed his current circumstances. After six years of marriage, Gretchen had recently returned to her native Germany, with their children in tow, but not before bluntly informing him that she wouldn't be coming back to the States, and he had nothing to gain by following or otherwise contacting her.

He was finding it progressively more difficult to conceal his drinking. His immediate supervisor, Master Sergeant Kroll, was content to overlook it, but he warned Yost that he couldn't cover for him indefinitely. If their lieutenant ever found out, Yost could kiss his retirement goodbye. At this point, though, even the prospect of losing his monthly stipend was the least of his worries. Just over a year ago, Yost's poker habit had escalated to the point where he owed nearly ten thousand dollars in gambling debts.

To make matters worse, he had recently crawled to a loan shark to cover the debt, so now he was over twelve thousand dollars in the hole and in legitimate fear for his life. Fortunately, his current existence—spending his daylight hours in the back of this rusted-out van and his nights piloting his yellow Hyster forklift—kept him on the protective confines of the base for most of the day, where the loan shark's muscle-bound lackeys were loath to venture.

Concerned about the potential of frostbite, he massaged the tip of his nose. His head ached and his back hurt, and he regretted not buying another fifth of liquor. Dusk roughly coincided with the end of the duty day at Wright-Patt, when it would be safe for him to leave the parking lot and blend into the swell of traffic departing the base. He would cruise home to his ramshackle house, sneak in through the back door, take a shower, change into a clean uniform, grab a quick bite if the amphetamines hadn't completely erased his hunger, and then head back onto the base in time to report for his shift at the warehouse.

Startled by the sound of two fighters zooming off the adjacent runway, Yost ducked down momentarily, afraid that he had been detected. Assuring himself that he was safe, he sat upright and exhaled.

He examined his face in the rear-view mirror. Startled, he didn't recognize his own reflection; he looked at least a decade older than his thirty-six years. He had lost a considerable amount of weight since he had started gobbling speed, and now his ashen skin was drawn tightly over his cheekbones. His left eye was still purple from last week's encounter with the loan shark's enforcers, and his sparse hair was rapidly turning gray.

He turned away from the sad visage in the mirror and gazed through one of the peepholes. A flicker of motion caught his eye; he observed a blue station wagon pull away from a brick building to the west of the parking lot. He had witnessed this event several times, so he knew exactly where the car was bound. Reaching down to pick up his Kodak Instamatic camera, he studied the front of a snow-covered hangar at the opposite side of the parking lot.

Billowing a dense plume of steamy exhaust, the station wagon pulled up to the hangar, where it was met by two security policemen bearing submachine guns. Two men in white coveralls climbed out, went to the rear of the vehicle, carefully unloaded a blue box resembling a casket, and carried it into the hangar. Yost snapped pictures with his Instamatic as the men unloaded a second box, identical to the first. He scrawled down the time in his pocket notebook.

As he considered the potential contents of the boxes, he turned up the volume of the transistor radio jammed into his shirt pocket and listened to Credence Clearwater Revival's "Suzie Q" playing through the earphone.

Over the past few weeks, Yost had made quiet inquiries about the site but had yet to locate anyone who had any inkling of what occurred behind the walls of the bland-looking building or within the old hangar. Even though it was clearly apparent that a hundred or more people worked at the offices of the Aerospace Support Project, the official base phonebook listed just one phone number for the entire facility.

In any event, he strongly suspected that *someone* out there would be more than willing to shell out good money for the fruits of his sleuthing. And Yost was confident that the money would be more than sufficient to buy his way out of his current jam.

Simulator Facility, Aerospace Support Project
2:50 a.m., Thursday, December 12, 1968

Yawning, Carson rubbed the beard stubble on his chin as he scanned the instrument panel for aberrations. The scan took him almost two seconds longer than normal, since his eyes were bloodshot and sore, and it was becoming increasingly difficult to focus. He looked to his right to check on Ourecky. They were wearing full suits for this scenario run, and Ourecky was experiencing chronic problems with his. "You okay, Scott?" he asked. "How are you coping?"

"My back is killing me, but that's nothing new," answered Ourecky groggily, stowing a flashlight. "My right leg is overheating, though. I think there's a kink in the suit coolant line. I have a cold spot on the back of my upper right thigh, and everything below there is burning up."

"Ouch." Carson glanced up at the mission clock. The clock read GET 32:50:54; they were nearly thirty-three hours into a thirty-eight hour mission profile. He checked his watch: it was almost three o'clock in the morning, and they had been lying on their backs in the Box since early Tuesday evening, except for their fifteen-minute breaks at six-hour intervals. "Can you stick it out until the next break, or do I need to request a time-out?"

"I think I can make it. It's really aggravating, though. You want some coffee?"

"Pass." Wiping a film of sweat from his brow, Carson sniffed the air. Stale and humid, the cabin smelled of warm electrical wires and several weeks' accumulation of steeped-in body odor. He watched as Ourecky yanked a plastic bag out of his right thigh pocket. The bag was partially filled with instant coffee. Ourecky stuck an ungloved finger in his mouth to moisten it, dipped the digit into the bag, and then stuck a damp clump of brown granules into his mouth. He chased the dry coffee with a squirt from the water nozzle.

"CAPCOM, this is Scepter One," said Carson. "I have an admin request."

"This is CAPCOM, go ahead." Carson frowned. It was Russo's voice, so abrasive that it grated like coarse sandpaper in his ears. Russo apparently had just come on shift and would likely cover the Capsule Communicator desk for the remainder of the simulation.

"It's getting a mite stale in here. Can you push up the air conditioner a notch, please?"

"Will do," replied Russo. "I'll expedite. Anything else?"

"Yeah. Be aware that my counterpart has a twisted coolant line in his suit. He plans to tough it out until the next scheduled break, but the techs will need to tend to it then."

"Good copy," answered Russo. "I'll let them know."

With the comms window completed, Ourecky closed his eyes for a moment, reviewing some of the information sent up from the simulated ARIA tracking aircraft. Carson noticed that he didn't immediately punch in the read-up data into the onboard computer, as he characteristically did, but instead rummaged in the storage pocket to his right. Ourecky extracted his slide rule and used it to verify the read-up data.

"Uh, Drew, we have a problem here," stated Ourecky quietly, continuing to deftly manipulate the slide rule. "Would you mind disabling the VOX, just in case?"

Carson toggled the Voice Control switches so that they could have a private conversation that couldn't be overheard outside the Box. "What's the problem, Scott?" he asked. He unwrapped his last stick of gum, sniffed it, and then stuck it in his mouth.

"They just read up the solution for the next maneuver. I have a different solution."

Carson's throat was parched, but the gum wasn't helping. As he squirted tepid water into his mouth, he thought about the maneuver; it entailed a relatively long burn of their four 100-pound aft thrusters to alter their orbit. The burn's attributes—to include its initiation, duration and several variables describing the spacecraft's orientation—were expressed as a series of numbers that would be manually fed into the onboard computer. The computer would largely control the burn and then provide feedback to indicate if the maneuver was successful or not.

"Drew, did you hear me say that I have a different solution for the burn?" asked Ourecky.

"I did. How much different?"

With no notes, relying entirely on memory, Ourecky articulated the ground's solution for the maneuver, then his own, and then pointed out the conflicting aspects of the two.

"Yeah, those *are* different," declared Carson. Drowsy, he whistled through dry lips. "Do you think it will really make that much difference which solution we fly? After all, we should still have some wiggle room to make some corrections later on, if need be."

"I don't think so," observed Ourecky. "The tolerances are so tight that we may not be able to recover if we fly the wrong solution. If we choose wrong, we might not make the intercept."

"Two paths diverged in a wood…" mused Carson, mulling the problem in his head. The simulated intercept "target" was actually an abstract series of calculations being run on a mainframe computer.

The ground controllers' maneuver solution was also generated by a computer. Surely it had to be more accurate than Ourecky's solution, given that the engineer had slept perhaps forty minutes in the past thirty hours. Beyond being sleep-deprived, Ourecky persisted with his annoying quirk of not writing anything down. As much as Carson had learned to trust him, this call was an easy one to make. "Look, Scott, this sounds like a no-brainer to me. We'll go with the ground's solution. Go ahead and punch it in."

"Drew, I think that would be a huge mistake," asserted Ourecky. "I've reworked this problem four times. I can't guarantee that my fix is perfect, but I think it's a lot closer than theirs."

"But they're running these equations on a computer outside, Scott," observed Carson. "You're working them in your head. That's a huge difference, in my book."

"Point taken, Drew," ceded Ourecky, arching his back to stretch. "But we're talking about *two* computers out there on the floor, not *one*."

"And what difference would that make?"

"Well, the target lives on one of the big mainframes, but their maneuver solution is generated by a computer not much more powerful than the Block One in here. It replicates the hardware they're going to fly on the ARIA tracking aircraft."

Ourecky yawned and continued. "That computer's doing the same thing that we're doing manually. It's making calculations based on where it thinks we are and then correlating that fix to where it thinks the target is. Usually, they're dead-on, but I just think our solution is more accurate this time. I can't explain why. My guess is that there are still some bugs in their computer program. Drew, computer programs aren't infallible."

Carson looked at the clock. Time was running out; he had to make a decision. He remembered the night in Alaska, when Ourecky saved the paraglider trainer—and possibly their lives in the process—and cemented his choice. "Punch in *your* solution, Scott."

Minutes later, Russo's voice rang out over the intercom: "Drew, this is Ed. I want you to switch your comms to Loop Four. I need to have a private conversation with you, and the guidance desk wants to review the maneuver calculations with Ourecky."

This was odd, thought Carson, yawning widely. *What could Russo possibly want to discuss in private?* "Switching to Loop Four," he stated, throwing the switches that would permit him and Ourecky to have independent conversations with the controllers outside on the floor.

"Do you hear me in there, Drew?" asked Russo. "Hey, babe, I hate to step out of role, but we need to keep this train centered on the track."

"What's the problem?" asked Carson curtly. He looked to his right; Ourecky was reworking the problem with his slide rule as he reviewed the various steps with the technicians outside.

"Our feedback display from your onboard computer shows that Ourecky didn't enter the maneuver solution we read up to him. Were you watching him when he keyed it in?"

Carson was becoming increasingly more aggravated. This conversation was detracting from the simulation. According to their current "orbit," they were several minutes away from the nearest communications window, so they really shouldn't be talking with anyone, even on an informal basis, and Russo should know that. Worse, this annoying chatter was eating up precious time, since they were less than five minutes away from executing the simulated maneuver burn, which he still had to prepare for. That would be difficult on any day, but at this point he was bone-tired. "Yes, I am very aware of what he entered, Russo. He had a different solution than what you read up. We discussed it in here, and we elected to fly his fix."

"Look, Drew, I'm just trying to make all of this a little less painful for everyone. I'm urging you to dump Ourecky's erroneous solution out of the machine and have him input our solution. I'm asking you this as a favor, Drew. Please."

"And why would I do that, Ed?"

"Because I discussed it with Gunter, and he's just about certain that you're not going to make the intercept insertion with Ourecky's solution."

"Well, I trust Ourecky, so we're flying his solution," replied Carson. "Case closed, Major."

"Not quite," answered Russo. "You are aware of the simulation rules, aren't you, Carson? If you don't make the rendezvous in the allotted time, we have to repeat this entire profile and we lose an entire week. Are you sure that's what you want? That doesn't seem too prudent to me."

Making sure that all was in readiness for the maneuver burn, Carson scanned his instrument panel as he pondered a reply. "Are you an idiot, Russo, or are you just trying to irritate me? I've been riding this Box for nearly a year now, so I don't need a lecture on the mission rules, particularly from *you*."

"I don't like your tone, Major. Order Ourecky to dump the computer and punch in the solution that we read up to him, and we won't need to discuss this later. You copy me in there?"

Carson felt something tapping him on the shoulder. Ourecky held up four fingers. Four minutes to the burn. "Yeah, I *copied* you, Russo, but we're riding what we have."

"Do I need to remind you that I'm senior to you, Carson?"

"Not in here you're not. We're *done* talking. Switching comms back to Loop One."

"What was that about?" asked Ourecky, stowing his slide rule.

"Nothing. Russo just wanted to know who I was taking to the prom. Ready for the burn?"

"Ready up."

"Then let's go."

Although over an hour had elapsed since the burn, the moment of truth had arrived. If Ourecky's solution was correct, then they would soon pick up the target in the optical reticle and the acquisition radar. If he was wrong, then there was little to be accomplished by remaining in the Box, since there was virtually no possibility they could execute the intercept in time.

Carson stared through the optical reticle. They were well into the catch-up phase of the intercept, trailing their target in a slightly lower orbit. This phase was carefully orchestrated so that it would commence right after orbital sunset, so that their target should appear as a faint star against a background of other stars.

The visual sighting would allow them to lock on to the target with the radar and then complete their ascent to the higher orbit in time to make intercept roughly at orbital sunrise. It was critical that the final phase—a series of braking maneuvers to adjust their speed to coincide with the target's—take place in daylight conditions.

"Radar yet?" asked Carson, blinking his eyes several times to moisten them.

"*Nothing* yet," replied Ourecky. He yawned deeply. "Man, Drew, I sure hope that I didn't lead you astray. I wouldn't know how to live with myself."

"Scott, I don't take it that way. It was *our* decision to fly your solution, and we'll live with the consequences no matter how they play out."

"Powering down the radar," said Ourecky, throwing a pair of switches on the radar panel. "Hey, if you don't mind, I'd like to close my eyes for a couple of minutes. I've been feeling kind of nauseous. Something I ate isn't agreeing with me. One of those cheese sandwiches had kind of a funky aftertaste."

"Okay. I have the controls, buddy." Carson adjusted the brightness of the illuminated reticle to its dimmest setting and continued to gaze at the current star field, remaining oriented so he could quickly detect any "star" that looked out of place. A few seconds passed before a pale dot, something like a sixth-magnitude star, floated in blackness near the center of the reticle.

He pulled his head back, blinked his eyes, squinted, and then looked again. The faint object was still there, and it didn't line up with any known stars, so it had to be it. "Scott?"

Ourecky awoke with a start. "Yes?"

"Power up the radar. I think I have a visual here."

It took a few moments for the radar to energize and for Ourecky to make adjustments, but then he observed, "Drew, I'm seeing some flicker on the acquisition light. Something's out there, all right. Stand by, stand by, stand by...."

Watching the barely perceptible speck in his reticle, Carson held his breath.

"*Solid* continuous light. Got it painted," declared Ourecky, fine-tuning the radar controls. "Target acquisition. Range to target one hundred-seventy-two miles and closing."

Both men breathed a sigh of relief. There was still plenty of work yet to be done, but this part was behind them. And best of all, thought Carson, the mission clock showed that it was time for their long awaited fifteen-minute break.

"Break time in two minutes," announced Russo. "Are you kids ready to climb out of there?"

"Yeah," answered Carson. "Yeah, Russo, we're both *more* than ready."

As he waited for the hatches to crack open, Carson heard Russo's voice, in an absolutely deadpan monologue: "Let me introduce myself, Major Carson. I am an Ourecky 9000. The Ourecky 9000 series is the *most* reliable computer ever made. *No* Ourecky 9000 computer has *ever* made a mistake or distorted information. We are all, by any practical definition of the words, foolproof and incapable of error."

Taken aback, Carson reached up and switched off the VOX loop. He turned to Ourecky and asked, "Do you have any idea what the *hell* Russo is yakking about?"

"*2001: A Space Odyssey*. You know: *HAL 9000*. He must think that's funny. I don't."

"Hey, Carson, I think I've figured out the solution to our computer dilemma," said Russo, laughing. "If MIT can't deliver the Block Two in time for our first flight, maybe we can figure some way to stuff Ourecky's head in a box and fly it in lieu of the computer."

Carson could see that Russo's attempts at humor were exactly what Ourecky didn't need at this point. The engineer's face was pale from exhaustion, and now he was visibly upset. Yet again, by sheer brainpower and the tenacity to stick to his guns, he had salvaged a mission. Granted, it wasn't as momentous as saving the machine in Alaska, but Ourecky's actions spared them the ordeal of re-running the entire scenario next week.

Instead of Russo lavishing Ourecky with the accolades he deserved, he ladled out ridicule and mocking scorn. Since he was so new to Blue Gemini, maybe his actions could be excused because he just didn't understand the tremendous mental and physical stresses of flying the Box, but still there was no defense for Russo being such an insensitive jackass.

"Drew," said Ourecky meekly, in a quavering voice. "Help me, please. I'm going to be sick."

Since Ourecky was fully suited up—including helmet and gloves—for the initial proximity operations, Carson knew that this was no laughing matter. Moving as fast as he could, he got Ourecky's helmet off. Jamming it down into the footwell, he was suddenly stricken with dread; no one had ever thought to put any airsick bags in the Box for this eventuality. If Ourecky puked on the suit, he would never be able to live it down and the suit techs would never forgive him. Then Carson remembered the coffee and grabbed the plastic bag out of Ourecky's leg pocket.

Cradling Ourecky's head as he held the bag to his mouth, he said, "In here, buddy...I promise I'll buy you another jar of Nescafe if you can just keep it all in this sack." Thankfully, in one massive retch, Ourecky did just that. Carson did his best to seal up the bag before stowing it away in his food locker. Then he used the squirt nozzle to rinse out Ourecky's mouth.

"Thanks, Drew," murmured Ourecky. "I owe you."

"You owe me nothing," replied Carson, flipping the comms switches to their normal positions. Thankfully, everything was back to normal, or relatively so, before the technicians unlocked and swung open the hatches.

Ourecky was as close to being physically spent as any man could possibly be; limp as a washrag, he allowed two technicians to extricate him from the cockpit and help him to his feet. Carson swung out of the cockpit and bounded to his feet without assistance, as if he had only spent thirty minutes in the Box instead of over thirty hours. Carson spotted Russo, still seated at the CAPCOM console, chatting with a controller. He waved casually at Carson and smiled.

Seemingly not paying attention to Russo, Carson calmly handed his helmet and gloves to a waiting suit technician. With stiff legs, he gingerly negotiated the metal steps from the elevated platform to the hangar floor. But when his feet hit the floor, he immediately picked up his pace, making a fast beeline toward the first row of consoles. Glancing up from his SIMSUP console, a panicked expression on his face, Heydrich dropped his clipboard and grabbed a phone.

Without breaking stride, Carson walked directly to the CAPCOM console and snatched Russo by his narrow black tie. He roughly jerked Russo out of his chair and swiftly escorted him to an isolated corner of the hangar, behind one of the mainframe consoles. Floundering, Russo didn't even have the presence of mind to remove his communications headset; the earphones were abruptly yanked from his head as their cable was pulled tight.

Still bearing Carson's helmet and gloves, the suit technician trailed the pair. He had obviously seen the pilot in action and knew what was about to ensue. He pleaded, "The suit, Drew, the *suit*! Please don't get any blood on your suit!"

Carson slammed Russo against the wall. Still grasping the tie with his left hand, he clenched his right hand into a tight fist and reeled it back to his shoulder, poised to let fly into the Russo's distraught face.

"Let go of him!" exclaimed Heydrich. "Haven't you been warned about fighting, Carson?"

"This ain't a fight, Gunter," said Carson. "Because a fight has two sides. This is simply a beating. I'm going to do all the delivering and Ed here is going to do all the receiving."

"I said stop it, Major Carson," said Heydrich. "*Now*. Let him go."

"Sorry, but no, Gunter. This guy's due a pounding, and I'm going to issue it to him."

Carson did not relent. Russo's face was turning red as an overripe tomato, and his eyes were starting to bulge out. His attention was obviously focused on the knuckles of Carson's fist that was still aimed squarely at his nose.

"Let go of him, Major Carson," Heydrich repeated. "You need to cease and desist, or I'll…"

"Or you'll *what*, Gunter?" snapped Carson defiantly. "Wait until Sheriff Wolcott rides back into town and tattle on me? Do you think I'm worried that Virgil might pull me off the flight roster? That might concern me, if I actually believed we had a chance of ever leaving the ground."

Hearing the commotion, Ourecky slowly walked up and stood next to Heydrich. "He's not worth it, Drew," he said quietly, in a hoarse voice. "Why don't you let him go? We still have a few minutes left on break, then we go back in the Box for just two more hours, and then we'll be *done*. Let's not lose that, okay? Go drink your Coke and then let's finish this."

Carson looked at Ourecky and then relinquished his grip on Russo's tie. He sighed, turned away from Russo, and said, "Yeah, Scott. You're right. He's not worth it."

Pacific Departure Facility, Johnston Atoll
9:38 a.m., Friday, December 12, 1968

Offering intercom headsets to Tew and Wolcott, the officer yelled over the noise of the four turboprop engines, "Since it's your first visit to the PDF, I've asked the pilot to do a fly-by at a thousand feet. We're a few minutes away from the island right now."

The officer was Lieutenant Colonel Ted Cook, the site manager for the Pacific Departure Facility launch complex on Johnston Island. He had spent the past several months overseeing the various construction projects on the island. The hectic schedule, strenuous labor, and Spartan living conditions had left him deeply tanned and spare of frame; Cook could easily pass for a South Pacific castaway in any Hollywood pirate movie.

Donning his intercom headset, Tew nodded. Following Cook, he and Wolcott pushed themselves out of the red webbing troop seats and went aft so they could look out the round window of the paratroop door on the left side of the C-130. There wasn't much room to maneuver; the cargo plane was jammed with aluminum pallets stacked high with supplies.

"Okay, here we are," said Cook, pointing at a nautical chart with a mechanical pencil and then gesturing out the window. "Those first three spits of land are Hikins, Akau, and Sand Islands. Johnston Island is coming up on our right. We're roughly seven hundred miles west/southwest of

Hawaii. These are coral reef islands, with a total landmass of slightly more than a square mile between the four of them, or roughly 660 acres. Johnston Island itself is roughly 210 acres. The atoll was used as a seaplane and submarine base during World War II.

"The island was retained as an emergency landing site after the war, and then there were several nuclear tests conducted here during the forties and fifties. Just so you're aware, they had a bit of a foul-up on a test conducted in 1958, where they launched a couple of Redstone rockets off the island. Each test involved a four megaton warhead that they detonated in space right over the island, so the fallout ended up coming right back down on the island."

Cook continued: "They launched several Thor ballistic missiles with nuclear warheads starting in 1962. They had a few mishaps. The worst one involved a Thor that blew up on the pad in July 1962. It scattered radioactive debris everywhere." He pointed out the window. "That's the Project 437 Thor launch site right there, further up the island from our site. I suppose you know that it's still operational."

As they passed to the north of Johnston Island, Tew could see that it was roughly rectangular, with its long axis oriented southwest to northeast. A 9000-foot runway ran down its center. He could see the ongoing construction at the PDF complex at the southwest end of the island. The site was dominated by the launch pad, a massive square slab of concrete measuring five feet tall above the surface and approximately a hundred yards on each side. Except for a partially buried concrete blockhouse that would serve as the control station for the launches, the pad was the only permanent visible fixture for the launch site. There was no stationary launch gantry jutting up from the launch pad, like a lonely derrick poking up from a Texas oilfield, nor anything else that could be spotted from a distance.

Extending into the water from the pad on either side were two steel-framed piers, each perpendicular to the concrete structure. Presently, a World War II vintage Navy LST—Landing Ship, Tank—was docked at each pier, oriented so that their bows pointed at one another. Tew could see that a stevedore party was in the process of unloading a large cylindrical object through the bow gates of one of the landing craft. The metal tube contained a dummy launch vehicle, which simulated a completed Titan II/Gemini-I stack that would have been hermetically sealed in a weatherproof shipping container at San Diego. Once the container was opened, the bottom half—mated into a pivot fixture permanently bolted to the launch pad—would serve as a cradle/gantry to elevate the Titan II to its vertical launch position. The rocket-bearing cradle would be erected by cables connected to a powerful winch on the second LST, which also carried the Titan II's hypergolic propellants and the pumps for fueling the rocket. The two highly modified LSTs, which also contained temporary berths and dining facilities for the launch crew, were the Navy's contribution to the Blue Gemini effort.

"That's actually the *second* concrete pad we've built," noted Cook, gesturing at the window. "If you're not aware, we had to scrap the first pad because the concrete didn't cure correctly."

"We're aware, Ted," observed Wolcott, slipping on aviator's sunglasses. "But this pad's up to spec, right?"

"It is, Virgil," answered Cook. He placed his hand on his heart in an effort to convey grave sincerity. "I would stake my career on it."

"I reckon you already have, hoss," said Wolcott.

Cook frowned and then continued. "We've completed the blockhouse. Most of the guts of it came out of Titan ICBM silos we're decommissioning. Our launch team is validating the equipment now, and I guess you can see we're also in the middle of an unloading exercise."

"It looks like your drill is going smoothly," commented Tew.

"Very much so, General," replied Cook. "The crew has it all down to a science. Our objective is to unload the stack, break it out of encapsulation, erect it, fuel it, do final preparations, and launch a payload entirely under the cover of darkness. Right now, the guys can safely accomplish all that in less than six hours. It's amazing to watch them work."

"Sounds great," said Tew. "Excellent work, Cook."

"Ditto for me, Ted," added Wolcott. "Very impressive."

"Well, gents, that completes our aerial tour," Cook announced. "The pilot informs me that he's preparing to go on final for landing, so you gentlemen need to return to your seats."

27

CHRISTMAS IN NEBRASKA

Aerospace Support Project
8:20 a.m., Monday, December 16, 1968

S canning a batch of last-minute expenditures, Tew glanced up to observe his staff trudge in for the morning staff meeting. To a man, they looked both weary and tense, like long-distance truckers aimlessly wandering through an all-night truck stop, simultaneously exhausted and yet wired on No-Doz pills and coffee. With Blue Gemini so close to culmination, their crushing workloads had escalated almost beyond the capacity of mere mortals.

As for himself, last week's excursion had nearly done him in. Besides visiting the Pacific Departure Facility, he and Wolcott had toured the Horizontal Assembly Facility at San Diego and scouted some prospective new pilots at Edwards Air Force Base.

He had even spent a night at home with his wife, although in retrospect he wished that he had just stayed away. Instead of the relaxing evening that he longed for, he was subjected to a withering barrage of questions about his steadily declining health. Every line of inquiry circled back around to one question: *Just how long did he expect to maintain this pace?*

As much as he hated to hear her ask it, it was a valid question, borne of her genuine concern for his well-being, and it could be answered on this very day. Tew had an appointment with his cardiologist this morning, who assured him that if there was no improvement since his last visit, his only remaining option was open heart surgery. The doctor had even scheduled the procedure in advance; based on the results from today's exam, Tew would report to Walter Reed tomorrow and remain there until the end of the year. Except for Wolcott, no one on his staff was aware that he might be going under the knife. Even his wife didn't know.

Although he could not consciously acknowledge it to himself, last week's foray was more or less a pre-emptive farewell tour, to finally witness the things that he had wrought. While his cardiologist was upbeat about his prognosis if surgery was necessary, Tew didn't want to inadvertently slip away from this world without actually witnessing the launch vehicles, spacecraft, and secret launch facility that were the fruits of his relentless labors.

Wolcott convened the meeting. "Gents, the boss and I are still a tad jet-lagged, so the swifter we gallop through this, the better." He turned to the personnel officer and said, "All yours, pardner. Make it as painless as possible."

Thompson, the disheveled personnel officer, stated flatly, "General, there have been no changes in the headcount and I don't anticipate any significant changes in the coming weeks. We stand a good chance of gaining the four pilots you interviewed last week, but not until they finish Phase Two of ARPS, so we won't see them for at least another six months."

"Thanks, bub," Wolcott said, nodding. "Intelligence?"

"No noteworthy developments on the Soviet side," asserted Colonel Seibert, the intelligence officer. "We're aware that they're preparing for another sizeable nuclear test. On the friendly side, NASA is still planning to launch Apollo 8 next Sunday, for a circumlunar mission. If it goes well, they'll be orbiting the moon at Christmas."

"Men headin' to the moon. Amazin'," noted Wolcott. "I'll cover operations and training. So that all you gentlemen are aware, General Tew and

I dropped in at the HAF in San Diego last week. We have two complete mission-ready systems in the quiver, ready to launch. Two more launch vehicles are being processed, and two more Gemini-I's are within eight weeks of completion. In February we'll launch a boilerplate Gemini-I to validate the Titan II and PDF. If we're successful, we fire a manned test shot in June."

"And the crew training?" asked Tew. "Are we making progress to the forty-eight hour mark?"

"Dead on, boss," Wolcott assured him enthusiastically. "And once we hit the mark in January, we can release Ourecky as per your guidance. My plan is to yoke up Carson with Ed Russo, so we'll have three full-up crews ready for contingency missions."

Tew noticed the rest of the staff exchanging nervous glances. He looked directly at Heydrich, who was clearly distraught at the mention of Carson and Russo, and immediately knew that something was seriously amiss. "Do you have something to add, Gunter?" he asked. "I don't like things being concealed from me. If there's bad news, I want it *now*, not later."

"It's *nothin'*, Mark," declared Wolcott. "There was a little dust-up in the simulator facility last week while we were out west. Nothing significant. Just a slight misunderstandin'."

"*Another* incident? What exactly happened?"

Wolcott started to answer, but Tew held up his hand. "What happened last week, Gunter?"

Staring at the table, Heydrich took off his black-framed glasses and loosened his black tie. In a faltering voice, he concisely recounted last week's incident.

"Carson *again*?" bemoaned Tew, getting red in the face. "You're telling me that Carson went berserk and came close to decking *another* officer?"

"Sir, I don't think we can lay all the blame at Carson's feet," Heydrich said.

"And just how could that *possibly* be, Gunter? If what you're describing is accurate, then who else could be at fault?"

"Well, sir, I suppose I'm at fault, probably more so than anyone else," Heydrich confessed sheepishly. "With the amount of stress that he's under in the Box, it's no surprise that Carson flew off the handle. The fact is that

Russo had been acting very unprofessionally up to that point, and Carson just reacted to it. I should have been monitoring Russo more closely. And to Carson's credit, sir, he didn't exactly run amok; he was calm and collected through the whole episode, short of throttling Russo and threatening to pound his nose flat."

Tew looked at his watch and drummed his fingers on the table while staring up at the ceiling. "Virgil, I expect you to resolve this conflict between Carson and Russo, particularly if we expect them to fly together. Make it abundantly clear to them that they won't be flying until they learn to behave like proper officers."

"I'll ride herd on them, Mark," asserted Wolcott. "There'll be no more scuffles."

Heydrich loudly cleared his throat.

"You have something to add, Gunter?" asked Tew.

"I do, sir," replied Heydrich. "I feel obligated to inform you that my personnel are exhausted and that most of them are angry about working over the holidays."

"Tarnations! We're *all* workin' over the dadburned holidays, Gunter," Wolcott said, shaking his head. "I don't know if you missed the telegraph, but we're facin' a hard deadline in January immediately followed by a launch in February."

"But that's only part of the problem," said Heydrich. "My men are hearing rumors that Blue Gemini will be cancelled anyway, so they're questioning the logic of being separated from their families at Christmas since it's almost a sure bet that they will be out of work in January."

Tew nodded. The men gazed toward him, as if he would confirm the prevailing scuttlebutt, but he was silent. He had heard the rumors, but was also aware that the leading Presidential candidates—Nixon and Humphrey—had recently received a series of classified briefings concerning ongoing Air Force activities, including a status report on military space operations. Blue Gemini wasn't specifically discussed, but was lumped in under the MOL program.

Afterwards, commenting on the briefings, both candidates made it explicitly clear that they only intended to launch men in space for peaceful purposes. They both believed that military space programs detracted from

NASA's efforts, so the MOL and all associated programs were expected to be cancelled early next year.

While Tew was disappointed that Blue Gemini might be cancelled, he was just as angry that the MOL program would be scrubbed, even though he was no longer involved with that effort. Right now, thousands of workers were employed by the MOL program—including almost a thousand with the Aerospace Corporation in California—and if these rumors were true, most of them would be unemployed by this time next year. Granted, some would be absorbed into other aerospace contractors or NASA, but the drive to land men on the moon was well underway, with most of the heavy lifting already accomplished.

On the other hand, cancellation of Tew's secret program would barely be a blip on the budget radar; most of the personnel involved in Blue Gemini were Air Force blue-suiters, so they could readily be transferred to other endeavors. *If the ceiling would soon collapse, what sense did it make to take these men away from their families during the holidays?* "Gunter, what's on your books until the end of the year?" asked Tew.

"Carson and Ourecky go into the Box tomorrow afternoon for a forty-hour full-up run. They come out on Thursday morning. As for next week, they go into the Box on Christmas Eve for a forty-two hour run and come out the day after Christmas. And the following week…"

"*Stop*," said Tew emphatically. He looked around the table at the weary faces of his subordinates. "Gunter, proceed with this week's plan, but we're standing down for Christmas. No training or activities until the first of the year, except for absolutely essential functions."

Perplexed, Wolcott's jaw suddenly dropped as if he had been gut-shot. He muttered, "But…"

"No buts. No later than close of business on Friday, I want everyone out of here. I want them to rest and spend time with their families. *All* hands, every man Jack, no exceptions."

"But, pardner," Wolcott quietly coaxed, "we're in the home stretch. We don't want to abandon what we've accomplished, do we? Can't we—"

"No," declared Tew, angrily glaring at Wolcott. "We'll still have the first half of January. If we don't make it, then we don't make it. We'll just accept the circumstances."

"Thank you, sir," said Heydrich. "I'm sure that the men will be most appreciative."

"I'm not done," said Tew, turning to the personnel officer. "Thompson, assuming we might indeed be cancelled, I want you to develop contingency plans to protect our people. I want to know what we can do to get them transferred or shifted over to gainful employment elsewhere."

Suddenly energized, Thompson nodded and made notes. "Will do, sir."

"I also want you to make some quiet inquiries with aerospace contractors, to see what sorts of openings they're anticipating in the coming months. It sounds like we have some lead time to cobble together a lifeboat for our civilian workers, so let's do the best we can." Tew put his hands flat on the table, took a deep breath, and bowed his head momentarily. "Unless anyone has anything tremendously pressing, we'll end this meeting. Anyone?"

No one spoke.

"Then, gentlemen, if I don't see you all before next week, Merry Christmas. And so there's no question, your New Year will be much less than happy if I catch any of your personnel lurking in these halls next week. Understood?"

All present nodded their assent and gradually started filtering out.

"Virgil," said Tew. "Stay."

"Not much choice, pardner. It's *my* office, too."

"Then all the more reason." Tew stood up and escorted Heydrich out.

"Merry Christmas to you, gentlemen," said Heydrich, pulling on a heavy gray wool Bavarian coat and fastening its bone toggle buttons. "And on behalf of my men, thank you again, Mark."

"Think nothing of it. Thank you for everything, Gunter. Merry Christmas," said Tew, closing the door behind the German engineer.

"Thunderation! What the hell has come over you, Mark?" demanded Wolcott. "These men didn't sign on for a dadburned quilting bee! How *could* you?"

Glowering, Tew shook his head. "Virgil, don't you dare raise your voice at me. Of all days, I don't need any more excitement today."

"Sorry. You're right, amigo," replied Wolcott, lowering his voice. "I apologize. But Mark, we're so close. So danged close. Can't you reconsider this?"

"You know, Virgil, we've worked so closely over the years, but I think that our perspectives have changed. To you, this whole endeavor is like a religious crusade. You won't be content until you set foot in the Promised Land or until everyone has perished in the process."

Tew continued: "I appreciate your ambition, but for me, this Project is not unlike any other mission. We do our utmost to achieve our goal, but we're also obligated to take care of our troops. If this mission is so critical, then we'll be allowed to proceed to its fruition, but we need to accept the fact that we may never leave the ground, regardless of how diligently we've worked."

"I s'pose you're right, Mark," muttered Wolcott.

"I *hope* I'm right. Would you mind giving me a lift to the hospital, Virgil?"

"As you wish."

Dayton, Ohio; 9:05 p.m., Tuesday, December 17, 1968

Yost switched off the van's lights, cut the engine, and coasted the last few hundred feet into the narrow alley. The brakes creaked slightly, but otherwise he was able to quietly park the vehicle behind his house. He was sure that if the loan shark's thugs were staking out his home, they would be watching the front, where he habitually parked on the oak-lined street.

He looked forward to a hot shower and a restful night in his own bed. He was on furlough until the warehouse re-opened after New Year's, so he planned to stay here tonight, pack up a few supplies, and then hit the open road. Since he had evaded the enforcers for the past week, he managed to squirrel away a few hundred bucks, mostly from his last paycheck, augmented by loans from the few friends still willing to front him some cash.

He intended to head east before the sun rose, toward the horse tracks of New Jersey. By carefully studying the racing forms and placing small bets, he would gradually amass enough cash to repay the loan shark and then build up a stake toward his future. *Stay disciplined. Stay sober. Read the forms. Study the horses. Stick with the plan. Nothing risky.* Just a little bit at a time, and he would be flush again before the champagne flowed on New Year's Eve.

Climbing down from the panel van, Yost smelled the pungent odor of stale urine. He had fumbled the milk can when emptying it earlier, completely soaking his shoes and his pants legs below his knees. He undid his belt as he considered taking the trousers off and sticking them in the trash before entering the house.

The grime-covered crust of old snow crunched under his feet as he walked toward the back door. He heard leaves rustling. A hulking thug materialized out of the shadows, tapping a baseball bat in his palm. "Mr. Yost? We've been waiting on you."

Yost panicked, scuttling backwards like a crab seeking the refuge of a submerged rock. He made it all the way to the van, groping for his key ring in his pockets, before he heard someone loudly clearing their throat behind him. He glanced back over his shoulder, glimpsing a second enforcer stepping out from behind a large tree.

The first thug paced forward, closing the gap. "Man, what the hell is that smell? Did you tinkle in your pants? Look, Yost, there's nothing to be afraid of. Just pay us what's due, and we'll be on our way." He gestured for the other man to search the van.

Within a few moments, the second enforcer emerged from the van, whistled, and tossed a small packet to his cohort. "Taped under the dashboard," he commented, in a high-pitched squeaky voice. "How original! If only I hadn't seen that damned gimmick a thousand times, I might have overlooked it. Hey, man, it's freezing out here! Can't we speed this up?"

Yost trembled uncontrollably, partly from the cold and partly from abject terror. His heart pounded like a struggling steam engine on the verge of blowing apart at the seams. An expanding stain of fresh urine supplemented the old. A sudden gust of wind tore through his clothes, chilling him to the marrow. "That's all I have," he pleaded, barely coherent in his fear, pressing his hands tightly together in a praying gesture. "If you let me keep it, I'll roll it up and be able to pay you in full. I know it's not much, but it's really all I have."

The thug removed his woolen mittens, wetted his fingertips, and quickly counted the cash. "Well, maybe you're right. Not much here. It doesn't seem right to deprive you of it," he noted, stashing the envelope in his jacket pocket. "But I can't grant you any leeway, Yost. Sorry, but I just can't budge. It's money owed." He directed the other man to continue searching the van.

As he searched, the thug heaved the milk crates out onto the driveway; Yost's binoculars, notebook, camera, prized collection of spy novels, and other sundry belongings plopped into the mud at his feet. Minutes later, the man re-emerged, shaking his head. "Nothing up here, boss, and if I stay in here a minute longer I'm going to barf."

The first thug signaled for him to climb down, and then turned his attention back to the cowering Yost. "You do know that we'll be back, right? You still owe a lot, and the interest is growing daily. And do me a favor. I'm not too fond of hiding out in the dark, so don't try sneaking in here again. Got me?"

Yost nodded fervently. "Look, I'll get your money. Just be patient with me, please."

"Oh, sure, Yost. I can be patient. But the problem is that my boss isn't so willing to be patient, and I have to answer to *him*, not to you."

"But I'll have the money! All of it. Just give me some more time."

"We'll be back, Yost. In the meantime, you probably want to get that foot looked at."

"Foot? There's nothing wrong with my…" muttered Yost. And then he felt the baseball bat thwack against his ankle from behind, brutally smashing tiny bones and ripping connective tissue. Toppling to his side in musty leaves and dirty snow, gasping in agony, he watched as the two men casually strolled away into the darkness. Before they disappeared from view, one turned back and hissed, "Merry Christmas, schmuck."

Waffle n' Egg Diner, Dayton, Ohio
7:18 p.m., Thursday, December 19, 1968

"Thanks for the lift, Drew," said Ourecky. "I'm sure that Bea can drive me back on base in the morning, and I'll look at my car then. I'm guessing that the battery crapped out, or maybe the alternator. Not too hard to fix, but I just don't want to fool with it right now."

"That's Bea's car, isn't it? Didn't you say she had a red convertible?" Bea's Kharmann Ghia was parked in front of the diner. The restaurant was decorated for Christmas, with strings of garish colored bulbs encircling the window frames. A painted silhouette of Santa Claus stood by the entrance, its plywood feet buried in dirty packed snow.

"Yeah, that's hers. She just flew in from Atlanta. We're going to grab a burger here and then head back to her place."

"Hey, Scott, if you do go back on base, you do know to stay out of the building, don't you? Ol' Cowboy Virg told us if he caught us in there before January that he would skin us alive."

"I'll steer clear." Ourecky scratched the two-day growth of beard on his chin. "I appreciate the break, but how will we make forty-eight hours? We're going to miss two entire full-up runs. I just don't see how we can recover the lost ground."

Carson cut off the lights and let the Corvette idle. "To be honest, Scott, it may be the best thing that ever happened. Lord knows Gunter and his guys need a break, probably much more so than us. Anyway, I don't think it matters how many times we go into the Box between now and the last run. We already have the procedures down absolutely cold. Now, we just knuckle down and grunt through it. We either make it on the last go or not."

"So tell me, Drew, how do you do it? It kills me to even think about climbing back into the Box, but you've been doing this forever, and it just doesn't seem to faze you."

Carson smiled. "Scott, when it starts wearing on me, I just do what I learned back in Beast Barracks at the Academy when they would stick us against a wall and make us brace for hours on end. In my mind, I put myself on a submarine. I walk through it from one end to the other, from one watertight compartment to the next, and I shut every hatch and dog them down tight, until I'm sealed up on the bridge. Then, when I'm isolated from everything, I just look at the world through the periscope. I can still function and react to the world outside the submarine, but I shut out all the pain and exhaustion and frustration. Does that make sense to you?"

"I suppose," replied Ourecky, snugging his wool scarf. "I just suspected that you might be popping speed or pain pills, especially since the docs keep offering them to us."

"Nope. It's all a matter of conditioning, Scott, just like going to the gym. You condition your mind and body to endure whatever comes. Drugs can be a temporary fix, but once you become even slightly dependent on them, you lose the edge and you can never get it back."

"I suppose that you're right." Seeing Bea wave through a window of the diner, Ourecky smiled and waved back. "So what are your plans for next week? Big date? Several big dates?"

Adjusting the Corvette's balky heater, Carson laughed. "No plans. I might fly up to Vermont to do some skiing. You? Do you and Bea have plans?"

"Not yet. If she's not flying next week, I'm thinking about asking her to come home with me to Nebraska, so she can meet my folks."

"Really? Sounds like you two are getting pretty serious. Are you sure you're ready for that?"

"I think so," replied Ourecky. "Hey, why don't you come in and grab a bite with us?"

Carson looked at Bea through the window and answered, "No, Scott. I don't think that's such a hot idea. You haven't seen her all week. I don't want to be a fifth wheel."

"You wouldn't be," Ourecky said. "Come on, Drew. I insist."

"Well, isn't this a pleasant surprise," said Bea, watching as Ourecky and Carson walked into the diner and approached the table. She slid out of the booth to hug them both. "Ouch! Have you lost your razor?" she exclaimed, stroking Ourecky's chin. "Have you not shaved all week?"

"Just a couple of days," said Ourecky. "We had an all-nighter."

"If you say so. But that's coming off when you get home, mister, or you'll be sleeping on the couch." Bea noticed that Carson seemed uncomfortable in her presence; it was like he couldn't bring himself to look at her.

"If you don't mind, I'm going to step away for a minute," said Ourecky. "Honey, can you order me a burger and fries? And a chocolate malt? Drew, can you keep Bea company?"

Carson nodded. Ourecky walked away, steering toward the restroom past the counter.

"You look beautiful, Bea," commented Carson, slipping into the booth opposite her.

"Thank you. You're very sweet, but I sure don't feel beautiful. This dry air is just doing me in. My hair feels like it's nothing but a big collection of split ends." She took a sip of hot tea. "So did you and Scott have a good week? Did you boys do any flying?"

"No flying. We were cooped up indoors all week. Procedures evaluations. *Very* boring."

"Oh. Sorry to hear that," replied Bea. "Listen, can I tell you something, Drew?"

Appraising the menu, he nodded and said, "Sure, Bea. What's on your mind?"

"I don't want you to take this the wrong way, but I watch you and Scott becoming friends, and it confounds me. I just don't understand it. To be honest, it *scares* me."

Apparently surprised at her concern, he laughed quietly. "I suppose I don't understand it, either. I didn't think much of Scott when he first came here, but I've really grown to respect him."

Bea watched their waitress at the counter; she had been staring at Carson for the past two minutes. She was a slim brunette in her late twenties. She straightened her apron, undid the top two buttons of her blouse, and then approached their table, order pad in hand.

"Are you ready to order?" asked the waitress, still intently focused on Carson. "Wasn't there another man that came in with you? Should I wait for him to come back?"

"I'll order for him," answered Bea. "Hamburger, French fries, and a chocolate malt for my friend. Grilled cheese sandwich, chips, and another cup of tea for me, please."

"How does your boyfriend like his hamburger? They usually cook them medium. It comes with lettuce, tomato, pickles, mayonnaise and onions."

Wrinkling her nose, Bea grimaced. "No onions, please."

"But Scott likes onions on his burgers," interjected Carson.

"Maybe I don't like onions on Scott, dear," replied Bea.

"No onions it is. And what would you like, sir?" asked the waitress, grinning broadly and batting her eyelashes.

"That pie looks good," observed Carson, pointing at a display on the counter. "Is it fresh?"

"It is. Just like homemade."

"Okay. Just coffee and a slice of apple pie. And let me have the check. My turn to splurge."

"Certainly. Oh, is that your Corvette out there?" asked the waitress, gazing through the partially frosted window. "That's really groovy."

"That's mine," replied Carson. "You like Corvettes?"

"Sure. Who wouldn't dig a hot car like that?" She scrawled her name and phone number on a blank ticket, handed it to Carson and winked. "Maybe we can go out for a spin sometime."

The waitress sashayed back to the counter. The cook glanced at the ticket and then returned to an argument with a trio of hippies, dressed in filthy clothes and scruffy field jackets, who apparently had an issue with their bill.

"She's cute," noted Bea, thinking how much the waitress resembled her friend Jill. "I'm sure she'll be very entertaining for a week or so, if that long. At least until you're bored with her."

"Bea, she gave me her number. It wasn't like I asked for it."

"Oh, yes, you did. You just don't realize it," Bea said, looking in her purse for a fingernail file. She had a chipped nail that was driving her crazy. She glanced at Carson's nails, and noticed how well groomed they were. For whatever reason, Scott had recently taken to fastidiously trimming his fingernails, and it appeared that Carson shared the same habit.

"So, Drew, how long do you intend to keep living your life like you do? How many pretty girls, shiny cars, and fancy watches do you need to accumulate before you're finally satisfied?"

"It's not like that, Bea," said Carson quietly, looking down at the table and self-consciously pushing his jacket sleeve over a newly acquired Breitling chronograph. "I'd like to have something more meaningful with someone, but I just haven't met the right girl yet."

The waitress brought his coffee and pie. "Don't lose my number, honey," she said, smiling at him as she filled Bea's cup with hot water.

"I won't," he replied, smiling back. Watching the waitress walk away, he cut into the pie with his fork. The apple filling was partially frozen.

"So whatever happened to your friend, Jill? The one you asked me about when we first met. She was really pretty nice. I was thinking…"

"Jill? Jill Osborn? Sorry, Drew, she's off the market."

"Married? Engaged?" He took a bite of pie and sipped his coffee.

"No. She's just out of town right now. She's visiting her sister in Columbus."

"Visiting her sister? I just thought…"

"She might be back someday, but I kind of doubt it, Drew. And I don't want to break your heart, but I don't think she would be too interested in seeing you." Bea dipped her tea bag into her cup and watched as the steaming water slowly turned brownish orange.

Apparently, Carson's conversations with Jill hadn't been too deep; otherwise he would know that Jill was an only child. And he was obviously too dense to comprehend what "visiting her sister" implied. Jill was pregnant, and she would die of embarrassment if Bea told Carson, since it was extremely likely that the baby was his. Extremely likely, perhaps, but not a certainty. Besides, Jill had already decided to keep the baby and had no need for Carson or any other man to interfere with her parenting, so it really made little difference who the father was.

"Oh. Bea, I have to apologize to you," Carson confided, scratching his unshaven chin and smoothing his moustache. "I think we got off to a rough start, and I want to make amends."

"So why this change of heart, Drew?"

"To be honest, when we first met that night at the Falcon Club, you struck me as someone who was entirely out of Scott's league," he stammered. "So, I…I…"

"So you thought I was more in *your* league," she said curtly, interrupting him. "Because you're the handsome virile fighter pilot, right? Well, Drew, it may shock you to learn that not all women fawn and swoon over handsome fighter pilots."

"I've learned that," noted Carson. "Anyway, I see you and Scott drawing closer all the time, and I would prefer that things not be so awkward between us. Like I said, I was a jerk that night, and I want to apologize."

"Okay. How thoughtful of you. So now you're forgiven. Anything else?"

"Yes. Bea, I have a favor to ask of you," he said. "A request."

"Go on."

"I don't know if this is just a passing fling for you, like some kind of temporary phase you're going through, but as much as Scott doesn't talk about it, I do know that it's very serious for him. He's clearly smitten with you."

Spooning the teabag out of her cup, she nodded. "It's serious for me, too, Drew. Obviously much more than you believe."

"I don't doubt it. But hear me out. Bea, Scott's my friend. Please don't hurt him."

"Okay, fair enough," she answered, reaching out to touch his forearm. "But I also have a favor to ask, Drew Carson. A request."

"Okay. What is it?"

"Drew, Scott's my friend. Please don't hurt him."

Both turned to see Ourecky walking past the counter, a silly half-grin on his face. He stopped to look at the jukebox, dropped in a quarter, and selected some songs. As if on cue, the brunette waitress brought their food.

"Wow," commented Ourecky, sliding in beside Bea. "That's a mighty good looking burger! I am absolutely famished. Bea, I suppose you were right about this place." He flipped open the bun, squirted on catsup, and noted, "Hey, no onions!"

Bea and Carson looked at each other and shared a smile.

"Time for me to shove off," said Carson, zipping up his leather flight jacket and picking up the check. "Scott, I suppose I won't see you until next year."

"Aren't you hungry?" asked Ourecky. "You haven't even finished your pie."

"I lost my appetite. I'm going back to the Q and stack some Z's before I head to Vermont."

Ourecky stood up and shook Carson's hand. "Be careful skiing. Keep an eye out for those ski bunnies."

"Oh, that I'll do," replied Carson, grinning. "No question there. You be careful yourself."

Bea studied them; it seemed like they were anxious to embrace each other, but could not bring themselves to do something so unmanly. As they stood close together, she also noticed that they shared similar reddish marks on their wrists and necks, as if something had been chafing their skin in exactly the same locations. Contemplating what that might be, she stood up, hugged Carson, and planted a peck on his cheek. Wiping away lipstick with her thumb, she leaned toward him and whispered, "Remember, Major Carson. We have a deal."

"Indeed we do, Bea. Merry Christmas to you."

"And Merry Christmas to you as well, Carson."

Carson went to the counter, paid the check, and made small talk with the waitress as she counted out change. As he went outside and started his Corvette, Bea watched the waitress pointing out the window and commenting to her friends. *She was probably in for quite a ride*, thought Bea, *but likely a short one, with a painful ending.*

"So, Bea, I have an idea," said Ourecky tentatively, dipping a French fry into catsup. "How would you like to go home to Nebraska with me for Christmas? You could meet my parents."

Bea was surprised, but pleasantly so; at the frenzied pace Scott had been working, she was shocked that he was getting any time off, except perhaps Christmas Day itself. For a brief moment she thought of reasons why she should say no, but then she realized that there were more reasons why she should say yes.

But there was a more practical matter to contend with: she was already scheduled to fly next week. On the other hand, she had sacrificed her holidays for the past three years so that Angie—a married co-worker—could be at home with her family. Considering the circumstances, surely Angie or another stewardess would be willing to take her place next week.

"Scott, I would love to, but I've got to go to the airport and make some calls tomorrow. And then I'll let you know. But yes, Scott, I would love to meet your parents."

Ourecky Homestead, Wilber, Nebraska
5:55 p.m., December 24, 1968

Bea marveled at the massive feast. They had been eating for at least thirty minutes, but had barely made a dent in all the food. Among the dumplings, sauerkraut, cheese, potato salad, mushrooms and other fixings was the centerpiece of the traditional Czech Christmas Eve meal: a platter bearing a large carp. His parents had bought the fish in town this morning; it was swimming in a wooden barrel when they selected it.

This was a Christmas unlike any that Bea had ever experienced. In her entire life, she had never once been in a single house with so many people; the big farmhouse was absolutely jammed to capacity with Ourecky's parents, grandparents, siblings, various in-laws, nieces, nephews, and a raft of grandchildren. It was nothing like the tiny but happy Christmas celebrations with her parents before her father was killed in Korea, and certainly did not resemble—thankfully—the sullen Christmases after her mum had married her stepfather.

"Some more carp, dear?" asked Ourecky's mother, a heavyset woman in her early fifties.

"Please, but just a smidgen, Mrs. Ourecky."

"Please call me Mama. I insist. There's no need to be formal; you're in our house, so you're in our family. And Bea, you're just so beautiful. Scott's letter just didn't do you justice."

"Thank you, Mama," replied Bea. She noticed that a corner of the living room, next to the broad hearth, was isolated by a clothesline and several hanging sheets. "What's that?" she asked in a low voice, leaning towards Ourecky. "Why is that part screened off?"

"It's a Christmas tradition from the Old Country," he replied. "This morning, before anyone woke up, my father put up the tree. No one but him is allowed to see it until after dinner tonight. He'll pull down the curtain, and then the kids will get their presents. Later, my mother will make pallets of straw and blankets under the tree, and the children will sleep there tonight, so they'll feel like they're sleeping in the stables, just like the night when Baby Jezisek arrived."

"*Baby Jezisek?*" asked Bea.

"Baby Jesus." He leaned toward her and quietly asked, "Can you finish that carp?"

"I'm not sure," she whispered. "I didn't want to be impolite when your mother offered it."

"Well, if you can't eat it, sneak it to me. It's bad luck to leave anything on your plate on Christmas Eve."

"I would have never pictured myself eating fish for dinner on Christmas Eve," she declared to Mama Ourecky, willing herself to finish the last morsel. "But this is *scrumptious*. Thank you so much for making me a part of your family."

As Bea chewed on the fried carp, she remembered that although she had grown up eating a fattened goose on Christmas Eve, her holiday repasts in recent years had mostly been in-flight meals or lonely dinners at near-empty diners. She glanced around the table and realized that all the plates were absolutely devoid of scraps or crumbs. She also noticed that everyone was quiet, gazing at each other with anxious eyes. Even the children, seated at three small tables in the dining room, looked apprehensive.

"Listen to me," cautioned Ourecky quietly. "When my father stands up, make sure you jump up at the same time, and we'll all go into the living room together as one, as quickly as we can."

"Why?" asked Bea, turning her head to diligently watch Papa Ourecky. "I'll tell you later."

Papa Ourecky pushed his chair away from the table and slowly stood up. Just as Ourecky said, the entire family—adults and children alike—followed his lead; in unison, as if connected by invisible threads, they rose from the tables and surged en masse toward the living room.

Papa Ourecky was almost a carbon copy of Scott Ourecky, except he was about an inch shorter and roughly thirty pounds heavier, with gray hair instead of black. As the family assembled in the living room, he held a finger to his lips and shushed the children.

Standing before the white sheets that divided the room, he declared, "*Stastne a Vesele Vanoce!*" Then he embraced Mama Ourecky and then hugged Bea. "*Vesele Vanoce!*"

"What does that mean?" she asked, turning to Ourecky. "What is he saying?"

"It means 'Joyous Christmas!'" he replied, embracing and kissing her. "*Vesele Vanoce!*"

"*Vesele Vanoce!*" she replied, fumbling the pronunciation and kissing him back. Around them, his family applauded and chanted, "*Stastne a Vesele Vanoce!*"

"Vesele Vanoce!" roared Papa Ourecky again, yanking down the sheets that concealed the Christmas tree. Squealing with glee, the children scampered to find their gifts. Bea marveled at the simple beauty of the festive tree; it was decorated with handmade ornaments—some many decades old—including delicately painted eggshells and walnut shells, pinwheels, and snowflakes fastidiously cut out of paper. A hand-carved Nativity scene was set at the tree's base. Mama Ourecky rushed forward and thrust a carefully wrapped gift into Bea's hands.

"Thank you, thank you," said Bea, hugging Mama Ourecky. "Thank you so much."

As she tore open the bright paper, she turned to Ourecky and asked "So why did we all have to get up from the table at once?"

"Tradition," he answered. "No one gets up before dinner is finished, and we all get up together because the first person to leave the table will be the first to die in the coming year."

"*Oh.* Good reason."

Hours later, they gathered in the darkened living room. Straw and sleeping children were strewn everywhere. "It's about to start!" whispered Papa Ourecky excitedly.

Mama Ourecky waved at Bea and patted the faded green fabric of the living room couch. "Bea, Scott, you young people sit here. The place of honor!"

"Another tradition?" asked Bea quietly, nudging Ourecky. "Please, please let your mother know that I can't possibly eat another cookie or any more fruit cake. I am absolutely stuffed!"

"Relax, dear. It's not a Czech Christmas tradition," replied Ourecky. "Apollo 8 is in orbit around the moon. They're going to have a live broadcast."

"Apollo 8? The moon?" she asked, sitting beside him. Mama Ourecky refilled Bea's wine glass as Papa Ourecky adjusted the rabbit ears antenna on the television. A low fire crackled in the hearth, warming the large room and decorating the walls with an undulating red glow. As pajama-clad children snored in the background, the adults sat silently, watching the grainy images of the stark lunar surface passing by seventy miles beneath the Apollo spacecraft.

They leaned toward the television, cupped their ears, and listened to the words of Astronaut Frank Borman, the Apollo 8 mission commander: *"The moon is a different thing to each one of us. I think that each one of us—each one carries his own impression of what he's seen today. I know my own impression is that it's a vast, lonely, forbidding type existence, or expanse of nothing, that looks rather like clouds and clouds of pumice stone, and it certainly would not appear to be a very inviting place to live or work."*

The broadcast went on for several minutes, with the distant astronauts describing various aspects of the foreign surface. At the conclusion of the live broadcast, they heard the voice of Astronaut Bill Anders: *"For all the people on Earth the crew of Apollo 8 has a message we would like to send you: In the beginning God created the heaven and the earth. And the earth was without form, and void; and darkness was upon the face of the deep. And the Spirit of God moved upon the face of the waters. And God said, Let there be light: and there was light. And God saw the light, that it was good: and God divided the light from the darkness."*

And then Astronaut Jim Lovell's voice continued the reading: *"And God called the light Day, and the darkness he called Night. And the evening and the morning were the first day…"*

After reading the final passage, Astronaut Frank Borman, the Apollo 8 mission commander, concluded the broadcast by saying, *"And from the crew of Apollo 8, we close with good night, good luck, a Merry Christmas, and God bless all of you—all of you on the good Earth."*

Obviously moved by the event, an awestruck Walter Cronkite wiped a tear from his eye and offered a few brief comments before the channel returned to regular programming.

"Men in orbit around the moon," commented Papa Ourecky, daubing a tear from his own eye as he switched off the television. He tucked his handkerchief into the bib pocket of his Liberty Brand denim overalls. "And they're supposed to land there soon."

"True. But there's still a lot of work yet to be done, Papa. They still have to rendezvous, and that's not an easy thing to do, trust me. And they've never flown the Lunar Module in space."

"No, Scott, NASA *has* flown the Lunar Module in space," interjected Papa Ourecky. "Back in January, on Apollo 5. It was unmanned, of course, but they did fly it."

"Right, Papa, but it's a giant leap from there to landing it on the moon with men in it."

"Your father sure seems to know what he's talking about," commented Bea. She drained the last of her wine; Mama Ourecky rushed forward to refill her glass. "How in the world do you know so much about the space program, Papa Ourecky?"

"Oh, we all used to follow it very closely while Scott was still at home. I've just kept up with it. It's all very exciting to me. Did you know that Borman and Lovell flew together on Gemini Seven?" he asked. "Can you believe that those two spent fourteen days in orbit together?"

Shuddering, like a giant icicle had been traced slowly along his bare spine, Ourecky closed his eyes tightly and muttered, "*No*. I couldn't possibly imagine that."

"So, Bea, what do you make of all this?" asked Mama Ourecky. "Isn't it *so* exciting that men are going to the moon?"

"I know we're going to the moon because Kennedy wanted us to," replied Bea, taking a sip from her wine. "Beyond that, I really don't know much. I just haven't followed the space program too closely. The first launches were exciting, I'll admit, but it's just gotten so boring after that."

"You don't follow it?" said Papa Ourecky. "I just assumed that since you and Scott were so close, that you must be a big NASA fan. After all, that's why he joined the Air Force, to be an astronaut. He wanted to be an astronaut ever since he was knee-high to a grasshopper."

"Really? He wanted to be an *astronaut*? That's funny. He's never mentioned that to me."

"Oh, Bea, he was deadly serious about it. He and some friends of his even built a mock-up of a Mercury space capsule. They even wrote to NASA to ask for information. On the weekends, they used to do practice missions. They would take turns; a few would be Mission Control while one would be the astronaut in the capsule. They took first place in the science fair that year."

Ourecky jumped in. "Papa…*please*. Please! Bea doesn't want to hear any of this."

"Oh, but I *do*," said Bea, laughing. "Please, please…tell me more."

"Bea, this boy was just head over heels crazy about outer space," related Papa Ourecky. "If he had drunk any more Tang, he would probably be orange tod

"Oh, this is just *too* much," blurted Bea. "Scott, I hope you aren't too offended, but I just have a really hard time picturing you as an astronaut. Your own space capsule? I'm sure it was something to behold. I just wish I could have seen it."

"But you *can*, Bea," exclaimed Papa Ourecky. "It's still out in the barn."

"Papa, I'm sure that Bea has *no* interest in seeing…"

"Oh, but I *insist*," exclaimed Bea. "I wouldn't miss this for the world."

"Well, it can wait until the morning, can't it?" asked Ourecky.

"But I want to see it *now*, Scott," she urged. "Lead the way, dear, please."

"If you insist. But grab your scarf and coat. It's cold out there."

Drawing in the familiar musty smells of the barn, Ourecky tugged a dangling cord and a single 60-watt bulb flickered on. The spacecraft mock-up was exactly as he left it, next to several neatly stacked bales of hay. Constructed primarily of scrap plywood and roofing tin, it was a very accurate model. Perhaps not worthy to be featured in Gunter's hangar of horrors, but certainly an engineering accomplishment for a bunch of high school kids.

"Wow," exclaimed Bea, looking at it in awe. "You really were serious!" She peered in the rectangular hatchway, admiring the precisely painted instrument panel. "But it's so *tiny*."

"It's life-size, true to scale," stated Ourecky. "That's why the original seven Mercury astronauts could be no taller than five feet eleven inches. They had to squeeze in there."

"Really? So you could fit in there? Why don't you climb in so I can see?"

"Uh, Bea…I'd rather not," he replied, looking into the cockpit. He shuddered as a chill passed over him, like a brisk wind had somehow blown through the barn's rough-hewn walls.

"*Please*, just this once. Just climb in so I can see what you would look like as an astronaut."

"*No*," he said firmly, shaking his head vigorously.

"Sorry," she said, cinching her scarf and blowing into her fingers. "I suppose I didn't know that it might dredge up bad memories. I guess if you were so keen to be an astronaut, it must have been quite a disappointment not to make it. Scott, I'm really sorry for dragging you out here. I was just goofing around."

"It's nothing," he said, embracing her closely. "Maybe it's just as well, though. Bea, I've wanted to get you alone ever since we've been here."

"Well, I want to be alone with you, too, Scott, but don't you think it would be a little uncomfortable out here? And it's *so* cold."

"That's not exactly what I meant, Bea," he said, laughing quietly. He took off his gloves, then turned away and dug in his pockets. Kneeling in the straw before the mock-up capsule, he held out an engagement ring and solemnly asked, "Beatrice Harper, will you marry me?"

"I *will*," she answered. "I will marry you, Scott. But there's one thing I have to ask."

Surprised that she answered so quickly, as if she had already contemplated the question, he stood up and kissed her. Gazing into her eyes, he said, "*Anything*. What is it?"

"Scott, you do know that I love you, don't you?"

"Well, since you've just agreed to marry me, I kind of suspected as much."

"Funny. Don't make me change my mind. Anyway, Scott, I've come to realize why I was attracted to you in the first place. You reminded me of my father, when we were all together as a family and we were *happy*."

Bea continued: "Scott, I don't know what you do at the base and I'm not going to ask, but I do know that you spend a lot of time with Drew Carson and it's not just a working relationship. It seems like you're always hanging out with him, at least when you're not with me."

"I know, Bea," he blurted, shivering against the cold. He bent over to brush straw from the knees of his corduroy trousers. "But..."

"Hear me out, Scott, please. I'm familiar with men like Drew because I grew up watching them. I don't know if you see him as just a friend or maybe some sort of hero, but I don't want to lose you like I lost my father. You probably think that I sound like a broken record, but I want you to keep your feet planted on the ground. Yeah, I know you have to fly sometimes as part of your job, but I don't want you going to flight school just so you can be more like Drew. Fair enough?"

Ourecky was tempted to laugh. With all the insanity and uncertainty in his life right now, she was asking for the one promise he could actually deliver. "Fair enough, Bea. No flight school. I *promise*. Now, are you still sure you want to marry me?"

"I do, Scott."

"Then let's get out of this cold and tell my folks the good news." He took a last look at the mock space capsule he had built as a teenager, not too many years distant, and then tugged the cord to switch the light off. He and Bea kissed in the cold darkness, then strolled back to the farmhouse, arm in arm.

As Ourecky reached to open the screen door, Bea said, "Hey, Scott Ourecky."

"What?"

"*Vesele Vanoce.*"

28

ENDURANCE

Simulator Facility, Aerospace Support Project
10:15 a.m., Tuesday, January 21, 1969

The day had finally arrived for the forty-eight-hour simulation, the acid test that would make or break the entire project. Carson and Ourecky stood at the base of the simulator platform as the suit technicians made their final adjustments.

Sipping coffee from a blue-enameled camp cup, Wolcott stood off to the side, quietly assaying the pair to determine their readiness. Carson looked as cocksure as ever, like a high school track star casually stretching before a race that he was sure to dominate, but Ourecky looked apprehensive, perhaps even more so than usual.

The next two days will be plenty busy, mused Wolcott. First, he would watch as the two boys were buttoned up in the Box, then stay on to ensure that things progressed smoothly. On Thursday, he would fly to D.C. to meet Tew at Walter Reed, where he was still recuperating from open heart surgery, and the two of them would go on to the big meeting at the Pentagon.

Ever a stickler for potential contingencies, Tew had insisted that two briefing charts be prepared for their progress report: one chart reflecting that they were successful in achieving the forty-eight-hour simulation requirement, and the other showing that they had missed the mark.

Wolcott finished his coffee and placed the metal cup atop a console behind him. He watched Heydrich conferring with his simulation controllers, reviewing final details before the Box was sealed and the scenario commenced.

Trying his best to stack the deck to favor Carson and Ourecky, Heydrich had agonized over the scripted profile for the mission. The exercise wasn't entirely canned; some key aspects would be randomly decided, so there could be no question later that the participants didn't have an unfair advantage or "cheat sheet" to skew the odds. Despite the random aspects of the exercise, Heydrich had painstakingly tuned the playbook so that the exercise would conclude almost exactly at the forty-eight-hour mark, and hopefully not a minute more, since he was certain that they would be treading on the fringes of Ourecky's endurance.

As if the forty-eight hours weren't enough, there were yet other unwelcome twists to the simulated mission. To ensure an unquestionably accurate replication of the pressures the men would endure in flight, an edict had come down decreeing that there would be no fifteen-minute breaks every six hours to stretch cramped muscles or pay a visit to the latrine. Carson and Ourecky would remain locked in the Box for the entire duration, hatches latched securely, until the final few minutes when they transferred over to the Paraglider Landing Simulator for reentry and landing. Additionally, the "loiter period," in which the spacecraft was powered down as they waited for an appropriate reentry window, would not be simulated. Consequently, the pair would execute the extremely stressful reentry and landing sequence without benefit of their customary twenty-minute nap.

Unlike most previous Box runs, not only were they wearing their full space suits, but they had also been fitted with the special super-absorbent diapers worn under the bulky ensemble. As in a real mission, where it was unlikely they would have adequate time to work through the elaborate procedures to capture and manage bodily wastes, *everything* would remain in the suits until the men emerged at the end.

Last but not least, the instructions dictated that the only foodstuffs authorized in the cabin were those items cleared for actual missions. That was a hugely significant deviation from previous Box runs, since it meant no instant coffee granules to energize Ourecky when his endurance started to flag. But it was only fair; with no fifteen-minute breaks, there would also be no Cokes to jump-start Carson.

Wolcott frowned as he observed Russo trying to chat up some of the simulation controllers. The situation with Russo had placed him in an uncomfortable position, particularly since it was now causing some dissension in the ranks. When he first raised the prospect that Russo might be a viable candidate to fly with Carson, Wolcott had inadvertently hatched a monster.

First, it was becoming painfully apparent that Russo *was* destined to fly, simply because there was no one else to pair with Carson. Second, although he was assigned here as a liaison officer, Russo now seemed intent on inserting himself in all aspects of Blue Gemini, even to the extent of usurping Heydrich's authority in the simulator hangar. And if he stood to fail at some juncture, that was the most likely, because the simulation crew was fiercely loyal to Gunter.

Beyond their devotion to Heydrich, the simulation crew—as a group—didn't appreciate that Russo had provoked the incident with Carson last month. Although Carson could often be abrasive, the controllers admired him for his competence and tenacity. Even more so, they respected that he had defended Ourecky; the simulation crew was an oddball bunch, and they probably saw much of themselves in the engineer, so anyone who attacked Ourecky—even verbally, even in a joking manner, as Russo had—was their sworn enemy.

Obviously steeling himself for the long haul, Russo wore the dark blue flight suit issued to the MOL "Can Men." A shiny stopwatch dangled from an orange cord around his neck. Trying his best to blend into the crowd, as if he were an active participant and not a caustic emissary, Russo stood out like a rodeo clown at a monastery. The more he tried to pal up to Heydrich and his white-shirted controllers, the more that they shunned him like an unwelcome leper. They had little time or patience for an inter-loper who had upset their otherwise happy family.

In wearing the MOL flight suit, Russo was effectively displaying his true colors; it was an outward display of his true allegiance. Although the MOL program and Blue Gemini were parallel efforts with wholly different objectives, they both drew from the same pool of dwindling resources, and it was only a matter of time before cooperation gave way to competition.

The suit technicians had finally completed their tinkering, and the two men were ready to enter the Box. Wolcott walked over and joined them. "Ready, hoss?" he asked Carson.

"Ready as we'll ever be, Virgil," answered Carson. Flexing his fingers in his stiff pressure suit gloves. Ourecky nodded in assent.

"I guess it goes without sayin' that we're counting on you two hombres to close this deal," said Wolcott, patting each man on the shoulder. "General Tew sends his regards, and he wishes you the best of luck. No matter how this thing shakes out, we're beholden to you both, and we'll make danged sure that you both get what you rightfully deserve."

"Thank you, sir," said Ourecky.

"Well, gents, I don't have anything particularly insightful or inspirational to say. Let's saddle up and ride. I'll see you two stalwarts in a couple of days."

Carson smiled. "See you then, Virg."

With Heydrich standing beside him, Wolcott watched as the two men ascended the metal steps to the Box. Technicians assisted them into their seats and connected various hoses and cables before closing the hatches. He turned toward Heydrich and asked, "Ready, pardner?"

"*Ja.* I think so," answered Heydrich, using his hand to slick back his greasy shock of black hair. "If nothing else, Virgil, we'll go out with a bang and not a whimper."

"I know it, Gunter. I'm confident in you and your boys. Do us proud."

Heydrich glanced around the hangar, stepped closer, and asked, "Is Mark Tew not coming down to lend those two a word of encouragement?"

"No, pardner. He's still in D.C. and plans to remain there until after the Thursday meeting. Besides, he ain't much for mixin' with the troops before battle, if you know what I mean."

"He's *still* in D.C.?" asked Heydrich, quickly signing a document handed to him by a subordinate. "Virgil, hasn't he been up there since…"

Wolcott nodded and then spat a stream of tobacco juice into a wastebasket.

"Virgil, is everything all right with Mark? He's been looking rough these past few months."

"He's fine," drawled Wolcott, tapping the metal lid of a puck-sized can of smokeless tobacco. "He's just been dealin' with budget and administrative issues, and it's a lot simpler for him to contend with them down there than up here. Closer to the flagpole, you know."

"They've finished their pre-launch," announced a controller. "We're ready to start."

Heydrich nodded, took his seat, and adjusted his headset. The countdown proceeded through the last few minutes—and then it was time.

"Stand by for engine gimballing," announced the CAPCOM.

"We are standing by for gimballing," replied Carson over the intercom. For a man who would be lying on his back for the next forty-eight hours, he sounded upbeat, even cheerful. Mounted near the consoles, a large speaker piped the simulated pre-launch noises—the whirring of pumps, creaking, groans, and squeaks—that Carson and Ourecky currently heard through their headsets in the Box. They were the sounds of a dormant Titan II rocket gradually coming to life, preparing to convert its potential energy into barely bridled explosive force.

"T-minus one minute," stated the CAPCOM.

"T-Minus one minute," echoed Carson.

"Stage 2-P valves opening in five… T-minus forty seconds… T-minus ten, nine, eight, seven, six, five, four …first stage ignition…two, one, zero. Hold-down bolts are fired. Lift-off."

"Hold-downs are fired and we have lift-off," chimed Carson. "The clock is started."

Standing behind Heydrich, Wolcott could not help but notice Russo punch the start button on his stopwatch as Carson called lift-off. *If that's how you're going to play it,* thought Wolcott, *you've got a hard two days ahead of you.* Knowing that Blue Gemini's fate now rested in Heydrich's hands—or perhaps in Ourecky's head—Wolcott closed his eyes and sighed.

Walter Reed Army Medical Center, Washington, D.C.
8:35 a.m., Thursday, January 23, 1969

"So, Mark, you look like hell," observed Wolcott, walking into the private hospital room and setting down a large gym bag. "Do you feel any better?"

"I do," replied Tew, sitting up in his bed. "Good to see you, Virgil. Did you bring my things?"

"I did, pardner," answered Wolcott. "Your street clothes and loafers are in the bag. Your uniform is hangin' in my VOQ room at Andrews. A courier is totin' the briefing charts and books; we'll meet him at the Pentagon. We need to scoot, though. Time's a'wasting."

Tew slowly stood up and untied his blue hospital gown. Wolcott gasped as he glimpsed the massive surgical scar that ran from the top of Tew's breastbone down to his navel. The swollen incision was closed with black sutures as thick as kite strings.

"It looks a lot worse than it feels," noted Tew weakly. "Although it does feel pretty damned bad. Kind of like getting ripped open with a chainsaw. It's funny, but hugging a pillow helps a lot with the pain, but I don't suppose that would go over very big across the river."

"Can't argue that, pardner." Wolcott reached into the gym bag and pulled out a knit shirt and khaki trousers. "And you're absolutely positive that you're cleared to leave the hospital?"

"Why, *certainly*, Virgil. I've been vegetating here for three weeks. I asked my doc if I could go visit some friends, and he said it was okay. We still have friends in the Pentagon, don't we?" Tew tugged the shirt over his head and winced as the fabric brushed over the protruding stitches in his chest.

"Friends at the Pentagon? I guess we do," replied Wolcott, handing Tew his folded khaki trousers. "I s'pose that we'll know by the end of the day, pardner."

Simulator Facility, Aerospace Support Project
11:05 a.m., Thursday, January 23, 1969

"Okay," said Russo, yawning as he checked his stopwatch. "We're drawing to a close here. Here's what I think we should see on the reentry sequence.

Start with a delayed retro firing, so they'll be forced to divert to an alternate landing site, and then…"

"*No,*" said Heydrich emphatically, shaking his head. "That's not how we do business here. We designed the scenario to plug in random events, but those events are generated strictly according to input from a random numbers generator."

"Random numbers generator?" asked Russo. "*Seriously?*"

"*Seriously.* Chris, hand me the random numbers generator."

"Here you go, boss," replied a sleepy controller, solemnly passing Heydrich a small black cylindrical object.

In a time-honored ritual, Heydrich shook the cylinder and then flicked it toward an open area on the table. Three dice rattled out, spun, and then came to rest between a chipped ceramic coffee cup and an overflowing ashtray. "Okay, Chris, tell the back row to load Post-Retro Scenario Five-Three-Two."

"Five-Three-Two," mumbled the controller, opening a binder and flipping through pages. "Five-Three-Two…here it is. Five-Three-Two. *Uh oh.* Gunter, you might want to roll again."

"Why?" asked Russo, attempting to look over the controller's shoulder. "What's wrong?"

"Among other things, Five-Three-Two calls for the command pilot to be incapacitated," answered the controller, slowly shaking his head. "I don't think we should…"

"Play it as it lies," stated Heydrich firmly, looking squarely at Russo. "The dice have spoken, and we're not going to argue. I am confident that Captain Ourecky can handle it."

The controller nodded, and turned to face the third row of consoles, the domain of the computer specialists. "Five-Three-Two," he announced.

"*Retrofire,*" declared Carson, punching the manual retro fire button and scanning the sequencing lights in the center console. The four retro rockets were programmed to "ripple fire," igniting one at a time, rather than "salvo fire," where they came to life at the same time. "Retro 3 is firing. Retro 2 is firing. Four is firing. All retros firing."

"I copy that all retros are firing," answered the CAPCOM. "Do you have your IVIs?"

Carson read the indicated numbers from the center of his panel: ""IVI read-outs are 269 aft, zero-one-zero left and 181 down. Everything is nominal."

"I copied 269 aft, zero-one-zero left and one-eight-one down, Scepter One."

"That's a good copy," replied Carson. Shuffling his feet in the narrow footwell, trying to improve his circulation, he looked to his right. "How are you feeling over there, buddy?"

"Not great," groaned Ourecky. "I'm mostly just numb at this point. My skull feels like it's packed with sand. I hope I don't offend you, Drew, but I really don't want to ever sit inside this thing again, *ever*, once today is over."

"Understood. I think you've paid your dues, Scott, and I owe you." Watching the sequencing lights, Carson announced: "Retro pack jettison."

"Copy retro pack jettison," replied the CAPCOM. "You guys ready to make the switch?"

"More than ready," answered Carson. He stifled a yawn. "We'll start disconnecting in here."

At this point, the clock would temporarily stop as he and Ourecky would be transferred from the Box to the Paraglider Landing Simulator. Although the swap heralded that their ordeal was nearly over, riding the paraglider simulator, which moved in all axes, sometimes felt like sitting in an out-of-balance washing machine. He hoped that Ourecky—who was as close to exhaustion as any man could be—could handle the few minutes that remained.

After the transfer was completed, the clock was restarted; the cabin rocked back and forth slightly, simulating their descent into the atmosphere. Gazing through his window, Carson watched the simulated drogue chute unfurl and blossom. "Drogue deployed," he stated. "Go to Rate Command."

"Drogue is out," confirmed Ourecky sleepily. "Switching to Rate Command."

"Minor oscillations," noted Carson. "Install your D-Ring."

"D-Ring." Ourecky fumbled the D-Ring into position and then locked it in place with a pip pin.

"Check your straps." Carson pushed himself down in his seat and cinched his shoulder straps. "Sixty thousand feet. Thirty seconds to main chute."

"Thirty seconds to main," confirmed Ourecky.

At the appropriate altitude, Carson punched the button to deploy the paraglider. Both men studied the wing as it gradually took shape, recognizing that it was taking several seconds longer than normal. "We have a burble up there," he observed. "It's not deploying cleanly. Stand by to cutaway and eject."

"Standing by to eject," replied Ourecky, poising his finger over the switch to arm the pyrotechnic charges that would jettison the paraglider.

"Paraglider is out," said Carson. "Looks like a good deployment, but we lost at least a thousand or more feet in the process. I think we still have enough glide slope to make it to the strip, though. Stand by to rotate to two-point landing attitude." He *hated* the next move. He punched a button and quickly swung his forearms up to guard from smacking face first into his instrument panel. The cabin jarringly lurched nose over, replicating the change to landing attitude. Almost immediately, he heard a buzzing tone in his headset. "*Damn*," he mumbled.

"What?" asked Ourecky.

Sitting quietly, Carson placed his hands in his lap, a subtle signal to Ourecky that he had been "incapacitated."

"Oh," replied Ourecky. Not missing a beat, he clutched the hand controller and immediately transitioned into the paraglider controllability checklist. "Stall check," he noted, gradually tugging back on the controller. "Stall check is good." He continued through the sequence and then switched on the TACAN receiver. Almost immediately, the TACAN acquisition light blinked on, indicating that they were locked on to a transmitter located at the landing site.

"Control, this is Scepter One, receiving TACAN on Channel Six," said Ourecky.

"Scepter One, this is Control. I copy that you are receiving TACAN on Channel Six. Field elevation is two-one feet, current altimeter is two-eight-nine-three. Winds from Two-Seven-Five, three knots gusting to five. You are cleared for Runway One-Eight. Runway surface is packed earth. Recommend you form a right-hand pattern."

With his hands still folded in his lap, Carson noticed that Ourecky seemed concerned about something. Finally, the engineer spoke: "Control, Scepter One, be aware that we had a delayed opening. I don't think we have sufficient altitude to safely fly a pattern and land upwind on One-Eight. Since the surface winds are negligible, I would rather shoot a straight-in approach and land downwind on Runway Three-Six. Can you advise clearance?"

Several seconds of silence passed, as the controllers were obviously discussing Ourecky's improvisation to resolve his shortage of operating altitude. "Scepter One, Control, Roger. You are cleared for a straight-in approach to Runway Three-Six. No change to winds or conditions."

"Control, I copy that I am cleared to land downwind on Three-Six," said Ourecky calmly. "Be advised that my command pilot is unconscious and I will need medic on field."

"Scepter One, Control, I copy that you will need a medic after landing."

Proud of Ourecky's ability to make a tough choice in a moment of intense stress, Carson grinned. *We're going to make it after all*, he thought, *the kid is going to put her down safely*.

Then, just by chance, Carson glanced at the mission clock on the upper right corner of his instrument panel, and was momentarily overwhelmed by panic. Ourecky had made a solid decision, but not flying the pattern would shave at least two or more minutes off the elapsed time at the end of the mission. Running the numbers in his fogged thoughts, Carson realized that it was very unlikely that they would actually achieve the requisite forty-eight hours.

Transfixed on the mission clock, frantically hoping that Ourecky will notice the time and recognize the problem, Carson watched the seconds painfully click by. But Ourecky didn't look at the clock, because he was doing precisely what any good pilot should be doing at this stage of the mission: he was focused entirely on safely landing the ship, and the elapsed time was not a variable that he need be concerned with at this moment.

"Control, this is Scepter One," announced Ourecky. "Field in sight, on short final."

Moments later, Carson heard a loud scraping sound—the simulated noise of the three skids making contact—in his earphones. Ourecky had pulled off the landing. He looked to the right to congratulate Ourecky, but heard snoring and realized that the engineer had fallen asleep. He suspected that he might be the only one aware of the time glitch, so he decided to keep his mouth shut and not broach the issue. This long-dreaded ordeal was *over*, and what happened afterwards was entirely out of their hands.

The Pentagon, 11:45 a.m.

Wanting desperately to rip off his uniform so he could scratch the relentless itch of the sutures poking out through his skin, Tew pretended to listen intently as a budget analyst pontificated about various current and proposed anti-satellite programs.

Droning on, the analyst asserted that deploying an unmanned anti-satellite system was more cost-effective than following through with Blue Gemini's satellite inspector/interceptor. Striving to present an aloof persona, Tew looked at Wolcott seated to his left. Wolcott was clearly fuming, seemingly ready to explode.

Tew looked to the head of the table, where General Hugh Kittredge was seated. Kittredge would ultimately make the recommendation on Blue Gemini's fate. To his right was Admiral Leon Tarbox, who headed up some secretive aerospace projects for the Navy. Currently, Tarbox was assigned to the MOL program, where he oversaw the Navy's participation in the effort. Coincidentally, as the MOL liaison to Blue Gemini, Russo worked directly for Tarbox.

The budget officer finally completed his pompous speech. Tarbox thanked him and then moved to stand at the head of the table. "So, gentlemen, I think it's clear that continuing with this project is far too expensive and too risky, and it's an opportune time to cut our losses. We've certainly learned a lot, and for that we're indebted to Mark Tew, Virgil Wolcott and their men."

Tarbox continued: "So, General Kittredge, our recommendations are as follows. First, shut down this program as swiftly as possible, but not before mothballing the launch vehicles and spacecraft to allow their use in short order, should a contingency arise. Second, reallocate Blue Gemini's remaining funds, hardware, and personnel resources to the MOL program. And third, transfer control of the 116th Aerospace Operations Support Wing to the Aerospace Rescue and Recovery Service, with the specific mission of direct support to the MOL program."

"Good points, Leon, and I concur with most of them," noted Kittredge. "But in fairness, let's make sure we hear both sides of the argument. Mark?"

Tew rose slowly out of his chair. As he did, Wolcott stood up to place a briefing chart on a tripod. Seeing that the chart was the version that reflected a successful completion of the forty-eight hour intercept simulation, Tew slowly shook his head, almost imperceptibly. In response, Wolcott grudgingly swapped out the briefing charts.

"Gentlemen," he said. The word was barely out of his mouth when a phone quietly buzzed on the table. One of Kittredge's assistants picked up the phone, listened, and then handed it to Tew. Tew put the receiver to his ear, listened, smiled broadly, and then motioned for Wolcott to swap out the briefing charts yet again.

"Good news?" asked Kittredge.

Handing the receiver back to the Major, Tew nodded. "*Excellent* news. I'll cover it shortly."

Just as the assistant replaced the phone, it buzzed again. Like a déjà vu scene, he picked it up and then handed the receiver to Tarbox. Smiling slightly, Tarbox listened and then declared, "No need to continue, Mark. It's a dead issue. Your guys just failed to hit a critical benchmark."

"*What?*" demanded Wolcott. "We've hit them all, including—"

"The forty-eight-hour intercept simulation?" smirked Tarbox. "Sorry, Virgil. Your guys came up short. I'm sure that they gave it their best, though."

"Wait," demanded Tew. He tasted bile rising in his throat, and suppressed the urge to throw up. Covering his options, he slid two steps closer to Tarbox just in the event that he couldn't hold it back indefinitely. "I just talked to Gunter Heydrich. He clearly said that they *just* completed that requirement."

"Maybe that's what Heydrich's claims," countered Tarbox. "But our auditor—Major Russo—monitored the whole show from start to finish. He kept precise track of time, and he reported that your guys missed the mark."

"I should have left you high and dry north of the Yalu, Leon," growled Wolcott, tapping the table with a wooden pointer. "Cryin' and whinin' and pissin' your britches, begging for someone to chase that MiG off your tail. Here's the durned thanks I get."

"Enough," said Kittredge, holding up his hand. "We will be *civil*, gentlemen." One of his technical advisors, a civilian aerospace expert, leaned toward him and whispered through a cupped hand. Kittredge frowned, then said, "Dr. Rutledge, would you mind conveying that bit of information to our esteemed colleagues? Hopefully before this tiff devolves into a fistfight?"

Retaining his seat, Rutledge said, "Gentlemen, there's no sense splitting hairs over the forty-eight-hour goal. It's not germane; the actual number is really inconsequential."

He continued. "When we wrote the benchmark requirements, we were concerned with your crews' capacity to complete a long-duration intercept mission with minimal assistance. We had to quantify it, put a number to it, so we punted it around and arrived at forty-eight hours. So hitting that *exact* number wasn't really important, since it was just an arbitrary figure."

"*Punted*? Arbitrary figure? Are you kidding me, pardner?" asked Wolcott. "Thunderation! We've been killing those boys for the past few weeks, and you're telling me that you could have blessed off on any long simulation with the wave of your hand?"

Awkwardly smiling, as if he could somehow deflect some of Wolcott's hostility, Rutledge nodded. "Forty-eight hours really means nothing. We just used it as a placeholder."

"Bub, I should use *you* as a placeholder," snarled Wolcott. "In a shallow grave."

Tew glowered at Wolcott, hoping that he could silently persuade his grizzled friend to just shut up. From his perspective, even though Kittredge seemed to have come into the meeting pre-disposed to agree with Tarbox, it looked like all the chips were still on the table and no one had drawn a pat hand.

"So the forty-eight-hour requirement was always subject to interpretation?" asked Tew. "Why didn't you clarify that?"

"Why didn't you *ask* for clarification?" replied Rutledge smugly.

For a long moment, there was silence. At an impasse, both parties looked betrayed and disappointed. Then the phone rang yet again. Kittredge's assistant picked up the receiver, listened, and then sprang up as if a powerful electric current had suddenly jolted through his body. Handing the receiver to Kittredge, he announced, "It's the White House, sir. Secure line."

Clasping the receiver to his ear, Kittredge listened intently. "Meet with *who?*" he asked. "*When?*" His face turned pale as he listened for a few more seconds and then hung up the phone. "Gentlemen, I've just been called away," he explained, gathering up his loose papers and stuffing them into an attaché. "We'll reconvene in an hour. If I'm not back by then, wait here until I return."

"So do we have your decision, Ed?" asked Tarbox glibly.

"When I return. Mark, you and Virgil will come with me. Bring that chart. Leave the other one; my aide will see to its destruction."

12:15 p.m.

Like a conquering hero of sorts, Ourecky was carried from the hangar floor into the suit-up room, hoisted on the shoulders of four technicians. Of course, triumphant heroes rarely doze in the midst of their victory parades. The men gently laid Ourecky in his recliner, and the suit technicians immediately started the intricate process of removing and storing the suit.

Unlike his groggy counterpart, Carson strolled into the suit-up room under his own power, to the applause of all gathered. Like a featured performer taking a bow at curtain call—selflessly acknowledging the contributions of the director, orchestra, lighting crew and other actors—the tired but euphoric Carson pointed to the snoring Ourecky, Heydrich, and the rest of the men present, and lightly clapped his hands.

Ourecky momentarily regained consciousness and gestured for Carson. "So we did it, right?" he asked.

"Yeah, buddy," replied Carson, leaning over Ourecky's recliner. "*You did it.*"

"Good," mumbled Ourecky. He yawned widely. "Hey, Drew. Can you do me a big favor?"

"Anything for you."

"Call Bea and tell her I'm not coming over tonight. I think I'll just stay right here and snooze." And with that, Ourecky lapsed back into deep unconsciousness.

Carson slouched into his waiting recliner and said, "Please shuck this thing off me." A technician went straightaway to work, unzipping the suit and carefully pulling the top portion over the pilot's head and shoulders.

"Sorry to be the bearer of bad news, gentlemen," gloated Russo, strutting into the suit-up room and waving his silver stopwatch. "But you didn't make the forty-eight-hour mark. It's official; I verified my time against the master elapsed time clock on the computer. You came up exactly one minute and thirteen seconds short. I admire your effort, but it looks like that hard work was all for naught."

"What the hell are you talking about?" demanded Carson. Yanking the recliner's handle and standing up, he was half out of his suit; his arms and upper body were free, but he was still encased in the bulky garment from the waist down.

Heydrich stood beside Carson. "It's just a minor misunderstanding, Drew. You gentlemen finished the forty-eight-hour scenario. That you came in under the clock is just not relevant."

Russo snorted, wagging the stopwatch. "I beg to differ. Rules are rules, and the requirement was for forty-eight hours, not a minute less. Besides, I've already called my boss, Admiral Tarbox. He's in the middle of a status briefing, as we speak, and he said Blue Gemini was already on the verge of being cancelled, so this is just icing on the cake."

"Icing on the cake?" asked Carson angrily. "Cancelled?"

"That's right," answered Russo, a quizzical expression on his face. "I thought you would be happy, Drew. After all, didn't you want to be released from Blue Gemini so you could fly overseas? I thought we were doing you a big favor."

Carson's face began to turn red. "Yeah, Russo, I do want to fly overseas, but not at the expense of all the people who worked so hard to make this happen. They don't deserve this."

"Well, Drew, it doesn't much matter, since this project was on the chopping block regardless of this simulation's outcome. This just makes it easier to swing the axe."

"That's *it*," hissed Carson, throwing up his hands and moving toward Russo. With the pressure suit half-hanging off his torso and gradually sliding lower, he slid his feet in an awkward shuffle. "That's it. Since you've already done your damage here, maybe you can fly back to California on the weekly medevac run."

"The suit… Carson… the suit," murmured one of the suit technicians. His voice was so soft that it was barely audible. He acted as if he sincerely intended to intercede, but he moved with the speed and intensity of a "B" movie zombie. "Be careful…not to get any blood…on the suit. It's really hard to sponge out."

Heydrich threw up his arm, momentarily halting Carson. "No!" he cried. "Carson, we've warned you about this. No more. You can't solve all your problems with your fists."

"This isn't just for me, Gunter. This is for *everyone*," Carson snarled. "Especially Scott and your guys."

"Oh, in *that* case," replied Heydrich, slipping off his glasses and handing them to Carson. "Why don't you just let *me* handle this, since I'll be out of a job tomorrow anyway?"

Raising his hands in panic, Russo was dumbfounded, not knowing what to make of the belligerent Carson, with his space suit now falling down around his knees, or the corpulent but uber-calm German engineer.

"*Lichter aus, schiesskopf!*" growled Heydrich, stepping forward and unloading a lightning-fast haymaker onto Russo's chin. Reeling from the impact, Russo collapsed like a debutante swooning at a midsummer cotillion. Shaking his fist as he blew on his reddening knuckles, Heydrich proudly declared, "*Technische Universität München*, 1939. You weren't the only one who boxed in college, Carson. You Americans had your heroes, but every good German boy ached to be Max Schmeling."

29

OUT OF ASHES

Aerospace Support Project
7:55 a.m., Friday, January 24, 1969

Heydrich looked over his notes, making ready to report on the grisly aftermath of yesterday's simulation debacle. Battling a pounding headache, courtesy of too much apple schnapps last night, he mentally prepared his case for the argument that was sure to come, that they had in fact accomplished the forty-eight-hour benchmark even though Ourecky had "landed" the simulator over a minute ahead of the requisite time. Heydrich sighed; he was prepared to leap on his sword if necessary, since he had been the one responsible for shaving the time so closely.

Knowing Tew and Wolcott as he did, he also expected to be publicly drawn and quartered for punching out Russo. The story had obviously made the rounds; several staff officers knowingly patted him on the back as they entered the room. Self-conscious of his bruised knuckles, he slipped his right hand under the table.

Despite his personal gloom, it seemed like there were some grounds for optimism. For the first time in several weeks, everyone at the table seemed upbeat and galvanized. Certainly, the mandatory holiday break had boosted everyone's morale, but Heydrich suspected that some momentous news was forthcoming, since Tew and Wolcott had flown back from D.C. last night instead of remaining over the weekend, as they were wont to do, and they had called for this uncustomary "all hands" staff meeting to convene on a Friday morning.

Searching for subtle clues, he looked toward the head of the table. Despite the pervasive rumors about his health, Tew looked none the worse for wear. Sipping coffee, which he had not done in months, the general flipped through a binder, pencil in hand, painstakingly checking columns of numbers. Wolcott offered no hints, either. With his pearl-buttoned gingham shirt and black bolo tie, wearing an expressionless face worthy of the most cutthroat riverboat gambler, he acted no different than usual.

Tew closed his binder, cleared his throat and spoke. "Without delving into minutiae of what was discussed in Washington yesterday, Blue Gemini has been deemed to be of great strategic importance. With some changes, we'll continue on the path we've established. If anything, we'll have more tasks to accomplish, but we're going to receive additional resources commensurate with our priority and the extra work that's expected of us."

Wolcott grinned, thumped his Stetson, and added, "Gents, this train has left the station and it *ain't* slowin' down. We're finally goin' to space."

Most smiling, the collected staff officers looked excitedly at each other and started shaking hands. Although he was thrilled with the news, Heydrich momentarily glimpsed into his future, seeing the massively increased workload that would certainly be his lot.

Tew loudly cleared his throat. "Settle down. In the coming weeks, our schedule will accelerate exponentially. Colonel Porter, can you update on activities at the HAF?"

"General," said Porter, beaming, "I spoke with San Diego yesterday afternoon. The boilerplate Gemini-I mock-up has been mated with the Titan II launch vehicle. They will perform their final diagnostics tests early next week, then encapsulate the complete vehicle in plenty of time to ship it to the PDF in time to launch next month."

"Any issues with the launch vehicle?"

"Nothing significant, sir. The testers kept bumping into a minor glitch with the feedback from the gimbal angle sensors, but they're addressing it. It sounds like a faulty module."

"Good work, Colonel. I want you to call the HAF straightaway after we're done here and instruct them to immediately *stop* what they're doing." Tew reached for his coffee and said, "Virgil, would you convey the rest of the colonel's instructions, please?"

"Gladly, boss. Porter, on Monday, you are to proceed to the HAF, where you will tell them to remove the boilerplate vehicle and place it in permanent storage and to then prepare the first Gemini-I for mating. I'm dispatchin' the first flight crew out there at the end of next week for their initial work-up on the vehicle, along with weight and balance tests. Any questions, pard?"

"Flight crew, Virgil?" asked Porter as he scrawled notes. "The first shot is unmanned."

"Not anymore," interjected Tew. "The February shot will be our one and only pure training mission. The crew will intercept an Agena-D target vehicle left over from the NASA missions. They'll test all the procedures, to include deploying a multi-function Disruptor."

"Disruptor, sir?" asked a newly assigned officer.

"The Disruptor is a little surprise package that we intend to leave on the Soviet platforms we target," replied Wolcott. "It can either blow the hostile satellite to smithereens, cause it to fall out of orbit, or just make it so unstable and jittery that the Russians will spend all their time trying to stabilize it. Plus they'll be yankin' their hair out trying to hash out the defects in their design."

"Again, gentlemen, next month will be our one and only dry run," said Tew. "We will launch missions at ninety-day intervals after that, with all subsequent flights to intercept live targets. Our primary focus will be to identify and disrupt their OBS platforms, but we are also authorized to pursue reconnaissance and military communications satellites."

The staff officers gasped.

"Intelligence," stated Tew succinctly. "Our first training shot was supposed to launch in June. Using our current catalog of targets, I want your people to determine our next three available flight windows. I want a preliminary assessment this afternoon and a complete report by mid-week."

"Will do, sir," replied Seibert. "A question, sir, if I might?"

"Go ahead."

"I assume that you're aware that peace talks are scheduled for tomorrow in Paris," stated Seibert. "The US is supposed to meet with the North Vietnamese and the Viet Cong. I have to assume that the war is rapidly drawing to a conclusion, and our troops will be coming home, most likely by next summer. How will this impact our project?"

"Not in the least," answered Tew.

"Well, pardner, that ain't absolutely true," interjected Wolcott. "If the war in Vietnam ends that quickly, then I won't have to listen to Carson and the other pilots continually whinin' about not being able to fly over there. Maybe for once, Carson will focus on the matters at hand."

"And finally," said Tew. "Virgil and I have a belated Christmas gift to present. Virgil, if you would do the honors."

"Gunter, considering recent events in your hangar, the boss and I were planning to install a boxing ring over yonder. But, pardner, there won't be enough room in the barn, because…" Like a television game show host awarding the grand prize, Wolcott declared: "NASA is transferring a Gemini procedures simulator—complete with all the bells and whistles—which we should start receivin' next week."

Heydrich somberly stared at the table, rubbed his throbbing temples, and sighed.

"So, Gunter, why so glum?" asked Wolcott. "After you've whined for months about your hand-me-down equipment and lack of resources, I expected you to be thrilled, but you look like a kid who just dropped his lollipop in the mud."

"Well, borrowing from your parlance, Virg, I hate to look a gift horse in the mouth. You can jam my hangar with simulators, but I still don't have sufficient manpower to keep them running."

"Not to worry, Herr Gunter. As part of the new personnel package we've been authorized, you're slated to receive a minimum of thirty bodies, split between blue-suiters and civilian contractors. Ain't that what you wanted, pardner?"

Later, after the staff officers filtered out, Wolcott leaned back in his chair, cleared his throat, and said, "Mark, we have another issue to discuss: our young stalwart, Captain Ourecky. With all these new developments upon us, friend, I want to keep him here. We need him."

"No, Virgil," replied Tew, shaking his head. "We'll follow through on our commitment to Ourecky. He goes out to the MOL office initially, and then to Cambridge for his PhD."

"But Mark, the MOL office is just plain saturated with eggheads as it is. Wouldn't it make more sense to keep him here? He's sure proven himself."

"That's precisely the point, Virgil. He's proven himself, time and again, and it's time that we pay what's owed. The MOL program has the resources, funds, and wherewithal to send him to grad school; we don't. So, just this once, let's do what's right. Okay?"

"Done, pardner."

"So where's Ourecky right now?" asked Tew, checking his watch.

Heydrich answered: "Sir, he stayed in the simulator hangar until about midnight, according to my guards, and then left the base. We don't expect him to come back until Monday morning."

"Fine. Where's Carson right now?"

"He's probably in his VOQ room, still asleep," observed Heydrich. "He was pretty beat yesterday. If he's not sleeping, he's probably working out."

"Locate Carson," said Tew quietly, turning to Wolcott. "I want to talk to him, as soon as possible. And just in case Ourecky shows up, tell the guard at the front desk to send him directly up here. Put an escort on him if necessary; I don't want him wandering around anywhere else in this building. And I don't want him talking to anyone. Period."

"As you wish, *jefe*," replied Wolcott. "Gunter, you probably want to pass the same guidance over to your guards in the hangar."

"Good point," noted Tew. "I want to see Carson now, and I want Ourecky in here this afternoon. Make it happen, gentlemen."

An hour later, Tew caught up with Carson at the base boxing gymnasium, where the pugilistic pilot was doing his utmost to pulverize a canvas punching bag. Pausing to admire his footwork, Tew was impressed with Carson's compact but powerful physique. For a relatively small man,

Carson's shoulders were broad and his arms heavily muscled; his chest and abdomen seemed chiseled from the hardest stone.

In the ring at the middle of the gym, two welterweights of the base boxing team sparred, exchanging less-than-enthusiastic blows as a frustrated trainer yelled at them to be more aggressive. Several other boxers lifted weights, thrashed speed bags, or skipped rope. The place smelled of sweat, bleach, and Ben-Gay analgesic balm.

Glimpsing Tew, Carson lowered his fists and removed his rubber mouth guard. "Sir?" he asked, seemingly not knowing what to make of Tew visiting this hallowed shrine to combatives.

"We need to talk, Major. I have a favor to ask of you."

"Yes, General?" replied Carson, removing his bag gloves before adjusting the white cotton tapes that encircled his hands and wrists. He used a towel to mop sweat from his forehead.

Tew glanced around before quietly speaking. "We're sending Ourecky away. He's finished what he came here for, and I've made arrangements for him to study for his doctorate."

"That's great, sir. I hate to see him go, but I'm happy for him. I know that's something he wants to do." Bending over at the waist, Carson tightened the laces on his high-top sneakers.

"You've grown sort of close to him, haven't you? You two have become friends, right?"

Carson stood up and said, "That's true, sir. I wasn't very fond of him at first, but…"

"That's where the favor comes in, Carson. You've probably heard rumors that we're shutting down Blue Gemini."

"I have, sir. Is it true?"

"*No*," replied Tew emphatically. "Quite the contrary. Carson, I want Ourecky out of here for his own good. That said, I want you to personally walk him through the out-processing. I want him to believe the rumors and to believe that we're closing up the shop for good."

With his arms hanging at his sides, looking perplexed, Carson asked, "Why, sir?"

"I don't want him looking back. *Ever.* I want him to focus on earning his doctorate, because that's what's going to be most beneficial to him and the Air Force in the long run. Also, just from a security perspective, we want the fewest number of people knowing what we're doing. It's need to know. Ourecky did his job here, and now he's leaving, so there's no need for him to know what we're doing in the future."

"Sir, I'm a bit confused here. You're going to tell him that Blue Gemini has been cancelled?"

"I'll do nothing of the sort," snapped Tew. "We're just going to tell Ourecky that we didn't achieve the requirement for the forty-eight-hour simulation, as it's written in black and white, and he can form his own conclusions from there. We're not going to *lie* to him. Moreover, you will not tell him any different, not if you intend to eventually fly. Do you understand me, Major?"

"Understood, sir," said Carson.

"Carson!" yelled the trainer from the ring. "Want a workout? I need someone to wake up Mister Shum. Maybe you and your fists could help."

"Give me a minute, coach," answered Carson, holding up his arm. "I'll be right there."

"So we understand each other? Good," said Tew. "Bring Ourecky around this afternoon, and we'll break the news to him. Sorry to disturb your workout, Major."

1:30 p.m.

Carson escorted Ourecky into the generals' office. With bloodshot eyes, looking much like he had just rolled out of bed, which he had, he was unshaven and pungently in need of a shower.

"Relax. Have a seat, gentlemen," said Tew cordially. "How about some coffee?"

"No, thanks, sir," croaked Ourecky hoarsely. "I'm fine."

"Thanks for coming back today, Captain. I know that you were supposed to be off, and you've certainly earned a break," said Tew. He nodded towards Wolcott.

"Okay, here it is, skins off," announced Wolcott. "Bad news ain't like cheese, so it won't get any more palatable as it ages."

"*Bad* news, sir?" asked Ourecky.

"Yeah, pardner. I'm sorry, Ourecky," said Wolcott, sliding a red-bordered paper across the table. "I know you're seein' this for the first time, but these are the benchmarks that we had to achieve in order to move on to the flight phase. Item number twelve states the requirement for the forty-eight-hour simulation. It's right there, neatly spelled out in black and white."

Perusing the paper, Ourecky yawned audibly. "I don't understand, sir. I thought that we…"

"We were one minute and thirteen seconds short, Scott," mourned Heydrich, hanging his head as if someone had just shot his favorite dachshund. "It was my fault. I didn't adequately anticipate the impact of the random drop-in events, like the delayed paraglider deployment. It's not your fault, Scott. You couldn't have handled the situation any better than you did."

"One minute and thirteen seconds?" asked Ourecky.

Wolcott nodded. "One minute and thirteen seconds. Doesn't seem like much, does it?"

Ourecky swallowed, grimaced, and asked, "And Blue Gemini?"

"Not your concern, son," replied Tew.

Ourecky looked like he had been gored with a poleaxe. Silent, Carson gritted his teeth and clenched his fists under the table.

"And now, the good news," declared Tew. "Assuming the likelihood that things might take a turn for the worse, I've been working on some contingencies on your behalf."

"Remember, hombre," drawled Wolcott. "Before you climbed into the Box on Tuesday, I told you that we would make danged sure that you got what you deserved."

But he doesn't deserve this, thought Carson.

Tew handed Ourecky a thin manila envelope. "On Tuesday morning, you're to report to the Manned Orbiting Laboratory project offices in El Segundo, California. You'll remain there through May, and then you'll proceed to the Massachusetts Institute of Technology, where you'll be enrolled as a doctoral candidate."

"MIT, sir?" asked Ourecky. "Doctoral candidate? PhD?"

"That's correct, Captain. Of course, you'll go as an Air Force officer, so you'll incur an additional service obligation, but I think that's a small price to pay."

"I agree, sir," said Ourecky, obviously still coming to grips with the situation. "It's quite an honor. But what will happen to the people here? Major Carson? Gunter? All the others?"

"Again, Ourecky, that's not *your* concern," said Tew, extending his hand. "You worked hard while you were here, and we want to do right by you. As for Blue Gemini, son, things don't always go the way we plan. In any event, it was an honor to serve with you."

Solemnly shaking Tew's hand, then standing at attention and briskly saluting, Ourecky said, "The honor was all mine, sir. Truly."

"Thanks for everything," said Wolcott, sticking out his hand as well. "You gave it your best, buster, and that's all that matters. I wish the best for you in California and at MIT."

Crestfallen, Ourecky shook Wolcott's hand. "Thanks for the opportunity to serve, sir."

"It was a pleasure to work with you, Scott," said Heydrich, hugging Ourecky and then wiping a tear from his eye. "I know that you'll do well at Cambridge."

As Ourecky turned to walk out the door, Wolcott said, "Just hold fast out there in the reception area, pardner. Carson will follow shortly, and he'll help you pack up today and process out on Monday. He'll be with you every step of the way. Shut that door behind you, please."

Carson shot Tew and Wolcott an angry glance and said quietly, "Are you sure you had to do it that way? Now, Ourecky's going to go through life with this weighing on his conscience, thinking that he failed us at a critical moment. It's not right."

"This was *my* decision, Major," snapped Tew. "And I would caution you never to question my judgment again. Have a seat. We still have another matter to discuss with you."

Still tired from the last two days, Carson plopped down in a chair. "Sir, I'll do as you ask and get Ourecky on his way to California. I'll stick to him like glue. He won't know anything."

"That ain't what this is about, pardner," said Wolcott, lighting a cigarette.

"We need to review crew assignments," declared Tew bluntly. "Of the two crews in the bullpen, which do you consider best prepared—currently—to fly a mission?"

"Howard and Riddle, by far, sir," replied Carson. "No question in my mind."

"That's our assessment as well," noted Wolcott. He puffed out a large smoke ring, which hovered lazily in the air before impacting the tiled ceiling.

Tew scratched his chest through his shirt and said, "Carson, I'm going to tell you something, and you are not to discuss it with the other crews, until I instruct you otherwise."

Carson nodded. "Yes, sir."

"Our unmanned shot in February has been scrubbed. We're still launching in February, but it will be a *manned* shot. Granted, it's only going to be a practice mission against a simulated target, but we're sending up a crew. After *that* mission, we will send up operational missions at three-month intervals to intercept, inspect, and destroy hostile satellites."

"We're goin' to *orbit*, Carson," added Wolcott excitedly.

Carson took a moment to absorb the new information, realizing that he had finally earned his ticket to space, and then smiled. "Well, sir, that's excellent news."

"You are *not* goin' on the first mission, pardner," said Wolcott, as if he could divine Carson's thoughts. "You're far and away our best pilot, but I won't tear up another crew just so we can pair you up with a right-seater who may or may not be able to work with you. Besides, since this is just goin' to be a low-pressure practice run, we don't necessarily want to play our first string quarterback."

"But it's the *first* mission, sir."

"Granted," interjected Tew. "But it won't be *your* first mission, Major."

"Here's what we're offering, pardner, on a take-it-or-leave-it basis," said Wolcott. "No later than the end of the day, Mark and I will make the call on who flies first. We're leaning towards Howard and Riddle, but regardless of our pick, that first crew will have top priority on all training resources. We want you to train them. Now, there's the matter of the next mission, which will be our first *operational* mission. We intend to pair Ed Russo with you."

Carson cringed at the mention of Russo. Just the notion of being jammed into a spacecraft with him was enough to make Carson nauseous.

"Now, pardner, this next part will probably be the stickiest burr under your saddle. We've determined that the critical place to fill is the right-hand seat. With your in-depth working knowledge of the intercept process, it would be much easier to shift you to the right seat, and put Russo on the left side. Besides, Russo is senior to you."

"And what are my other options, sir?" asked Carson.

"None, if you want to fly the first operational mission," said Tew. "You fly with Russo or not at all. If you won't, we can leave you in a trainer position, working under Gunter Heydrich, until our next crop of pilots comes in. So you can either voluntarily choose to fly with Russo or you can ride the bench for a year while preparing the other crews to fly. Your choice, Major."

"And what if I elect to leave Blue Gemini, sir? What if I choose to resign my commission?"

Wolcott snuffed out his cigarette and said, "Carson, I don't know if you read all the fine print before you signed, but you don't have the *option* to resign your commission. You belong to us for at least the next five years. If you want to quit Blue Gemini, feel free to do so, but you best be mindful that you won't be heading to Southeast Asia, if that's what you envision happening when you leave here. No. In fact, Carson, you'll be renewin' your friendship with Major Agnew."

"And where is Major Agnew, sir?" asked Carson.

"Can't exactly tell you, hoss," replied Wolcott. "But I'll give you a hint. Before you board *that* plane, you'll be drawin' two sets of extreme cold weather gear, much like the stuff you used at Eielson. The second set is for when the first batch eventually wears out."

"I'm sorry, Carson, but you didn't leave us with a lot of options," said Tew. "You're an excellent pilot, but you're fortunate that you're not sitting in Leavenworth right now, given your childish antics. Assaulting a senior officer is a serious court martial offense. Be back in here at sixteen hundred, and let us know your decision. And if you breathe a word of this to Ourecky, then you won't have a decision to make, because I will make it for you."

Carson stood up, saluted Tew, and left. Wolcott chuckled and said, "Good thing I didn't mention that we've both assaulted plenty of senior officers over the course of our careers, huh?"

"You just had to remind me, Virgil, didn't you?"

Wright Arms Apartments, Dayton, Ohio; 6:20 p.m.

As he trudged up the stairs, Ourecky's head swirled with all the things that had to be accomplished. He wanted to ask Drew to join Bea and him for dinner sometime over the weekend, but Carson just seemed angry and distant most of the afternoon. Ourecky couldn't decide if Carson's ire was something that he had caused. *One minute and thirteen seconds*, he thought, *literally months of hard work undone in one minute and thirteen seconds.*

Bea met him at the door. "Let's go dancing, Scott," she said seductively, slipping her arms around his waist as she kissed him. "It's been a dreadful week. I just want to do something fun."

"It's been a horrible week for me, too," muttered Ourecky, sliding out of her grasp. He slipped out of his coat and hung it on a peg by the door. He shuffled toward the couch, took a seat, and patted the worn upholstery. "Bea, I have something to tell you, and you're not going to be very happy about it."

"Scott Ernst Ourecky!" she exclaimed, wagging her finger at him. "Don't you *dare* tell me you're going to flight school. You *promised*."

"Nothing like that, Bea," he responded. "Please, please sit down. Something happened at work today, while you were out shopping, and I had to go back to the base. General Tew wants me to go to California for a while. Bea, I'm really sorry, but I have to leave on Tuesday."

"California?" she asked, furrowing her brow and frowning. Then she smiled, ever so slightly, licked her upper lip, and said, "California… *California*. So, Scott, how long will you be there?"

"A few months. At least through the end of May."

"Where in California, darling?" she asked calmly, picking up a Delta Airlines in-flight magazine from the coffee table. She opened it and quickly flipped through the pages to a map in the back of the magazine, which depicted Delta domestic routes and cities in the United States.

"El Segundo. It's near Los Angeles. Right by the airport, in fact." He looked at her, surprised at her unruffled demeanor. "Bea, aren't you even the least bit upset? I'm going to California for a few months. I'm leaving on *Tuesday*."

Studying the map, she pursed her lips and smiled. "Oh, Scott, you *are* in the Air Force so it's not a shock that they would send you somewhere on temporary duty. At least you're not going overseas, to Vietnam or Korea or someplace horrible like that. California is *nice*. We have a route to LA from Atlanta, so I can just catch a flight over, spend the weekend with you, and then come back in time to fly on Sunday afternoon. It'll be great, Scott. I'm looking forward to it."

Stunned by her response, Ourecky strained to make sense of what was happening. Maybe it was just because he was dog-tired, but everyone seemed at least one step ahead of him, or that there was something that they all shared that they weren't willing to share with him. He was sure that Bea would be distraught with his news, at least momentarily, but she was taking it all in stride. In fact, she was beyond being calm and collected; she seemed downright *happy*.

"That's not all, Bea. After I go to California, I won't come back here."

"*Really?* Where would you go?" asked Bea. She sipped some red wine from a glass she had left on the coffee table and then handed the glass to him.

He finished the wine. "Well, the Air Force wants to send me to MIT for my doctorate."

"Really? That's *great*, Scott. That's what you wanted to do after you left the Air Force, and now the Air Force is going to pay for it? That's excellent."

As tired as he was, something she said just didn't mesh. And then it came to him. "How would you know that the Air Force is paying for it?" he asked, setting the glass on the end table.

She looked at him for a moment, smiled hesitantly, and then answered, "Well, I just *assumed* that they would pay for it, dear, if they're sending you back to school. And I'm sure that they'll want something for their investment, won't they?"

"Yeah. I'll have to stay in for three more years."

"That's worth it, Scott. More than worth it."

"So do you think you might consider going to MIT with me? I know we haven't set a date, but maybe we could get married after I come back from California."

"That sounds like a plan," she replied. "A *good* plan."

"And you don't mind going to Massachusetts?" he asked.

"No, Scott. I don't mind at all. I just want to be with you, wherever you are."

Contingency Stocks Commodity Warehouse # 2
Wright-Patterson Air Force Base, Ohio
9:45 p.m., Friday, January 24, 1969

"Wow, man, cool," said Yost, grinning as he jingled the ring of keys around his finger. Providence had arrived; he just could not believe his good fortune. "Are you sure?"

"I am," asserted his supervisor, Master Sergeant Kroll, a large man of Estonian descent. "Really, Yost, you would be doing me a *huge* favor. I know you've fallen on some hard times lately, what with Gretchen leaving you and cleaning out your bank account and all, so I guess I would be doing you a favor, too, so let's just consider it a mutually beneficial arrangement."

Yost was so overjoyed that he was almost moved to tears. It was almost too much to ask for, a heaven-sent opportunity to reverse his present circumstances. "Okay, let's go over it one more time, just to make sure that I've got it all down. I don't want to let you down, Bob."

"It's simple. I need you to feed the cat and pay the bills while I'm in Thailand. Anna will kill me if anything happens to that damned cat. Don't let it out, no matter how much it cries."

Kroll continued. "There's a shoebox with pre-addressed envelopes for all the bills, along with a check for each bill for every month. The rent will stay the same; the other bills might fluctuate a little. You're welcome to stay over as often as you want, but make sure you turn down the thermostat and turn off the lights when you leave or my electric bill will be sky high. You're welcome to anything edible in the refrigerator or cabinets; any chow after that is on your dime. Got it?"

"I think I got it," replied Yost. "And you'll be TDY at Udorn for six months?"

"At least. And Anna will be at her folks' place in Oregon until I get back. Keep an eye on the mail; if you see anything that looks important, forward it to her. The address is taped up by the telephone. Okay, Eric, what's the most *important* thing to remember?"

"Feed the cat."

"Correct," said Kroll. "And the Mustang is yours to drive until I get back, but I would really appreciate it if you kept it to a minimum. Premium gas only. It burns oil, so expect to put a quart in every week, unless you want to spring for replacing the rings. Rinse it off at least once a week, or more often if it starts building up a salt film. Please take good care of it."

"And you're paying me fifty bucks a month on top of letting me stay in your place?"

"Yeah," replied Kroll. "Thanks. I almost forgot about that." He reached into his pocket and pulled out the thick wad of cash that had been disbursed this afternoon as an advance on his TDY pay. Peeling off five twenties, he added, "Here's the first two months in advance."

Wiping a tear from his eye, Yost said, "I can't tell you how much this means to me, Dan."

"Well, you should know that it's on the third floor and there's no elevator," replied Kroll, glancing down at the heavy cast that encased Yost's right leg below the knee. "I hope you feel the same way after you've been up and down the stairs a few hundred times."

Apex Minerals Exploration Inc., Dayton, Ohio
2:15 p.m., Monday, January 27, 1969

"Here," said Grau, handing Henson an envelope. "Plane tickets and instructions. I need you to go back to Gabon. You'll leave in the morning."

"Already?" asked Henson. He had just been to West Africa in November, immediately after matriculating from Grau's training course. "I thought that the earliest I would head back there would be in June. So I'm not scouting Brazil this month?"

"Not now, Matthew. We've received instructions to activate the staging site in Gabon for the last week in February. I want you to head back over there to complete the preparations."

"Last week in February? Why?"

"We don't ask *why*, Matthew. We just do." Grau reached into his desk and pulled out an old Florsheim shoebox. "Here are your operational funds. Same accounting rules as always."

Henson opened the box and pulled out a rubber-banded packet of 500-franc banknotes. The bills looked spanking fresh from the printing presses of the Banque de France. The grim face of famed mathematician-philosopher Blaise Pascal peered out from the front of every note.

"Any questions, Matthew?"

"No, sir." Henson stood up to shake his hand and glanced at the picture of Grau, his African wife, and baby in Paris. "So did you talk?" he asked. "I think I probably would have."

"What?" asked Grau, closing his desk drawer.

"Your eye," said Henson, gesturing toward the picture. "When they tortured you."

"What? My eye? Hah! So when did I tell you that I lost my eye when I was tortured?"

"Well, I just thought…"

"You thought wrong. If you must know, I lost my eye at the hands of my wife's family."

"You're kidding, right?"

"Not in the least," said Grau, self-consciously touching his black eye patch. "Her family didn't take too kindly to her becoming involved with a white man. Her brothers caught us together in my room one night. They took me out to the street and beat me. My eye ruptured, and it became infected later, and then I lost it. I suppose I'm lucky, though, because they probably would have beat me to death if Aminata had not stopped them."

"So after that, her family accepted you and then you were married?"

"No. We were *already* married. We had wed in secret a month prior to that. They stopped beating me because Aminata told them so and begged them to stop. Of course, they probably would have finished the job if they had known she was pregnant."

Grau continued. "Anyway, I lost my posting in Senegal and ended up in a hospital in Paris. By the way, French medicine leaves much to be desired. Maintaining sterile conditions is not their strong suit, not by a long shot."

"So this picture was taken after you got out of the hospital?"

"Yes, but not for my eye," replied Grau. He held up his left hand and wiggled the stump of his ring finger. "This was much later, after the Russians cut off the first knuckle of this finger and osteomyelitis set in. I ended up losing the whole digit. I was fortunate I didn't lose the whole hand. French medicine again."

Henson cringed. "So, Mr. Grau, what happened to your wife? Aminata?"

Grau closed his eye as a tear formed and ran down his cheek. "She died in childbirth, exactly two years after that picture was made. Our son died with her."

"I'm sorry, Mr. Grau. I shouldn't have asked."

A quiet moment passed, and then Grau opened his eye and looked at Henson. "No. It's painful, but I appreciate that you caused me to remember them. So I guess you're still curious about the baby girl in the picture. My daughter, Adja."

"Adja?"

"Yes. My brother and sister-in-law raised her in California since I was out of the country most of her life. She's a sophomore at Kent State. I would like you to meet her sometime, Matthew, but I should warn you: she's not a bit warm and cuddly like me."

"Well, that's certainly disappointing, Mr. Grau," replied Henson. "But I would like to meet her someday. Well, I suppose I have a plane to catch tomorrow, so I had better go pack."

"Bon voyage," replied Grau. "And be careful in Gabon, Matthew. Keep your wits about you."

James M. Cox Municipal Airport, Dayton, Ohio
7:30 a.m., Tuesday, January 28, 1969

Standing at the gate, Ourecky checked his tickets. He heard the boarding call and turned to face Carson. "I suppose this is it, Drew," he said. "Any idea where your next assignment is?"

Carson frowned. "Not exactly. We're not immediately closing down the office, Scott. There's still paperwork to do. I'll be at Wright-Patt for at least a couple of months, maybe even longer."

"Any chance that you'll make it to Vietnam?" asked Ourecky.

"I doubt it. Ol' Cowboy Virg made it abundantly clear that I won't be flying in combat anytime soon, if ever. I guess I had better just come to grips with that."

"Well, I would think they could make allowances for you. Look, Drew, I have something to ask you. Bea and I are planning to get married in June, in Nebraska. Our plans aren't absolutely firm yet, but I would really like for you to be my best man."

Carson laughed. "Best man? I'm honored, but have you discussed that idea with Bea? I wouldn't think she would be too enthused about it."

"I've talked to her about it. She's warming to the idea. Why don't you think about it, Drew?"

"I'll think about it, brother. It's really going to depend on where I'm at, though. It's kind of hard to see that far into the future, considering the circumstances."

The loudspeaker blared, announcing the last boarding call. "I guess I had better head out."

"Yeah." Carson started to shake Ourecky's hand and then awkwardly hugged his shoulders. "It's been an honor, Scott. Really. I would be happy to fly with you anytime. Sorry we had to do all of our flying in the Box, though."

"Not all of it," said Ourecky. A gate attendant waved frantically at him.

"We did have our moments, didn't we?" said Carson, laughing. "Take care of yourself, Scott Ourecky. Tell Bea I said hello. She is coming out this weekend, right?"

"Right. I'll tell her. Take care of yourself, Drew. We'll see each other again, soon enough."

"Yeah we will. Soon enough."

30

FEBRUARY

Simulator Facility, Aerospace Support Project
11:35 a.m., Wednesday, January 29, 1969

An additional procedures simulator had arrived from NASA, which required that the hangar door be opened so that Heydrich's technicians could ferry in the pieces from the three tractor-trailer trucks parked outside. A small forklift was used to convey most of the larger apparatus, while a steady stream of workers wheeled in cabinets, consoles, and other smaller components.

Despite all the bustling activity associated with the new arrival, a simulated mission was underway, with Crew Two preparing for next month's launch. With the wintery weather blowing in through the gaping entrance, the temperature inside the hangar hovered in the low teens.

Normally, the hangar's interior was all but sweltering, since the poorly ventilated structure retained the heat collectively generated by several large computers and a multitude of various electronics. Heydrich's men usually worked in shirtsleeves, but today they were clad in cold weather gear—bulky Arctic parkas, wool watch caps and gloves—on loan from the supply room.

When they weren't lugging gear or engaged in the ongoing simulation, the technicians clustered around several strategically placed electric space heaters. Hot coffee and cocoa were in great demand, and the usual stockpile of glazed doughnuts was exhausted in short order.

Pondering whether it was sufficiently cold to induce frostbite, Carson rubbed his tingling nose. This was the first time that he had worked on this side of the magic curtain, and he wasn't fond of it. Although it was cramped and miserable inside the Box, he had been constantly busy in there and—moreover—had been the center of attention. Right now, in his role as CAPCOM, he was just another cog in Gunter Heydrich's diabolical machine. Granted, he was a cog of slightly more importance than most, but a cog nonetheless.

Two of Carson's flight school classmates had been killed in Vietnam yesterday. The shoot-downs were so frequent now that he wondered how many months would elapse before his class was winnowed down to nothing. Of course, even when all of his classmates were sleeping in Arlington or cemeteries elsewhere, he would remain, since he could not be killed in combat, and it was unlikely that he would die atop a rocket. So he was stuck with the unenviable task of calling the wives and girlfriends of the dead and captured to offer his condolences. As uncomfortable as he was in attempting to assuage their grief, he most dreaded the inevitable moment when they asked when he would finally be going overseas himself.

As for his present circumstances, things were as bad as they possibly could be. Russo had received his promotion orders for lieutenant colonel yesterday, so the issue of seniority was no longer in question. To more effectively cement them together as a team, Wolcott had decreed that they would work together in the Simulator Facility, training Crew Two—Tom Howard and Pete Riddle—to fly Blue Gemini's first mission.

This morning, the insufferable Russo had blasted into the facility like the bitterly cold gusts now pouring in from the windswept parking lot. An unusually strange tension had set in between Russo and Heydrich. After last week's melodramatic episode, Heydrich was reluctant to be seen as provoking any form of confrontation, so he wandered around the hangar like a gun-shy pointer hound, afraid to openly challenge Russo on any issue. Russo, exploiting the vacuum, swiftly sought to establish dominance, sticking his nose everywhere that it didn't belong, and dictating things that he had no business dictating.

As the CAPCOM, Carson was obligated to remain at his station for the duration of his twelve-hour shift, unless a suitably qualified replacement could stand in for him. Presently, the only qualified replacement was Russo, and Carson was not about to subject his friends in the Box to even a second's worth of Russo's pestering voice, regardless of the circumstances.

Since there were no unfilled electric sockets anywhere near his console, the glowing red warmth of the nearest space heater was over fifty feet away, so Carson sat at his console, headset clamped on his skull, encased in a parka, with teeth chattering. Although his bladder felt ready to explode and he desperately wanted a hot cup of coffee, he was determined to keep at his station. At least the guys in the Box were warm right now, and they were allotted a latrine break every six hours, whether they needed it or not.

Deep in his control panel, a failing capacitor buzzed persistently, like a pesky fly that needed swatting. As he shivered, Carson looked to his left; Heydrich sat motionless at the next station, with his head and face hidden by the hood of his ancient but heavily insulated Luftwaffe parka. Only an occasional cloud of steamy breath, puffing from the luxuriant wolf fur that lined the hood, lent any tangible evidence that he was still animate.

To Carson's right sat Russo, who had talked almost non-stop for the past two hours. Paging through the "glitch book," a three-ring binder that contained the multitudes of assorted problems that could theoretically befall the Gemini-I as it orbited the Earth, Russo relentlessly called out malfunction after malfunction. Each simulated discrepancy resulted in the simulation crew twisting the knobs or throwing the switches that caused displays and instruments to subtly change within the Box, which in turn required immediate and diligent action on behalf of the two-man crew.

"Scenario item number 674," chimed Russo cheerily, like he was reading the name of a tow-haired tyke from Santa's "nice" list. "*Computer start circuit failure.*"

The controllers anxiously looked to Heydrich for guidance. Although he was supposed to set the tone and tempo for the entire exercise, Heydrich had the authority to veto Russo at every turn, but he said nothing. A silent puff of mist through wolf fur signaled his tacit consent.

Knowing that Russo could not see them, the bewildered workers shook their heads as they tweaked their dials and threw their switches. Seconds later, through the intercom speaker, they overheard Howard and Riddle curse as they grappled with the error.

"And now item 312," said Russo, placing a gloved finger on yet another discrepancy. "*Continuous yaw-left condition due to excessive steam venting.*"

"They're still working 674," observed Carson, nudging Heydrich's elbow.

"That's my point," replied Russo. "Item 312. *Continuous yaw-left condition. Now!*"

Minutes passed before a technician in the second row reported that the crew had successfully addressed both glitches. "Then hit them with item 1425, *Primary horizon sensor failure*," ordered Russo. "Then wait two minutes and throw them item 551, *OAMS helium pressure indication zero.*" He stood up from his console and yawned. "I'm going to grab some coffee and then go outside to stretch my legs for a minute or two. Can you cover for me, Gunt?"

A wisp of vapor puffed out of Heydrich's fur-lined hood. He was silent even as the frustrated technicians behind him scurried to comply with Russo's directives. As Russo walked away, with the heels of his insulated boots clicking on the concrete floor, Carson heard Heydrich's muffled voice, softer than even the quietest whisper, counting his steps: "*...acht, neun, zehn, elf, zwölf, dreizehn...*" Then the engineer held up a finger and asked, "He's by the coffee urn, *ja?*"

"Right," replied Carson.

Heydrich slowly flipped back his parka hood, adjusted his glasses, and smoothed his awry mop of black hair. "*Gunt!*" he cursed. "Where does that *verdammter dummkopf* come off thinking that he can call me *Gunt?*"

"Gunter, why on earth don't you say anything? Why are you letting him do this?"

"What good will come of challenging him?" responded Heydrich. "If he wants to believe that he walks on water, then let him, at least until he sinks in over his damned head. Maybe then Virgil will come to his senses and yank him out of here!"

"I guess you're right, but I don't know how much more of this I can stand," noted Carson. "Spending the next few years at some Arctic outpost is really starting to look like an appealing option. At least I'm dressed for it."

"*Ja,*" muttered Heydrich, nodding his head. "At least you have that option, Drew, but I don't. But I'll tell you this: if you suddenly hear a thunderclap, smell sulfur, and see that imp burst into flame, you'll know that Satan has granted my wish. I may spend eternity in hell afterwards, but at least I'll get some work done today."

A green light blinked on Carson's console; it was an indicator for the private intercom loop, a subtle signal that the men inside the Box wanted to discuss an administrative matter. Simultaneously, in his earphone, Carson heard Howard's voice. "Tom, I'm on Loop Four. Can you switch to the same so we can chat?"

"Give me a second, Tom," replied Carson. Pretending to stretch, he stood up and looked around. He glimpsed Russo standing at the hangar's open entrance, sipping coffee as he watched some of Gunter's technicians unloading one of the flatbed trailers. Waving his arms in sweeping gestures, as if to tell them the optimal way to carry each console and computer cabinet, Russo seemed intent on infringing on their business as well. "Okay, coast is clear. I'm on Loop Four. What's on your mind?"

"Drew, we're working this intercept scenario, but you guys keep chucking glitches at us like we don't already have enough crap to keep us busy. Between you and me, I think we have a pretty damned tight handle on glitch management. What we really need to practice is working together as a team to fly the intercept, and this damned malarkey isn't helping that process."

"Tom, I'll see what I can do." Carson looked at a sheet of paper on which he had tallied the number of malfunctions. Since the simulation had begun over three hours ago, Russo had thrown over a hundred different anomalies at the crew. Howard had every right to be infuriated.

"Yeah, Drew, you *do* that," groused Howard angrily. "If this were a new car and I had that many damned things go sour within a couple of hours of pulling out of the lot, I would just hang a lemon on it and park it in front of the dealership."

Raising his voice, Howard continued. "At this rate, I half-expect to look out the window and see a gremlin eating a doughnut out there. Realistically, if we go to orbit and start experiencing this many glitches, I'm calling for the first available Recovery Zone and then I'm lighting the retros. This is not training. This is just asinine harassment, pure and simple. Can you not ask Gunter to lighten up just a little bit? Just what the hell is he trying to prove?"

"It's not Gunter, Tom. He's sitting right here beside me, and he's just as unhappy as you are," replied Carson, glancing around to make sure that Russo had not drifted back into earshot. "It's your buddy Russo. He's calling the shots out here. *That's* the problem."

"Well, do something and do it quick," pleaded Howard. "Or I'm climbing out of here to handle it myself, and I'm not afraid to draw blood. If Virgil doesn't like that, then maybe he can fly this first damned mission himself."

2:10 p.m.

"Tarnation, it's freezin' in here," observed Wolcott. He looked over Carson's shoulder and noted that the next scheduled contact window wasn't for another fifteen minutes. "Son, don't you have the common sense to wangle a dadburned heater over here?"

Hearing the familiar voice, Heydrich flipped down the hood of his parka. "Virgil, what brings you here on this miserable day?"

"Just paying a visit. I thought I would mosey over and chew the fat. Got a minute to chat?"

"Of course," replied Heydrich, removing his headset. "We would probably be more comfortable in the locker room or the suit-up area." He turned to Russo and said, "Colonel, can you handle things while Virgil and I talk?"

"I've got it covered, Gunter," replied Russo, still intently studying the glitch book.

"*No*, pardner," said Wolcott. "I need to talk with you as well. Carson, you too. Gunter, shuffle some of your troops to cover these chairs. We won't be gone too long."

As the four men walked across the hangar, Wolcott paused to light a cigarette and asked "How's this new Box lookin', Gunter? How long until you have it up and running?"

"Some key pieces are missing," answered Heydrich. "If need be, I'll send some of my guys down to Cape Canaveral to track them down. Even then, it will take a few months to shake out all the bugs. These things are very complex monsters, Virgil. They're just not meant to be moved around. Instead of trying to get this one completely installed and operational, it would probably be more practical for us to cannibalize it as parts fail on the existing simulator. One of my original DDP-224's is on the verge of failing, so it would be nice to have a spare."

"DDP-224? What's that, pardner?"

Heydrich launched into a sneezing fit. Stopping to pour a cup of coffee, Carson looked over his shoulder and interceded for him. "A DDP-224 is a digital mainframe computer, Virg. Three of them run most of the functions on the Box."

They passed through the locker room on the way to the suit-up area. One wall was dominated by a row of twelve gray metal lockers. Seven were marked with strips of masking tape, each bearing a name: Carson, Agnew, Jackson, Sigler, Howard, Riddle and Ourecky. Two suit technicians played cards. Looking at Wolcott, one said, "You shouldn't smoke in here, Virg."

"You're right, Joel," replied Wolcott. He licked his fingers and then snuffed out the cigarette before discarding it in a wastebasket. "You boys mind lendin' us a bit of privacy?"

As the technicians left, Wolcott sat down and said, "I won't beat around the bush, because I know that you're busy. Carson has informed me that we have a significant problem over here."

"There is *no* problem, sir," stated Russo, sitting stiffly upright in one of the two recliners. "Major Carson obviously misinformed you."

Carson spoke up. "Sir, I didn't intend…"

"That will be *enough*, Major," snapped Russo.

"No, buster, *that* will be enough," countered Wolcott, removing his hat. He strummed his fingers on the Stetson's broad white brim. "You need to hush in the saddle until I let you know when to speak. You understand that, pardner? Are we square?"

Russo nodded sheepishly. "Yes, sir."

"Now, like I said, there's obviously a problem here, and I appreciate Carson bringin' it to my attention, particularly since it appears that much of it is my doin'. It looks like you and Gunter have somehow got sideways, and we're going to sort it out before we walk out of here. To get things out on the table, I want Carson here to describe what he related to me. Carson?"

Carson looked as if he had been unexpectedly called to testify at a high profile murder trial. "Well, Virgil," he said quietly. "I suppose it's mostly a question of training philosophy. When I was in the Box with Agnew and then Ourecky, Gunter did an excellent job of covering just about every contingency, and he made sure to hammer them home if he saw that we had a weak response, but he didn't overdo it. He placed more emphasis on making sure that we practiced the routines and procedures necessary to accomplish the mission."

"Okay, pard," said Wolcott, holding up his index finger. He reached into the pocket of his denim barn coat, pulled out his packet of Red Man, and stuffed a sizeable lump into his mouth. "And how does that differ from Ed's approach?"

Carson glanced quickly at Russo and then looked away. "He throws the crew one curve ball after another. I know that stress is part of the game, but I think that he's overdoing it."

"Duly noted," said Wolcott. "So, Carson, on which side of the fence do you land?

"I prefer Gunter's approach. In my opinion, having to respond to a constant string of malfunctions detracts from training. I think that the crew can become so conditioned to things going wrong that they can forget what to do when things go right."

"It *doesn't* matter if they can correctly perform a maneuver if they have a stuck thruster in the middle of executing it," Russo snapped angrily. "They have to be able to work under stress."

"I think that Carson still has the floor, Ed," interjected Wolcott. "At least until I say otherwise, and I'm pretty danged sure that I'm still the head honcho here. Carson, anything to add?"

"Yeah. Russo also doesn't throw the glitches in a logical sequence. It's all sort of erratic and scatterbrained, and that's a big issue. Gunter normally worked us through some sort of failure chain, where if we didn't detect and resolve the first issue quickly enough, then it cascaded to another related issue and so on. You have to understand that Tom Howard could probably recite the glitch book from memory at this point, so I'm sure that's he's not happy with things being chucked at him with no rhyme or reason."

"Anything else, pard?" asked Wolcott.

"One thing. I think that the strongest argument for Gunter's training methods is just how well Ourecky was trained. By the time he left here, he was as competent as *any* pilot on the roster."

"But he's *not* a pilot," sniffed Russo.

"Yeah, he's *not*," snapped Carson. "But if I had to pick between Ourecky or *you*, I would leave you on the ground. In an instant. As it is, you don't even rate sitting in that chair of his."

"Enough, children, enough!" growled Wolcott. He spat tobacco juice in a wastebasket. "Okay, gents, we're on a short fuse and it's burnin' fast. We don't need to be continually locked in an arm wrestling contest on who's the foreman in this shop. I'm going to ask you gentlemen one question, and one question only, and I'll make my decision on the basis of your answers."

"But just so it's absolutely clear: once I decide who will run this show, it's a done deal. I don't want to hear about any grousin' or grumblin' over here, and if I hear the slightest peep about one of you cowpokes takin' a swing at the other out behind the bunkhouse, I'm marchin' back over with my bullwhip and six-shooter, and that'll be the dadblamed end of it. Savvy?

The three men nodded solemnly.

"Okay," said Wolcott. "You first, Russo. If I put you in charge of this show, can you have those boys prepared to fly by the 25th?"

"Without question, General," replied Russo without hesitation. "I am absolutely confident that they will be ready, and that we will be able to launch on schedule."

"Gunter," said Wolcott, leveling his gaze at the German engineer. "Same question."

Heydrich looked up at the ceiling and fidgeted with his hands. "Virgil, in all sincerity, I am *fairly* sure that we can have them ready, but I'm not *absolutely* sure. Granted, it's a relatively simple mission, all things considered, and it's mostly canned, but you're asking us to accomplish a momentous task in a very short period of time."

"Can you do it or not, pardner?"

Removing his back-framed glasses, Heydrich stared Wolcott dead in the eye and answered, "I'm not sure, Virg, but we will do our utmost to see that they're ready."

Rolling his eyes and smirking, Russo covered his mouth with his hand.

"Well now, hombre," said Wolcott, looking toward Carson. "Quite a danged disparity we have here, so it falls to you to present the argument that will swing my decision. Assumin' that you're going to ride for whomever I choose to lead this drive, what say you, hoss?"

Carson took a sip of his coffee, glanced first at Russo and then at Heydrich, and then turned to face Wolcott. "Sir, I'm in agreement with Gunter. Honestly, I don't know if we can have them completely ready in time, but we'll damned sure do our best."

Like the Hindu god Shiva pondering the fate of the world, Wolcott was meditatively silent. He closed his eyes for a few moments, opened them, and observed, "Well howdy, ain't this confoundin'? I feel like I'm a big ranch owner at the rodeo, trying to decide which of my top hands is going to try for eight on the prize bull."

Wolcott continued. "On the one hand, I have a rider who's been on buckin' bulls all his life, and he's tellin' me that he ain't entirely positive he'll be able to stay on, but he'll do his damndest to try. On the other hand, I've got a dude rider chock full of gumption. He barely knows what a bull's ass smells like, but he insists that he can clamp on for the duration."

Listening to Wolcott, Heydrich frowned, Russo grinned, and Carson—silently awaiting his fate—showed no expression at all.

"Gunter," declared Wolcott flatly. "It's been your show and it will remain as such. Russo, Carson, so long as you're under this roof, you will take your cues from Herr Heydrich. *No* squabbling, *no* bickering, and *no* hesitation. Savvy?

Elated, Carson breathed a sigh of relief and nodded his head.

"Yes, sir," muttered Russo.

As the men left the room and went back to the hangar, Wolcott leaned toward Carson and asked, "So, pardner, it's a lot different to be workin' *outside* the Box, isn't it? Have you gleaned anything of value from this experience?"

"Yes, sir, I have indeed. I sure do have a much greater appreciation for Gunter and his guys now. This is not a fun job."

Forest Park Apartments, Dayton, Ohio
3:30 p.m., Sunday, February 2, 1969

Yost's right foot was still encased in a cumbersome cast. Back in December, the loan shark's goons had managed to break virtually all of the miniscule bones that comprised his right ankle and foot. He would be in the cast for at least another month, perhaps longer. Sighing, he used a length of coat hanger wire to scratch underneath the plaster shell.

He had finished composing the letter over an hour ago. He read it again for the tenth time, to be absolutely sure that the text said precisely what he wanted to say without revealing too much. He flipped through his stack of pictures and selected four snapshots that should serve to spark the reader's curiosity.

His plan was simple: He would make several copies of the letter by hand, and have multiple copies printed of the pictures he had chosen, and then he would send them to several major news magazines and newspapers tomorrow. He anticipated that he should get a bite in relatively short order, probably before the week was out, and then his pressing financial problems would be behind him.

Marvin Gaye sang "I Heard It Through the Grapevine" from the radio in the kitchen. Yost heard loud mewing from the bathroom. He had only been here slightly more than a week, but he was already tired of the noisy and perpetually unhappy cat, and wasn't sure if he could make it through the next few months without killing it. If nothing else, the ungrateful feline might "accidently" slip out the door, where it would undoubtedly be mauled by the small pack of stray dogs that frequented the parking lot. Not having to buy cat food and kitty litter would save him a couple of bucks a week, and he wouldn't have to tell Kroll until he had returned from Thailand.

He guzzled the dregs of a Schlitz and rummaged in the refrigerator for another bottle, but realized he had drunk the entire case. It was just as well; beer wasn't his beverage of choice, but since they were in the fridge, he felt obligated to polish them off. After all, that's exactly what Kroll had asked him to do. He walked back to the bedroom and collected his laundry basket.

Yost was still coming to grips with his astoundingly good fortune. It was like he was a snake who had shed an uncomfortable and restrictive skin, emerging as an entirely different creature. He somehow had dodged the loan shark's enforcers since his last run-in, and he was fairly certain that they had not become attuned to his new routine and environs.

Even the simplest things, like sleeping in a real bed with real sheets, were indescribable pleasures. He had taken so many hot baths that his flesh might be in danger of permanently wrinkling, but he still felt as if he hadn't cleansed away the uncomfortable grime of his recent past. Of course, it was a pain to keep his cast dry, but it was worth it.

And now he was going to experience the inexplicably simple joy of washing his clothes without having to constantly look over his shoulder at the Laundromat. He dropped a box of Rinso laundry detergent into his laundry basket and headed out the door.

Balancing the full basket against his right hip as he pulled the door closed, he heard a clicking noise. Instinctively throwing up a hand to guard his face, he furtively glanced to his left. Fully expecting to see a thug with a gun, he was pleasantly surprised instead. Four doors down, an attractive blonde was leaving her apartment. *Things just keep getting better and better*, he thought, *I must be living right.*

He strolled in her direction, and asked, "There's supposed to be a laundry room in this building. Do you know where it is?"

"In the basement," she replied, looking over her shoulder as she locked her deadbolt. "Fourth door on the right, past all those storage rooms. By the way, don't waste your dimes on the last dryer. It spins your clothes around, but it doesn't ever heat up." She glanced down at his cast, "So what happened to your foot?"

"Slipped on the ice."

"You're new. Did you just move in?" she asked, putting her apartment keys in her purse and slipping on some suede gloves. A page of newsprint fluttered by, carried by a frigid gust of wind. "I wasn't aware that there were any vacancies."

"I'm kind of house-sitting for a friend of mine. I'm in 38," said Yost.

"Married couple with a loud cat? No kids?"

"That's right. The Krolls. He's going to be away for a few months and she's staying with her mother in Oregon."

"Oh, yeah," she said. "I've seen them. I'll tell you, I could do without that cat's screeching. It keeps me up some nights. I'm surprised that the super lets them keep it, with all the complaints that people make."

"Sorry. I didn't realize that it was such a nuisance." Yost smiled at her, thinking that he would definitely do something to resolve the cat situation, sooner rather than later.

"So where do you work?"

"I'm in the Air Force," he said, balancing the wicker basket as he fumbled in his pockets for coins. "I'm on the base."

"Really? What do you do?"

"Uh, it's sort of classified. I could tell you, but then I would be in a lot of trouble."

"Pilot, huh?"

"Could be," he replied, grinning. *Pilot?* He thought about piloting his squeaky-wheeled forklift around the poorly lit and drafty warehouse. Captivated by her polite smile, he wished that he had thought to brush his teeth and put on some more presentable clothes.

"Well, look, I've enjoyed talking to you, but I'm running a bit late."

"Sorry I held you up," he answered. "Maybe we'll see each other again sometime."

"Maybe," she said, turning toward the stairwell. Her heels quietly clicked on the concrete as she walked away.

Ogooué-ivindo Province, Gabon
8:05 a.m., Friday, February 14, 1969

As they drove northeast along National Route N4, past thatch-roofed shanties and undernourished livestock, Henson was reminded that while Gabon was a prosperous nation, at least in comparison to most other

African countries, that affluence had not yet dwindled down to all the masses, many of whom still lived in destitution and squalor.

Flipping down the heat-cracked visor, Henson was relatively sure that his driver thought he was crazy, and today's junket probably reinforced that notion. He had hired the driver—Georges Essangui—and his truck two weeks ago, not long after he'd arrived in Gabon. For today's chores, Georges' cousin Aymar had tagged along to help with manual labor; Henson looked to the back of the truck, and saw Aymar, sound asleep, reclining atop the six wooden boxes that they had loaded earlier at his warehouse. Henson was especially fond of the Essangui cousins; they were hard workers and knew how to be discreet, so long as they were marginally sober.

The Renault truck—a durable GLR 8 Berliet diesel built in 1953—could readily haul eight tons, which was ideal for Henson's purposes. Georges was delighted that Henson took such a keen interest in ensuring that his truck was kept in tiptop shape, even to the extent that he was willing to make a fairly substantial investment—by Gabonese standards—for long overdue engine repairs, transmission work, and new tires. Henson's only stipulation was that the truck be available on short notice, with no questions asked.

"*Dis-moi*," said Georges, casually steering with one hand as he held out a small burlap sack with the other. "So tell me, *chef*, we drive around all over the countryside and hand out these bags. Then we go to the marketplace tomorrow morning and wait for people to bring them to us full of dirt and rocks, and you pay them for them. *Correct?*"

"*Exacte*," replied Henson. "But we only give the bags to people who legally own their property. If they don't have a deed to the property, then their dirt is of no value to us."

"*Insensé.*" Georges shook his head and rolled his eyes. "*Tres insensé.*"

Mayela, Gabon; 4:30 p.m.

After distributing sample bags most of the afternoon, they arrived at their campsite, a parcel of land owned by a farmer named Claude Biko. An old dirt airstrip, abandoned by the French when they left Gabon in 1960, ran through the center of Biko's property.

Back in November, Henson had tentatively surveyed the strip as a possible emergency landing site, as well as a potential jumping-off spot to accommodate rescue operations in adjacent countries. It was suitably remote, plenty long enough to accommodate a DC-3 or Twin Otter, and could probably serve as a field for a C-130, if some simple improvements were made.

Fortunately for Henson, the strip's hardpan surface was in good condition and free of any vegetation that would otherwise need to be cleared. Unfortunately for Biko, the French had apparently saturated the parcel with a potent defoliant to curtail weed growth. While Biko had paid only a seeming pittance for his land, he was unable to cultivate any crops of value, so he barely kept up with his mortgage payments. After once thinking he was a shrewd bargainer, Biko had since become the laughingstock of the district. Henson felt sorry for him, since the French had obviously neglected to mention that they had permanently poisoned the land.

"Here we are," said Georges, cutting off the engine and lighting an unfiltered Gitanes Brune cigarette. "Home sweet home."

Henson stepped out of the cab and looked around. "Georges, you and Aymar unload those boxes and then find us some firewood for the evening, *s'il vous plait.*"

"*Merde*, those boxes are heavy! We just about broke our backs loading them on the truck. Why can't we just leave them up there tonight?"

"I don't like them weighing on the springs," answered Henson.

Georges laughed. "You forget it's *my* truck. They are not going to hurt *my* springs."

"Please just do as I ask, Georges. Do you remember that I'm paying you extra?"

"As you wish, *chef*," replied Georges, using a faded red bandana to wipe dust from his broad face. He and Aymar could be twins; both in their late twenties, they were a slight shade lighter and a couple of inches taller than Henson, with their dense hair cut close to their scalps. "So what's in those crates, anyway?"

"My surveying tools and some ore comparison samples." Henson heard voices and looked up to see Biko and his family approaching. "Georges, can you translate for me, *s'il vous plait?*"

Georges met them by the truck, listened to their request and then translated from an obscure Bantu dialect into French. "Biko says that he would be honored if you would join him for dinner."

Henson looked at Biko and smiled hesitantly. Biko was small and gaunt, almost frail; his threadbare gray trousers were held up with a frayed piece of hemp rope. His emaciated wife, barely clothed in a tattered blue cotton dress, and six naked children looked like stick figures in an obscenely realistic cartoon.

Henson knew that if he dined with Biko, his family would probably go hungry tonight, and possibly for a day or two as well. But it was a tremendous insult to refuse such a heartfelt invitation, and he thought for some way to handle the dilemma without hurting Biko's feelings.

Then an idea came to him, and he said, "Tell him that I would be honored to share a meal, but I must confess that I crave chicken, like my mother cooks back home. I had intended to buy one in the market and roast it over our campfire. Would it be too much to ask that his wife cook my chicken tonight? And also, since I miss my family so, could his wife and children join us also?"

As Georges translated, Henson glanced at the Biko's spindly wife; trying hard not to smile outright, she was clearly salivating at the tantalizing notion of fresh chicken for supper.

The driver listened to the Biko's reply, and relayed it to Henson. "He said that chicken would be very nice, and that his wife would be *honored* to cook it for you."

Covering his heart with his hand, Henson nodded at Biko and smiled. "*Tres bon. Merci.*"

As Biko and his family padded away on bare feet, Henson peeled off francs and put them in Georges' hand. "I'll wait here," he said. "Go to the market and buy the two biggest chickens that you can find. And some vegetables: plantains, yams, avocadoes, whatever they have today."

Grinning, Georges smiled broadly and nodded his head.

"And since I know that you're going to do it anyway and try to hide it from me, buy Aymar and yourself a bottle to stay warm tonight. Buy yourself a chicken as well. And no women, Georges. Don't you dare bring any women around here. I need my sleep and don't need to be kept up all night with your partying. Understood?"

"*Sous-entendu.*"

11:23 p.m.

Passed out in the Renault's cab, Georges snored loudly. Aymar was curled up in a fetal position by the smoldering remnants of the fire, also sleeping soundly. Moving slowly so as not to disturb them, Henson unlocked one of the six boxes and removed five old coffee cans, a liter-sized fuel can, and a high-powered flashlight.

He walked out on the disused airstrip, carefully pacing out distances, and arranged the cans in the designated pattern. The bottom of each coffee can was lined with an inch of sand, into which he poured a small amount of oil mixed with gasoline.

Checking the luminous hands on his watch, he lit the cans at the appointed time. From high overhead, the smudge pots would stand out as bright markers against the dark landscape, but they were virtually invisible from the ground. Right on schedule, he heard the increasing drone of an approaching MC-130E Combat Talon's turboprop engines. Using a shrouded flashlight, he repeatedly blinked the Morse code letter "P"—dot-dash-dash-dot—to the unseen pilots.

As the plane roared overhead in the faint moonlight, Henson counted six parachutes. Over the next few hours, as his companions slept, he located each bundle and manhandled it to their campsite. For each package, he unlocked and opened one of the six boxes, removed the heavy sandbags that were contained within, scattered their contents in the undergrowth near the road, and then refilled the box with the contents of the supply bundle and rolled up parachute.

Henson was already awake at sunrise, tending to a pot of coffee over a small campfire, when Georges and Aymar came to life, both groaning from hangovers.

"*Mon dieux!*" exclaimed Georges, tugging on his plaid cotton shirt, not realizing that he was putting it on inside out. "They must have put something in that bottle last night."

"*Oui*," said Henson, pouring himself a mug of coffee. "It was a terrible poison called alcohol. Now, those boxes aren't going to jump back on that truck by themselves. We have to be at the market in an hour, so to work, you two!"

Doing as they were told, the two dusky men struggled to load the cumbersome boxes on the truck. "Have these things gotten heavier overnight?" grumbled bleary-eyed Aymar, massaging his temples and panting after loading the first two.

"More likely that you have grown weaker overnight," chided Henson. "Hustle! And push them all the way to the front, *s'il vous plait*. We have a lot of samples to collect today."

As he tore off a hunk of bread for breakfast, Henson noticed Biko approaching. "Georges," he said. "Please see what Monsieur Biko wants."

"Gladly," said Georges. He spoke to Biko, then to Henson. "He said he has heard that you are paying for soil samples. He wants to know if you want a sample from his land."

Henson laughed and replied, "But of course! I should have already asked."

Smiling, the unshod and hapless farmer accepted a sample bag, and then made a great show of strolling around the extent of his sterile acres, periodically stooping over to collect an interesting stone or choice clod. In a few minutes, he returned with the bag full.

Henson filled out a tag and tied it to the sample before digging some francs out of his wallet. He paid Biko three times what he normally paid for the samples, plus a generous amount for using his property as a campsite, and profusely thanked him for his kind hospitality.

Makokou, Gabon; 8:05 a.m., Saturday, February 15, 1969

Makokou's Saturday market was alive with activity. Throngs of people, apparently oblivious to the oppressive heat, milled through the maze of stands, shopping for the week's groceries. Ebony children frolicked noisily in the dust. A wizened old man, dressed in native garb of garish colors, roasted peanuts over glowing charcoals and sold them by the handful. A gaggle of elderly women jabbered about the poor quality of cassava this season. A meat vendor wielded a large knife to slice a rear quarter from a goat carcass. A swarm of black flies buzzed incessantly around the butchered goat.

As the morning drew on, the equatorial sun glared down in force, so Henson and his two cohorts sought refuge under the shade of an old French Army tarp, flimsily suspended by four lengths of bailing twine. A pair of comely Bantu lasses eyed Georges and Aymar, giggling coyly. Stripped to their waists, the cousins responded by enthusiastically flirting and flexing their ropy black muscles.

A steady stream of property owners formed a line to deliver their bags of soil. Henson methodically labeled every single bag and then denoted the corresponding property on an old French Survey topographic map before compensating the owner with a fair sum at least commensurate with walking to the market to deposit the sample. In accordance with Henson's instructions, Georges placed the bags in neatly ordered rows alongside the truck.

As he bought samples, Henson noticed two men talking in a dialect that he didn't recognize. Occasionally, the men smiled and laughed while pointing at him. "Georges," he said quietly. "What is so funny? Am I doing something wrong? Am I violating some sort of local custom?"

"Attendre, chef, s'il vous plait," replied Georges, casually flicking away an expended Gitanes. A curious child scampered to grab the discarded cigarette. "Wait, boss, I'll find out."

Georges went to speak to the boisterous men. The three chattered and laughed as if they were sharing a private joke. He returned to the table and quietly confided, *"Chef,* it's a cultural thing. Those men are brothers, and they have it in their heads that they're superior to you."

"Superior to me?" asked Henson. "Why on earth would they think that?"

Obviously uncomfortable, Georges looked down and scraped his foot in the soil. *"Chef,* they're directly descended from a clan that raided villages and sold other tribes to the Portuguese and Dutch, to be shipped to the Americas."

"What does that have to do with me?"

"They said that since you are American and black-skinned, then it was likely that their ancestors captured your ancestors and sold them into slavery. They were joking that since you had shown up back in Africa, they could claim title to you and compel you to till their land."

For a moment, Henson fumed with anger, but then just as quickly settled down. Turning to Georges, he subtly twisted his watch around on his wrist, so that the dial was on the inside of his forearm, and then said quietly, "Do me a favor. Bring my field pack from the truck cab, *s'il vous plait*. It's tucked under the seat."

Henson waited patiently until the smirking brothers reached the front of the line. Maintaining a deadpan expression, he treated them no differently than anyone else. He asked them to verify their property on his map, and then he opened their bag to examine the soil sample. Looking inside, he quietly exclaimed, *"Intéressant! Tres intéressant!"* Mumbling under his breath, he reached into his khaki canvas field bag, pulled out a large magnifying glass, and pretended to examine the soil even more closely.

Smiling, delighted that Henson would be so interested in their specimen, the men leaned forward, paying rapt attention. Staring through the looking glass, Henson creased his brow, frowned, and gasped. Then he reached into his field pack and extracted a small Geiger counter.

Switching the machine on, he held the sample bag open with his left hand and waved the radiation detector's probe over the dirt with his right. The little yellow box clicked ominously. *"Mon dieu! Radioactives!"* asserted Henson, with all the scientific authority of the esteemed Madame Curie herself.

Dismayed, the men stepped back from the table. Murmuring quietly. The other people in line drifted away from them, shunning the pair as if they were stricken with smallpox or the plague.

Feigning his grimmest expression, Henson stood up and walked over to one of the men. Clasping the man's white cotton shirt with his left hand, he swept the Geiger counter's wand over his chest. The machine ticked again, even faster than before. He repeated the gesture with his brother, with the same results.

"Radioactives!" hissed a toothless old crone in the crowd. She removed a purple kerchief from her head and swished it at the men like she was shooing away a litter of sickly hyena pups.

Translating for Henson, Georges politely told the pair that Apex Minerals Exploration wasn't interested in their sample and requested that they please take their bag away, preferably as far as possible. Grabbing their burlap sack, the slave traders' descendants did just that, scurrying out of the market as swiftly as they could.

Henson used a small whiskbroom to sweep off the surface of the table and then rinsed his hands with water from his canteen. Then he consulted his watch, noting the time on its radium-painted hands, and waved for the next property owner to come forward with his sample.

Libreville, Gabon; 5:48 p.m., Saturday, February 15, 1969

Driving into the outskirts of Libreville, the truck was stopped by a gendarme patrol. "What's in the truck?" demanded a corporal. He wore a faded French "lizard pattern" camouflage uniform and a moth-eaten maroon beret, as did his two subordinates. The three carried well-worn French MAT 49 submachine guns and looked very hot and very bored. In the distance, undulating heat waves roiled up from the road's scorching macadam surface.

"*Sol*," replied Henson, offering the corporal his canteen. "Dirt. Soil samples for ore analysis."

"*C'est du sol, d'accord*," confirmed a gendarme private, using a stick to prod the burgeoning heap of burlap bags in the wooden bed of the truck. "*Beaucoup de sol.*"

"Oh, so you're that crazy bastard who travels around buying dirt?" asked the gendarme corporal. He guzzled water from the aluminum canteen, replaced the cap, and then wiped his mouth with his sleeve before handing the flask back to Henson.

"*Oui*," replied Henson, with a grandiose flourish worthy of Marcel Marceau. "*C'est moi. J'achète du sol.*"

"Pass, *imbecile*," muttered the listless corporal, laughing, waving his hand like a traffic cop and pointing toward Libreville.

9:30 p.m.

Henson sat at a makeshift table, examining his map. Georges and Aymar had unloaded the boxes from the truck into his small warehouse and had departed for home over an hour ago.

The old sheet metal building retained most of the heat from the day and felt like the Devil's own sauna. The place stank like old tires, which had been stored here long before it fell into disuse. Henson heard the squeak of an unoiled door hinge and looked up to see four white men—all dressed in work clothes—emerging from a storeroom in the back of the building.

"Henson?" asked one of the men. In his late twenties, he was tall and thin, with blonde hair, a closely cropped beard, and spoke with a pronounced Midwestern accent. "Matt Henson? I thought I recognized your voice. You were in my assessment cycle back at Aux One-Oh."

"Aux One-Oh? Yeah, I do vaguely recall…"

"Remember? We did a search together. We always wondered what happened to you. You were doing so well, and then you just suddenly vanished off the face of the earth. The scuttlebutt was that you took a swing at Captain Lewis and were shipped off to Leavenworth."

"Oh…Finn," replied Henson, finally recognizing his former classmate. He removed his HF radio set and portable TACAN beacon from one of the boxes. "Good to see you again, Ulf. I've just been busy. Look, this box is all my stuff. Your gear is in those five boxes over there. Check it out and then we'll move it to a hide site for safekeeping. So you're one of the scouts?"

"Yeah. Me and Davis. Baker and Nicholson will stay here with the gear and monitor the radio. The other guys are scattered all over, and some won't be in until tomorrow night."

Henson nodded and pointed to some boxes toward the back of the warehouse. "There's food and water in there. There's an outhouse out back. Make yourselves at home."

As Henson studied his map and made notes, the men unpacked their boxes. Two held a variety of small arms—a couple of .30 caliber Browning machine guns, six .45 caliber Thompson submachine guns, and several venerable M1 Garand rifles. The guns weren't top-of-the-line modern hardware like the stuff they carried at Aux One-Oh, but plenty formidable to stand up to any firepower that they might face in this part of the world. Other boxes contained ammunition, tactical radios, field gear, batteries, and enough essential supplies to outfit a squad for a commando raid or rescue mission.

Chewing on the end of a pencil, Henson watched the men as he reflected on the day's activities. Collecting and lugging the soil samples was a labor-intensive ruse, but one well worth the effort since it accomplished several objectives. Besides providing reasonable cover for his own reconnaissance and last night's supply drop, the sampling operation also afforded ample justification for him to ferry the pair of scouts—posing as survey helpers from Apex—to reconnoiter Biko's airstrip and the routes to the frontier regions.

Pushing himself out of his chair, he walked to his hammock to grab some shuteye before his labors began anew in the morning. As he passed by the pile of sample bags in the corner, he noticed that one burlap sack had burst open at the seams. Henson knelt down and examined the tag; the sample was from Biko's property.

Even in the dim light of the warehouse, a sparkle caught Henson's eye, and he looked more closely; in the midst of the spilled black earth were six miniscule flecks of gold. Apparently, the French had missed something when they so swiftly sold their real estate to the unsuspecting and gullible Biko. He laughed quietly and promised himself that when he and the scouts stopped in Mayela to examine the airstrip, he would quietly tell his friend Biko that his land was probably very, very valuable after all.

Special Security Detachment, Aerospace Support Project
10:15 a.m., Thursday, February 20, 1969

Jimmy Hara looked up from a surveillance report to see one of his counter-intelligence agents, Dean Trask, enter the bullpen. Trask had just come from the weekly multi-agency counter-intelligence coordination meeting hosted by the Wright-Patterson Office of Special Investigations. "Anything new from our OSI brethren?" asked Hara.

"More of the usual crap," answered Trask, rummaging through his desk. He produced a Baby Ruth candy bar, tore open the wrapper, took a bite, and added, "Soviet spies lurking *everywhere*. News at eleven."

"Those paranoid OSI guys never change," sniffed Hara. "Hey, save me a bite!"

"Buy your own." Trask tugged a pocket notebook from his jacket and said, "Hey, there was one new wrinkle. The OSI said that a national news magazine was contacted by someone who has "indisputable evidence" that UFOs are being stored and studied at Wright-Patterson."

"Man, that damned rumor *never* dies," said Hara, laughing. "I know you're new here, Dean, but we hear that one almost like clockwork. Every six months or so, someone claims that they know where the flying saucers and alien corpses are stashed."

Trask finished his snack, crumpled the wrapper, and threw it into the trashcan on the other side of the office. He nodded and said, "Well, the OSI warned everyone to keep an ear out. They suspect it's someone on base shooting their mouth off. Probably some disgruntled civil service guy looking for a payday."

"Obviously," Hara said. "Let's be good, obedient spy chasers and mention it at our next counter-intelligence briefing for Blue Gemini personnel. Even though this UFO business doesn't pertain to our operation, it wouldn't hurt to keep everyone a little tuned up. It never hurts to pay attention, because sooner or later, there might be someone out there paying a little too much attention to *us*."

Aerospace Support Project
10:10 a.m., Saturday, February 22, 1969

Wolcott removed his white Stetson, placed it solemnly on the table, and said, "Okay, folks, this is just like that awkward moment when the blushing bride is standing at the altar and the preacher tells everyone to speak now or forever hold their peace. If this crew ain't ready, then you need to be forthcoming. As much as we're all chomping at the bit to launch, we're not going to place these two gentlemen at risk because there's been a shortfall in their training. So if you have something to say, then now's the time to spout. If we need more time to prepare, Mark will pick up the phone and request a stay of execution. Otherwise, amigos, we go to space. Mark, anything to add?"

Tew shook his head.

"Okay," said Wolcott. "Round the table, it is. Gunter? Do you sign off?"

"They're ready," replied Heydrich.

"Russo, do you concur?"

Russo nodded glumly.

Wolcott shifted his gaze to Carson. "Well, pard, do you have anything to add?"

"No, sir, I don't," answered Carson. "I wish that we could spend more time with them, but these men are ready."

"Tom, Pete, are you ready?"

"We've *been* ready," answered Howard.

"General Tew, I certify that this crew is ready for flight," averred Wolcott, lightly clapping his hand on the table.

Tew nodded. "Then we go."

31

PACIFIC DEPARTURE

Pacific Departure Facility, Johnston Island
12:12 a.m., Tuesday, February 25, 1969

Carson, Russo, Howard and Riddle had spent the night aboard the LST. The landing ship, built at the height of World War II, had transported the assembled Titan II and Gemini-I from San Diego. It also served as a staging base while the rocket was prepared for launch; the vessel's troop spaces, designed to accommodate a company of soldiers or Marines, had been converted to living quarters for the Air Force personnel and civilian contractors who stayed aboard during pre-launch preparations. A second LST was configured as a fueling and servicing platform. With the Titan II already fueled, the ship had moved several miles offshore, to be followed by the first LST once the last personnel of the launching party disembarked in less than an hour.

Even when securely lashed to their piers at port, Navy ships rarely slept, but typically hummed with vibrant activity night and day. It was not unusual for hungry sailors to sit down to "mid-rats," a meal traditionally served at midnight in order to make sure that all hands on all watches were adequately nourished.

With the launch just mere hours away, the four pilots didn't dine in the officers' wardroom, opting instead for mid-rats in the regular galley. Incognito in white coveralls, they looked no different than the fifty or so technicians involved in Blue Gemini. Sliding their brown plastic trays down stainless steel runners, they dutifully fell in line behind drowsy sailors trying to grab a quick bite before returning to their nocturnal chores.

Carson checked the time. As the de facto shepherd for the flight crew, he observed that they had less than an hour to eat and make last minute preparations before Howard and Riddle went ashore to don their space suits in a converted travel trailer.

"Want to jump ahead, sir?" asked the sailor ahead of him. "The skipper says that you civilians have important work to do." Gangly and pimple-faced, the teenager had probably been drafted, or at least was a draft-motivated volunteer seeking to avoid ground combat in Vietnam; on average, approximately thirty of his brethren died there every day.

"Thanks, but no," answered Carson, selecting silverware from a dispenser. "I appreciate it, but we'll wait our turn like everyone else."

"So what's on that rocket out there?" asked the sailor.

"Not sure. What do you think?"

"The scuttlebutt is that it's a nuke," answered the sailor, filling a brown plastic cup with apple juice. "I just hope we're far enough away when it goes off. I want to have kids someday."

"Good luck with that," replied Carson, filling a mug with coffee from a stainless steel urn.

On the bulkhead behind a sullen cook, a hand-lettered sign stated: *"Take what you want but don't bring it back."* After receiving their choices from the limited bill of fare, the four men carried their trays to a table in an isolated corner of the galley.

A half-conscious young sailor passed by, smoking a cigarette and absentmindedly humming Steppenwolf's "Born to be Wild" as he swabbed the deck with a nearly new mop. As he moved on, the pilots realized that they had nearly a quarter of the dining area to themselves.

Keeping their voices low, they talked quietly about the mission preparations; a casual observer would probably assume that they were mid-level technicians, trying to resolve a wiring malfunction or deciding where to go drinking during the planned two-day layover in Hawaii. Normally a voracious eater, Howard nibbled on toast and barely touched his eggs. On the other hand, Riddle gobbled down his food like a castaway who'd just been rescued from a desert island.

"Get any sleep?" asked Carson. He used his fork to cut his undercooked bacon and tried a bite. Chewing slowly, he decided that he would forgo the remainder.

"Slept like a log," noted Riddle. "I'm surprised that you guys were able to rouse me."

"I didn't sleep at all," confided Howard quietly. "Drew, it wasn't because I'm scared. I guess I'm just anxious to get on with it. We've worked so hard to make it here."

"That we have," Carson said. "Do you feel confident with everything?"

"Yeah. Really, this flight is just about as simple as they come. It's very basic stuff that we've been practicing for almost a year. To be honest, Drew, I regret that ol' gunslinger Wolcott didn't give you the first flight. You earned it, in spades."

"We've got *ours*," interjected Russo. He sipped his orange juice. "Yeah, maybe you guys are going up first, but ours is the shot everyone will be watching. First *operational* mission."

Picking up his fork, Russo continued. "When this stuff finally hits the history books, it'll be sort of like the relationship between Shepard and Glenn. Sure, Shepard was the first American to fly in space, but it seemed like everyone forgot who he was as soon as Glenn went into orbit. And poor ol' Gus Grissom, he went up between Shepard and Glenn, but no one remembered…"

Russo immediately fell quiet as he recognized that the other men were scowling at him. Test pilots had their own unique traditions and superstitions, and one steadfast and inviolable rule was that no one spoke of the dead—in any manner—on the day of a new aircraft's inaugural flight. "Sorry," he noted. "My mouth was moving faster than my brain."

"You seem to be rather stricken by that malady, Ed," quipped Riddle. "Maybe you should just stick your foot in your mouth and leave it there. It would be a lot more convenient."

"It's not '*Ed*,' Major. You can either call me Colonel or sir," snapped Russo, reverting quickly from momentary embarrassment to his normal insufferable self. "I've earned that."

"No, Ed, you haven't earned *crap*." Riddle doused his hash browns with hot sauce. "The problem is that you believe that this is some kind of contest that you're destined to win. Well, despite your superior rank and obvious brilliance, you're no different than the rest of us. The only reason you're here, *Ed*, is that a good man couldn't handle the pressure and flaked out."

"Okay," said Carson, looking up from his tray. "That's enough, Pete."

"No, Drew, it's not. I've had enough of this guy trying to lord over us, and on *this* morning, I am going to have *my* say." Riddle forked the last bit of scrambled eggs into his mouth and then pointed the utensil at Russo. "If *Ed* here wants to cry to Virgil, that's his prerogative, but only if he's willing to suit up and take my place on that firecracker out there." Riddle dropped the fork on his tray. "Oops, I'm sorry, *Ed*…I forgot, you *don't have* a suit yet."

Seething, Russo was silent.

Leaning over the table, Riddle quietly added, "Virgil may be saving you for the first operational mission, *Ed*, but here's something for you to mull over: In a few hours, when those hold-down bolts blow, Tom and I will ride our little chariot into orbit, and you'll be stuck on the ground. This stuff will never hit the history books, so even though I'll rise into the heavens on a blaze of obscurity, at least I will have gone there. So you may fly in a few months, but you can warmly bask in the knowledge that you will fly *after* me."

"Okay," said Howard. "That *is* enough, Pete."

"As you wish, boss. You *are* the mission commander, and I am obligated to listen to *you*." Looking at Howard's plate, he asked, "Are you eating the rest of that toast, Big Head?"

Howard shook his head, and Riddle flipped the cold bread slices onto his plate before slathering them with a dollop of apple butter.

"Aren't you a little concerned about accumulating too much residue in your gut?" asked Carson, trying to dispel some of the lingering awkwardness. "You guys will be up there for two days, remember? And with the rendezvous schedule as it is, you may not have time to yank out a waste bag and climb out of the suit to take a crap."

"Yeah," answered Riddle. "I figure we won't have that luxury, but I would rather have a full diaper than be hungry up there. I don't function well on an empty stomach, and those dainty little tea sandwiches and squirt tubes just don't do it for me. So I'm stoking up now, courtesy of our seafaring brethren." He devoured the toast in short order and then finished his coffee.

"Ready?" asked Howard. The four men rose together, carrying their trays to the dishwasher's window. Looking at the unfinished bacon on Carson's tray, a Filipino dishwasher glowered and pointed at a sign that declared "*Scrape all food waste off tray.*"

Carson obediently emptied the tray, placed it in the rack, and then spoke to Howard. "We have to go ashore in twenty minutes. That gives you just enough time to brush your teeth and hit the head. You had better enjoy that while you can, because you won't be seeing any porcelain for a couple of days."

"Drew," said Riddle sheepishly. "You mind holding on to something for me?"

"Sure," replied Carson, watching Howard disappear down a ladderwell. "What is it?"

Riddle handed him a white envelope. "Nothing. Just give it back to me when I get back to Wright-Patt. Uh, otherwise I want you to open it if…"

"That's not going to happen," Carson said, taking the envelope. "You guys are going to be celebrating back in Ohio by the weekend, so don't act so damned morose, Pete."

As the four pilots walked down the LST's narrow gangway, Carson looked toward the gantry where the Titan II was poised to launch. Small work lights illuminated those parts undergoing a final "once-over" inspection, but otherwise the rocket was a dark monolith, barely visible in the pale light of the moon, jutting up from the hardscrabble landscape of the coral reef island.

Designed as an ICBM to heave thermonuclear warheads at the Soviet Union, the Titan II could be launched at a moment's notice—less than a minute, actually—with minimal preparation. Unlike the massive Saturn moon rockets charged with highly refined kerosene and liquid oxygen, the Titan II required no igniter to spark its fuels into combustion. Its hypergolic fuels were relatively docile, so long as they were kept separated, but all they had to do was meet and the result was furious and instantaneous combustion. Additionally, the oxidizer component of the mix—nitrogen tetroxide—did not require cryogenic storage, like liquid oxygen.

Thus, unlike the previous generation of ICBMs, the Titan II could be maintained in a launch-ready state with its fuel stored on board. For such an exceptionally complex and capable piece of military hardware, the Titan II was remarkably "shelf-stable," mostly by virtue of its unique propellants and exquisite design. It was perfectly suited as an ICBM that could linger dormant in a silo, patiently awaiting a call to action. But it wasn't a perfect design. The hypergolic fuels—hydrazine and nitrogen tetroxide—were costly, toxic, corrosive, and dangerous to handle.

The PDF pad workers servicing the Titan II rocket had been carefully screened and selected from the Strategic Air Command ICBM crews. Although the work was hard, working with Blue Gemini was a plum assignment, since the workers labored above ground, instead of living a mole-like existence in cramped silos. Additionally, their normal duty station was at the HAF in San Diego, where they also participated in the launch vehicle assembly and pre-shipment preparations, far removed from the windswept and lonely plains of the Midwest.

Two hours later, Howard and Riddle emerged from the Airstream suit-up trailer and boarded the van that would transport them to the pad. The pad workers had already removed the fiberglass shroud that concealed the Gemini-I spacecraft, so all was in readiness for the crew to board. As the men arrived at the gantry, the handful of remaining workers paused momentarily to wish them a safe flight. Then they returned to their hectic labors, diligently checking off the last items on their pre-flight countdown checklists.

The gantry supporting the Titan II was not large enough to support an elaborate "white room" boarding structure, so an open-air platform, accessible by a painfully slow elevator, had to suffice. The precarious platform could only accommodate one pilot at a time, along with the three pad workers who would insert the men into the spacecraft and strap them in for the bone-shaking ride to orbit. Howard, as the command pilot, would board first.

Carson borrowed an intercom headset to speak to Howard as he entered the elevator cab. "See you back at Wright-Patt, Big Head," he said, shaking Howard's gloved hand. "It's funny, but you'll be landing at Edwards long before I make it back to California. You two do good up there. I envy you. Sure you don't want to swap out? It's not too late, you know."

Picking up his portable cooling unit, which was about the size and shape of a large briefcase, Howard laughed. "No, Drew, I'm riding this ticket all the way."

7:00 a.m.

Except for a short hold to verify some key communications links, the countdown had progressed smoothly. As Howard had asserted, the mission was a straightforward replication of tasks proven time and again during the ten manned flights of NASA's Gemini program. The first major objective was an "m=4" rendezvous with an Agena-D target vehicle left over from a NASA flight. The rendezvous, so named because the interception would be accomplished within the first four orbits or roughly six hours into the mission, would be assisted with radar tracking and calculations provided from the ground, tracking ships at sea, and EC-135E ARIA aircraft.

For the next twelve hours, the pilots would practice close-in maneuvers that would be used during later interception missions against suspect Soviet satellites. Then they would break away, gradually increasing separation, before attempting a second rendezvous, effectively an acid test of the procedures that they had refined during the past several months. It would be flown "closed loop," with minimal support from the ground, and would culminate with the deployment of a hoop-shaped "Disruptor" device that would encircle and snare the Agena.

Afterwards, the pilots would power down the Gemini-I spacecraft for a six-hour "loiter" period. After resting, they would power up, reenter the Earth's atmosphere and deploy their paraglider to execute a night landing at a remote dirt strip at Edwards Air Force Base, California.

The partially buried blastproof blockhouse wasn't spacious, just barely large enough to house the twenty men who would oversee the launch and control the initial flight until the Gemini-I was successfully injected into low Earth orbit. At that point, control of the mission would transfer to Blue Gemini's Mission Control Facility located at Wright-Patterson Air Force Base.

Carson manned the CAPCOM desk in the blockhouse. Russo served as the Range Safety Officer; one of his principal tasks was to destroy the Titan II rocket if it went astray. Wolcott and Tew were present, strictly to observe the momentous event.

The countdown went into its final seconds. Monitoring the status of the launch and taking cues from the various controllers over an intercom loop, Carson relayed critical information and instructions to the flight crew. "Mark. T-minus one minute. Stage Two fuel valves coming open in five seconds…T-minus thirty seconds…Everything is looking great…T-minus twenty seconds …T-minus ten, nine, eight, seven, six, five, four… first stage ignition…two, one, zero. Hold-down bolts *fired. Lift-off.*" He glanced up quickly to watch a television monitor; belching a thick cloud of billowing smoke, the rocket gradually rose on twin wisps of scarcely visible flame.

"Copy bolts! Lift-off! The clock has started and we're on our way!" yelled Howard. His voice was barely audible over the roar. "This horse has left the stable!"

"Good luck, guys," said Carson.

"Roll program started," announced Howard.

"Roger roll program," replied Carson. There was so much whooping and cheering in the room that he had to hold his hand over his earphone to hear Howard's barely audible voice.

"Roll program is…*complete*!" declared Howard moments later. "Initiating pitch program!"

"Copy pitch program started. Looking good," replied Carson. Several of the launch controllers were already peeling cellophane wrappers from fat cigars.

Twenty seconds into the flight, Riddle commented, "Clouds coming up. Beautiful! Wish you could see this, Drew."

Yeah, I wish I could, thought Carson. "What's your cabin pressure, Squeaky?"

"Cabin press is at five point nine," answered Riddle.

"Be aware that we're experiencing a lot of pitch oscillation," said Howard. "Getting a bit rough. Should smooth out momentarily when we go supersonic."

Over the intercom loop, the "FIDO"—Flight Dynamics Officer—announced, "We're showing some erratic readings on the gimbal sensors."

"How erratic, pard?" demanded Wolcott, stepping out of his role as a silent observer.

"*Very*," replied the FIDO, an Air Force major. "But we can't tell if we're losing the gyro platform. For some reason, it's overcompensating."

"Thirty seconds on the clock. Be advised we are experiencing a slight pitch over," reported Howard. "We'll watch to see if it corrects itself. Right now, it feels like we're porpoising." The gyro's overcompensation resulted in progressively increasing pitch oscillations, which caused the rocket to fly in a slowly undulating pattern through the air, like the rhythmic pattern of a porpoise breaking the surface of the water and then submerging, over and over and over.

"This stack is not going to hold together very long. Advise on abort," demanded Howard.

"They don't have sufficient altitude for a modified Mode II abort," declared Russo. "And there's no Mode I with this platform!"

For once, Russo was correct, thought Carson. Mode II—firing their retro rockets to escape the malfunctioning booster—was the crew's only abort option, but it wouldn't be viable for another twenty seconds. Mode I abort, which had been available to NASA Gemini astronauts, was not available on the Gemini-I since it wasn't equipped with standard Weber ejection seats.

"*Screw* the protocol! *Abort!*" yelled the launch control officer, punching the red abort button on his console. "*Now!*" The abort command coincided with an electronically transmitted tone signal that immediately shut off the first stage booster engines.

Barely a second passed before Howard calmly stated, "Abort Indicator and Engine I lights are showing *red*. Now attempting a modified Mode II abort. Abort handle to Shutdown."

Carson unconsciously groped for a phantom abort handle at his left side as he reviewed the checklist in his mind. After five seconds, Howard would shove the abort handle from the Shutdown setting to Abort, which would initiate a swift chain of events that would hopefully save the two pilots. Explosive charges would sever the Gemini-I from the wobbling booster. Immediately afterwards, the spacecraft's four retro rockets would fire in unison to blast the spacecraft from the doomed rocket. The purpose for the painfully long five-second delay was to allow the booster's upward velocity to sufficiently dissipate so the retros would be effective.

Transfixed on the television monitor, which was linked to a tele-scope-equipped camera tracking the rocket's ascent, the launch team anxiously watched for the flash of pyrotechnic line charges detonating. But the charges didn't fire and the spacecraft did not separate; the nose of the Titan rocket was already keeling over rapidly, and the resultant sheer-ing forces were too great to allow the Gemini-I to break away.

"We're pitching over sharply, about to fully nose over," reported Howard coolly, like he was sports commentator casually describing a line-up change in a baseball game. "Mode II abort is a No Go. This is *not* recoverable, gentlemen."

Suddenly, as if all their eyes were linked by a common thread, the launch crew's collective attention focused on a large red "Destruct" button on Russo's Range Safety console. After first swiveling a lock-out key, Russo would press the button, which would initiate a radio signal, which in turn would detonate a series of explosive charges built into the Titan II booster rocket. The intent was to prevent an errant booster from acci-dently injuring anyone on the ground.

Of all the contingencies that they had rehearsed ad infinitum, this was the one scenario that no one has seriously contemplated. Frozen in place, with a terrified expression on his face, Russo desperately looked to Wolcott, as if silently seeking guidance at this awful moment.

Without speaking, Wolcott answered. He reached past Russo, calmly turned the safety lock-out key, and then quickly and deliberately pressed the red plunger once, then twice more for good measure. On the television screen behind him, the pin-wheeling rocket was disintegrated by a massive explosion, instantly rendered into a plume of cascading fire and smoking debris. Speechless, the men watched the rocket's demise in stark black and white.

Picturing the faces of Howard and Riddle in his mind, Carson could barely breathe, but he continued to monitor the UHF channel as if they might have miraculously escaped their doom. *Maybe they separated just before the explosion,* he thought. Still watching the monitor, he studied the scattered chunks now spiraling down toward the sea, waiting for one to sprout a parachute. *If anyone could keep their cool and pull out of this, Big Head Howard could do it.*

Wolcott took his hand from the Destruct button and turned to face the others. Most were in shock, some looked ready to cry, two or three were already starting to sob. Wolcott's weather-beaten face was implacable, like a stoic cavalryman holstering his Colt after putting down a lame mule. "Is there some reason that you hesitated, pardner?" he asked Russo, speaking in a tone almost glacial in its detachment. "Surely you know the procedures."

"But they were…" Russo stammered. "…they had no way to…"

"Eject?" asked Wolcott, drawing an unfiltered Camel from a pack in his pocket. He lit it with his Zippo held in an unwavering hand. "Correct, but what would have been gained by delaying the inevitable by just a few more seconds? Did you forget your job, Colonel? Did you forget that we have ships out there on picket? What if that rocket had remained intact and crashed down on one of them? Or what if it had made it to Hawaii? Have you considered that, pard?"

Silently sobbing, Russo despondently shook his head.

"Next time, pardner, do *not* hesitate. Do what has to be done." Wolcott inhaled deeply from the cigarette, held it in for a moment, and exhaled through his nose. "Now, the rest of you, put down those damned cigars, snap out of your shock, and do your damned jobs. Let's figure out the point of impact so we can get out there pronto to start fishin' pieces out of the water."

6:10 p.m.

Carson stood by the pier, watching the sun going down in the Pacific sky. It was a beautiful evening, with a gentle breeze blowing in from the ocean. He heard a steady drone from the north and realized that it was a contract passenger plane, ferrying back the first contingent of the island's regular workers from Hawaii.

Wolcott walked up. Waving, he beckoned Carson. "Hey, Carson, would you mind giving me and Mark a lift to the other side? We're on the first shuttle back to Hickam Field."

"Sure, Virg." The two of them walked to the far side of the blockhouse, where Tew waited with two B-4 suitcases. Carson helped them load the bags into the back of the Willys jeep and then climbed behind the wheel. He awkwardly shifted the open-topped jeep into gear, grinding some metal in the process, and they jolted to a start and headed north along the airstrip. He thought Wolcott seemed amazingly calm and collected for someone who had lost a rocket and two pilots less than twelve hours before. But then he realized that the retired general had cut his teeth in a generation where death and destroyed aircraft were daily occurrences.

As for himself, he had personally witnessed ten fatal crashes, including two where he had been piloting a chase plane close enough to see the doomed pilots' faces when they died. But something about this morning's incident had unnerved Carson, and he hadn't realized what it was until late this afternoon: it was the very first time that he had ever witnessed an intentional act that had resulted in someone's death.

"So, pardner, I s'pose you're havin' some second thoughts after this morning, huh?"

"We wouldn't think any less of you, Carson," interjected Tew. "That's human nature."

"I'm not afraid, sir, if that's what you believe," replied Carson. Trying to keep this morning's events in proper perspective, he remembered that men climbed into cockpits every single day in Vietnam, not knowing what the day held. But despite the uncertainty of their circumstances, at least they had some limited means of escape if their aircraft was shot down or otherwise disabled. "I just wonder if it was really such a good idea to pull out those Weber seats."

"There's really no way of knowin'," replied Wolcott. "After all, NASA launched this same rocket ten times with men on top, and never once did they have to rely on the ejection seats. They came awful close that one time, and even then it's questionable if those boys would have made it through. So, pardner, you could say it all comes down to a toss of the coin."

"True," said Carson. *Yeah, but with the ejection seats Big Head and Squeaky would have at least had that coin to toss*, he thought. *But on the plus side, at least they died with three hundred pounds of additional mission gear inside the pressure vessel.*

"Pardner, what happened this morning was a tragic loss, but a harsh reality is that airplanes and rockets will never be an absolutely safe means of travel," commented Wolcott.

Carson pulled up to the terminal building and switched off the jeep's engine. He watched for a moment as the incoming passengers disembarked from the arriving plane. Dressed in gaudy Hawaiian shirts, they laughed and joked, oblivious to this morning's tragedy. Carson thought about the next launch, when he would be riding the rocket, and wondered if he could be nearly as calm and professional as Howard if he found himself in similar circumstances.

"Son, I would be lyin' to you if I told you that we weren't in a pickle," said Wolcott, swinging his B-4 bag out of the jeep. "I know that you and Russo ain't geehawing, but the fact is that we can't possibly get this next rocket off the ground without you sittin' on top of it. I want you to know that, Carson. The fate of this project is very much in your hands."

"Aptly stated," added Tew. "We are relying on you, Carson."

"Sirs, I appreciate your candor. I don't have any intention of backing down. I'm planning to ride the next rocket out of here, but you need to know that I can't go up with Russo."

"Major Carson," said Tew flatly. "We have *no* other options. Are you telling me that you would rather go up with one of the men from Crew Three?"

"No, sir. I just have to ask you both if you are men of your word."

"Pardner, that goes without sayin'," replied Wolcott.

"And you kept your word to Ourecky, right?"

"We did," answered Tew. Removing his Stetson, Wolcott nodded solemnly in concurrence.

"Then, Virgil, I'm going to ask you to recall a conversation that we had a few months back, when you asked me to work with Ourecky on the forty-eight-hour simulation. All I ask is that if you kept your word to Ourecky, then you keep your word to me as well."

"I know where you're going, pardner, but we ain't sending you to Vietnam, so don't ask."

"That's not it," said Carson. "Virgil, you told me that if we made the forty-eight hour mark, then I could have my choice of right-seaters."

"I do recall that," said Wolcott. "But it ain't particularly relevant right now, since we're runnin' short of options. But out of sheer curiosity, pardner, who exactly do you have in mind?"

"I want to fly with Ourecky. I trust him. I can't trust Russo."

"No!" barked Tew. "I won't allow it. Besides, Ourecky would never agree to it anyway. He's just a few weeks away from going to MIT. He's not a pilot, and he's not going to agree to some harebrained idea, particularly after what happened this morning."

"Well, you're probably right, sir," said Carson. "But would you at least grant me an opportunity to ask him? If he refuses, then what harm could come of it?"

"No!" reiterated Tew, pulling his B-4 bag from the jeep. "And besides, we don't even know if he's even physically qualified for the project. It makes no sense to invest a lot of thought on this if it eventually turns out that he can't pass the medical exam."

Looking out to sea, Wolcott was silent for a long moment. Then he looked back to Tew and said, "Mark, just by happenstance, I *do* know the answer to *that* particular question. When he went through his flight physical to fly back-seat with Brother Drew here, Ourecky got the full work-up. The bottom line is that he's medically qualified for space flight. I'm sure that the docs would want to give him another once-over just to be absolutely sure, but he's good to fly."

Tew shook his head. "I can't believe we're even entertaining this as a serious notion. It's a long trip back to Ohio, Major Carson. I won't promise you anything, but I'll think about it."

32

EL SEGUNDO

Los Angeles International Airport, California
2:15 p.m., Thursday, February 27, 1969

S itting in the pilot's lounge of LAX's general aviation facility, Carson perused the morning newspaper as he waited for Ourecky. Ironically, tomorrow's scheduled launch of Apollo 9 was delayed because the astronauts—McDivitt, Scott and Schweickart—had caught colds earlier in the week. If cleared by NASA physicians, the trio would launch on Monday. President Nixon, who would be just returning from his trip to Europe, would be in attendance to view the lift-off.

Closing his eyes, Carson wondered if the President would ever witness one of his launches. Certainly, the President was cognizant of Blue Gemini and its implications. He had to be aware of Tuesday's incident at the PDF, and perhaps that's what compelled him to travel to Cape Kennedy. It would be only the second time that men rode a Saturn V rocket into space, and even though the behemoth was designed by the brilliant von Braun and his stellar coterie of German rocket scientists, there was still a possibility that something could go awry.

Perhaps the President's media-savvy handlers surmised that there could be no better place for him to be. If events proceeded as planned, the President would be there to witness history in the making and assume ownership of it. But if the Saturn V faltered on lift-off, he would be perfectly positioned to call attention to the tremendous risks undertaken by previous Administrations—Democratic, no less—as he consoled the widows and asserted personal leadership in the aftermath of a national tragedy. Either side of the coin was a winning toss.

Carson's mouth felt dry as he reflected on why he had called Ourecky to meet. He unwrapped a stick of Juicy Fruit gum and stuck it in his mouth. This was a hastily planned stop on their trip back to Wright-Patt, so he didn't have an abundance of time. The Manned Orbiting Laboratory Project Office, where Ourecky worked, was located immediately south of the airport at Los Angeles Air Force Station in El Segundo.

Wolcott had called last night, granting Carson permission to invite Ourecky back into the fold. Without disclosing his true intent, Carson approached Russo to ask for the short layover. Much to his surprise, Russo had not resisted his plan; of course, although Russo worked at Wright-Patt, he was still assigned to the MOL office here, so he likely harbored his own agenda.

He set aside his paper and looked out the window. Los Angeles International was one of the busiest airports in the world; every day, tens of thousands of people came and went from here. Everything seemed so orderly and efficient on this sunny afternoon, but Carson was well aware that the airport was associated with two fatal air crashes just over a month ago.

On January 13, while on landing approach, a Scandinavian Air System DC-8 splashed into the Santa Monica Bay roughly eleven miles west of the airport; fifteen people died but thirty survived. Less than a week later, after losing an engine and electrical power following takeoff, a United Airlines 727, on its way to Milwaukee, plunged into the same bay, killing all twenty-eight people aboard. As Wolcott had implied, flying airplanes could often be a perilous business, but not all the risk was assumed by the men behind the controls.

Rather than avoid the pervasive stench of the massive oil refinery in nearby El Segundo—a town otherwise known as "El Stinko" to

LA residents—Russo remained out by the T-38, subjecting the aircraft to a cursory but superfluous inspection. He was clearly distraught after Tuesday's incident but seemed determined not to show any outward sign of his distress.

Carson sipped a Coke as he watched a man on the other side of the lounge. Wearing blue jeans and a white T-shirt, the man looked to be of college age. Likely a student pilot, he was engrossed in the busy work of assembling a flight plan. The table before him was covered with the brightly colored FAA sectional maps.

Measuring route segments, the man used an E-6B "whiz wheel" flight computer to calculate various elements—leg timing, estimated fuel expenditure, wind corrections—as he pieced the plan together. After a few minutes, he neatly folded his maps and tucked them into a binder, stood up, and walked over to Carson's table. "Is that your Talon out there?" he asked, gawking out the large window at the needle-nosed trainer.

"It is," replied Carson. "Actually, I'm just riding the back seat today."

"Cool. I'm still learning to fly. Just soloed three weeks ago. I'm in the middle of a cross-country run. Just wish that I hadn't landed here. I'll be lucky to get out of this place, after they stack up everyone else in front of me. I'll probably burn most of my fuel just taxiing and waiting."

"Just be patient," counseled Carson. "Are you flying that Piper Cub?"

"Yes, sir," replied the man, donning a leather flight jacket that was much too heavy for the warm day. "Don't NASA astronauts fly T-38's like yours to maintain proficiency?"

Carson nodded.

"Cool. Would you mind if I went out and looked at it?"

"No problem with me. Just talk to the man out there. His name's Ed Russo. It's his plane."

The man thanked him, pulled a baseball cap over his shoulder-length blonde hair, and walked out into the bright sunlight. Meeting him at planeside, Russo was obviously receptive to the interruption; the two walked around the T-38 for almost ten minutes, and Russo even let the neophyte pilot briefly climb into the cockpit.

After the man left to file his flight plan, Carson watched an official blue sedan pull into the parking lot. Three men—two wearing Air Force dress uniforms and a third in Navy whites—disembarked and walked toward the jet. Russo saluted one of the newcomers and then gestured toward the building. One man broke off and headed his way; spotting Ourecky's distinctive gait, Carson stepped outside and called him over.

"Man, it's good to see you, Drew!" exclaimed Ourecky. The two men shook hands and then embraced. Other than the flight suit the engineer had worn on hops or during training, Carson had never before seen Ourecky in uniform. The blue space above his left breast pocket was woefully bare, except for the red and yellow ribbon of the National Defense Service Medal, the "everyman award" given to all military personnel when they completed their basic training.

"Good to see you, too." Carson pointed at a picnic table beside the building. "Let's go over there, if you don't mind, so we can have some privacy." He looked back in the direction of the T-38. Russo's hands were tracing a trajectory through the air, an arc that stopped abruptly and then plummeted toward the ground, so he knew precisely what he was describing to the two visiting officers. "Scott, who are those guys?"

"Those two? Talking to Russo? That's General Stokes and Admiral Tarbox. They're both bigwigs with the MOL office."

"I know Stokes. What about the Navy guy? Tarbox?"

"Tarbox? Eventually, there will be some Navy-specific MOL missions. Admiral Tarbox oversees the preparations for those. He apparently has some potent political connections. Definitely not someone you want to cross."

"I'll make note of that," Carson said. "Thanks."

"So what brings you and Brother Ed to sunny California?" asked Ourecky, sitting down at the picnic table. "Seeking gainful employment with us Can Men? Are you guys preparing to lock the doors back at Wright-Patt?"

"I can't speak for Russo, but I'm definitely not here to look for a job," said Carson abruptly. In fact, he had been here just over two years ago, to interview for the third group of military astronauts who would eventually fly on the MOL. Four Air Force pilots were later selected; Carson did not make the cut.

"This isn't a casual visit, is it? You're not just passing through, are you?"

Shifting his eyes to the sun-warped redwood boards of the picnic table, Carson avoided Ourecky's ever-inquisitive gaze. "No, we're not just passing through, Scott. The truth is that I have some really bad news: Tom Howard and Pete Riddle are dead."

Ourecky swallowed. "Oh, man. Big Head? Squeaky? How did it happen? T-38 accident?"

"No, Scott. I feel terrible having to break it to you like this, but they died in a launch accident. Blue Gemini wasn't cancelled when you left. Tew just wanted you to believe that. Tom and Pete drew the first practice mission. Tuesday morning, they launched from the PDF, but they apparently had some sort of gyro failure. The rocket had to be destroyed, and we lost the guys in the process. I was there, on CAPCOM, so I saw and heard it all."

Ourecky suddenly looked like he was in shock. "They couldn't execute a Mode II?"

Carson shook his head solemnly. "Not enough altitude and the platform was already too unstable. At least it was quick for them."

"I suppose that there's something to be said for that."

"Well, they signed on for the job," observed Carson. "It's not like they were coerced. Look, Scott, I didn't come here just to dump bad news in your lap. I came here to ask you something."

"What is it, Drew?"

Carson started to open his mouth, but was waylaid by an overwhelming sense of trepidation. He knew plainly what he was here to do; the question was foremost in his mind, but his mouth was not willing to form the words. For a fleeting moment, it was if he was drawn up into the imaginary submarine he had invented to insulate himself from pain and stress. The hatches were sealed tightly behind him, and he was isolated in the conning tower, looking out through his periscope, separated from his present reality so that he could reflect on the potential consequences of his actions. He knew what he had to ask Ourecky, but he was very conscious that when he asked the question, both of their lives could be indelibly altered by the outcome.

"What is it, Drew? What do you want to ask?"

How can I do this? thought Carson, as if he was trying to reach out telepathically to Ourecky. *Scott, how do I ask you to do something that will likely kill both of us?* Finally, the words poured out of his mouth, in almost a jumbled stutter, like he was an awkward adolescent asking the prettiest girl in class to accompany him to the homecoming dance. "Scott, I came here to ask if you might want to fly. To orbit, that is. With me." At that moment, a heavily laden TWA 727 took off, punctuating Carson's question with the roar of its three engines.

Ourecky looked skyward and gasped. "Are you *serious*?"

"I am serious. More accurately, *we're* serious. Tew and Wolcott granted permission for me to ask you, but you're not under any obligation one way or the other."

"But I'm not a pilot," declared Ourecky.

"Scott, you probably never dreamed that you would hear these words from me, but that doesn't matter in the least. You and I both know that."

"But why *me*, Drew?" implored Ourecky, looking toward the T-38. "Even with Tom and Pete gone, there's still Parch Jackson and Mike Sigler on Crew Three. And last but definitely least, there's Ed Russo. Wasn't Wolcott going to team you with him?"

"Yeah, I was supposed to fly with Russo, but the generals finally realized that it wasn't going to pan out. They're allowing me to choose who I want to fly with, and I choose you. Of course, that's contingent on your decision. If you say no, then I'm back on square one."

"Drew, this is an awful lot to absorb at one sitting," Ourecky said. "To be honest, I'm still stunned that Blue Gemini wasn't discontinued."

Carson nodded. "If you decide to fly, you'll find out more about what had happened. Blue Gemini wasn't cancelled; it was accelerated. That's why Tew and Wolcott agreed to let me talk to you. Regardless of which crew goes next, they have to know the spacecraft and procedures absolutely cold. Just by circumstance, Scott, you may not be a pilot, but you're far more current and knowledgeable than anyone else, especially on the computer and calcs. Besides that, the two of us have more training as a crew than Jackson and Sigler. And with the launch schedule as it is, there's no possible way to bring them to the same level in the time that's available."

"And Russo? What about him?"

"I won't even discuss Russo," answered Carson flatly. "Flying with him is not an option that I'm willing to entertain. I would rather fly with an airsick chimpanzee."

"Gee, thanks, Drew." Ourecky pointed at his heart. "As Bea's fond of saying, that just makes me feel warm right here."

"That's not what I meant." An orange-painted Coast Guard HO-4S rescue helicopter clattered by, low overhead, on its way to the Coast Guard air station located on the airport.

"I know. Just kidding. So when is the next launch?"

"As early as April, but we still have to work around NASA, and their flight schedule is up in the air. They're gunning hard to land on the moon this summer, but in order for that to happen, everything has to run smoothly next week, provided those guys are healthy in time to launch."

"And after Apollo 9, then everything has to function on Apollo 10," interjected Ourecky. "What a huge gamble they're making."

"Right. In any event, Scott, I really can't discuss any more specifics unless I know that you're even slightly open to the possibility of flying. If you're not, then we need to end this conversation now, and you need to focus your energies on what you're currently doing and forget about Blue Gemini and forget about me. Are you even interested?"

At this point, Carson almost wished that he hadn't come here in the first place. He felt terrible about putting Ourecky on the spot and almost hoped that his friend would say no.

"I'm open to the idea, but I haven't made up my mind," replied Ourecky hesitantly. "It's a lot to absorb. So, how many missions?"

"Eleven more, at least. They plan to launch every three months, all against live targets."

Ourecky whistled quietly. "*Eleven* missions? What's the plan to divvy them up? Assuming that I fly with you until a replacement comes along, would we alternate missions with Crew Three?"

Exhaling slowly, Carson shook his head. "Nope. Scott, it's unlikely that the Block Two computer will be delivered before the fourth mission, as it's currently scheduled. Gunter and his guys are working hard with Crew Three, but it's doubtful that they would be able to fly a closed loop profile without the enhanced computer. And to be honest, everyone is more than a bit skittish about following Tom and Pete."

Carson pointed toward the three men standing by the T-38. "If you can believe it, Russo was jarred so bad he couldn't even speak for two days.

At least Parch and Mike weren't there to see it in person. They were working in Mission Control back at Wright-Patt, but that doesn't make them any less reluctant to jump on top of a rocket."

"No kidding."

"That said, if you choose to come back, it's virtually certain that the first four flights would be ours. Assuming that they adhere to a ninety-day cycle, that's a minimum of a year."

"How about MIT and my PhD?" asked Ourecky. "Did Tew say anything about that?"

"No, he didn't say anything specific, but obviously it'll be delayed. Scott, as much as I want you to fly with me, I have to be honest enough to tell you that if you decide to come back, you can probably kiss your PhD goodbye, particularly since the funding flows out of the MOL's spigot. I don't think the Can Men will be willing to invest any more money in you if you're no longer on their books, regardless of Tew's influence."

"Okay. I can live with that. But what am I supposed to say to Bea? Right now we're planning to get married in June. And regardless of what happens, I still want you there."

Carson rolled his eyes and groaned. "*Bea*? Oh man, I haven't even thought about Bea yet. I'm sorry, Scott, but I can't lend you any suggestions on that end. If you're looking for advice about women and how to deal with them, I'm the last guy you should ask. I can fly supersonic airplanes and rockets, but I'm absolutely mystified by women," he admitted, shrugging his shoulders. "Look, as much as we're going to be on the road in the coming months, does Bea really have to know that you're going to be back at Wright-Patt?"

Ourecky laughed nervously. "It's not quite that simple. First, she's been flying out here every other weekend, so it wouldn't take her too long to figure out that I'm not here."

"Ouch. Is she coming out this weekend?"

"No. She's staying back in Ohio. Besides, Drew, it's been hard enough on me to mislead her about what we were doing the past few months. I wouldn't feel right about going back to Wright-Patt without letting her know."

"Hopefully, she'll understand. It sounds like she's been around the military long enough to understand that things are always subject to change."

"She *definitely* knows that."

Carson glanced out at the T-38. Still chatting with the VIPs, Russo had already started his pre-flight inspection so they could make their scheduled takeoff window. "Scott, Russo and I have to zoom back to Ohio. I don't need an answer now, but I want you to think about it."

"Well, it's a lot to think about," Ourecky said.

"I know. Whatever you choose to do, I'll respect your decision," said Carson, scratching the left side of his moustache. "I don't want to get all weepy here, but I've realized something in the past few weeks since you've been gone. Just so you know, I generally don't share much about my personal history, for good reason."

Carson explained, "My dad died in a training crash in Idaho when I was three. My mom never recovered from that. She died in a car wreck when I was eight. She was drunk, cruised out over the white line, hit another car head-on and killed two other people besides herself. I was sent to live with my uncle. Uncle Jack was so busy climbing his career ladder that he didn't have much time for me, so he packed me off to a military boarding school in Alabama, then I went to West Point, then the Air Force, and now I'm here."

"I didn't know any of that," said Ourecky. "I'm sorry."

"There's nothing for you to be sorry about. Anyway, although I've been surrounded by people all my life, I've really been on my own. I think that's why boxing always appealed to me, because it was just me against some poor schmuck who was unfortunate enough to land on the same card. The coaches liked me because I was small, fast, and powerful. They always loved that I could pound an opponent into oblivion without mercy, but all I was really doing was unloading all the anger and frustration I felt."

Ourecky rubbed his left shoulder. "Gee, Drew, that explains a lot."

Carson smiled. "Scott, what I've come to recognize is that I've never really gotten close to anyone. I've never had a steady girlfriend, mostly because most women are smart enough to sense that I have no intention of becoming too attached, or because I punch out of the relationship when a woman starts pressuring me for a commitment.

"I have never learned to function as part of a team, because it's just not been necessary. When I climb into the ring, I know that the only person I can depend on is myself. It's the same with flying fighters; yeah, they beat it into us to always trust and rely on our wingman, but the reality is that I always fly by myself. But this business is entirely different, and the past few weeks have made me realize two things. The first is that I can't go up there by myself."

"And the second?"

Hesitant to answer, Carson was silent for a moment. "Scott, I can't go up there without you. I don't trust anyone else enough to make that trip. For my part, I *have* to go and I *want* to go. Please go with me."

33

THE BIG WHEEL

Aerospace Support Project
8:30 a.m., Wednesday, March 5, 1969

"I 'm sorry that we had to be so deceptive, Ourecky," said Tew. "I hope that you can understand the circumstances. We're just thrilled to have you back."

"Ditto for me, hoss," added Wolcott, grinning as he patted Ourecky on the back. "Welcome to the brethren. You have reached the inner circle, and now you are one of us."

"Gentlemen, if you will excuse me," said Tew. "I have an appointment at the hospital. I'm sure that we'll chat later."

"Anything...*new?*" asked Wolcott.

"Just a follow-up. I'll let you know how it comes out."

As Tew departed, Wolcott spread a large calendar on the table in front of Carson and Ourecky. Tapping his finger on a date in mid-June, he declared, "That's our tentative launch date, provided we put this danged investigation behind us in time."

Wolcott continued: "The remainder of March will be taken up almost entirely by mission-specific technical training here at Wright-Patt. At the end of April, you two will go down to Eglin for the full dose of SERE Contingency training. Anyway, the remainin' time will be mission-specific prep work. The good thing is that most of that will be what you were already doing in the Box before January, so you won't need too much catch-up work. The bad news is that you can plan on eighteen-hour days, seven days a week, for at least the two months before we launch."

"Sir, when will we get weightlessness training?" asked Ourecky. "Don't they do that with KC-135's flying parabolic profiles? I don't see it on the schedule. Am I missing something?"

"No, pardner. If we somehow slip ahead of schedule, we might let you jump on the Vomit Comet for a few Zero-G profiles, but otherwise we ain't wastin' time on it."

"Wasting time?" asked Ourecky.

"Correct, pard. You've been in the Box, Ourecky, so you know how much room there is. More accurately, you know how much room there *ain't*. Once you're on orbit, you ain't going to be doing any somersaults. If you have the time to snatch some Zero-G before you fly, great, but otherwise, the main thing that you need to know is that loose crap will float around in the cabin, and you have to manage the clutter. That's about it."

"Oh. I guess I was really looking forward to that, sir."

"Someday you'll have an opportunity. But speaking of *gravity*, there is something else."

"What's that, Virgil?" asked Ourecky.

Wolcott stubbed out his cigarette and grinned. "You have a ticket to ride the *Big Wheel*! Tomorrow, you and Carson will go to the Johnsville Naval Air Development Center in Pennsylvania. It's in Warminster, right outside Philadelphia. They have an unscheduled block on their human centrifuge on Friday, and we've arranged for you to take advantage of it."

"I thought there was a centrifuge here at Wright-Patt," said Ourecky.

"There is, but it's currently booked for physiological testing. Besides, it ain't the *Big Wheel*. Nothin' holds a candle to the Johnsville centrifuge, pard. It's the real McCoy."

"But what's the rush on the centrifuge, Virgil?" asked Ourecky. "It can't wait? Surely the Navy will have some more open blocks between now and our final pre-flight work-up." He thought of Bea; he hadn't seen her in two weeks and was anxiously looking forward to their reunion tomorrow night. Besides, she only thought he was in Ohio temporarily; he had yet to break the big news to her. Certainly, she would be thrilled that he would be back in town to stay.

Wolcott's face took on an expression like he was entering a corral to break a recalcitrant mustang. "No, Mr. Ourecky, it *can't* wait. We ain't hagglin' here. If you're going to ride a rocket, the Big Wheel is a rite of passage that you must endure. The flight docs have declared you medically certified and you may have the gumption to fly, but that's all for naught if you can't soak up heavy G's. Not everyone has the stomach for it, and it's far better that we find out now than several weeks down the road. The Wheel is available on Friday, so we strike while the iron is hot. I really ain't in the mood to hear any more danged balking or bellyachin' about it. That good with you, pardner? Are we square?"

"Yes, sir," replied Ourecky sheepishly. "That's good with me."

"Splendid," said Wolcott. "Now, if you two gents will excuse me, I'm going to mosey down the hall to the boys' room. With my internal plumbing gettin' as bad as it is, I might just start pourin' my coffee directly into the urinal instead of filtering it through my kidneys first. That would sure save me a heap of time every day." He stood up from the table and slowly ambled out the door. His knees and hips literally creaked as he walked.

After Wolcott departed, Carson punched Ourecky lightly on the shoulder. "In the future, try hard not to agitate Ol' Virg, if you don't mind. He *owns* you now. Before this, you had been a temporary worker, more or less, but now you're on a crew, and that's Virgil's domain."

"I understand that, but I just got back here, and I would like some time to adjust."

"Oh yeah, *sure* you need time to adjust." Carson laughed. "You can't yank the wool over my eyes. You're just trying to finagle some extra time with Bea. I feel really sorry for you, but we have to ratchet up the program. Every millisecond between now and June is precious. Remember when I asked you if it wouldn't be better to just not tell Bea that you're back at Wright-Patt? Now, maybe you understand that I was trying to do you a favor."

"If you say so," muttered Ourecky.

"And this trip doesn't have to be miserable. We'll stay at the Navy Yard. And I've got some hot numbers for a couple of girls in Philly."

"I don't think Bea would be too thrilled with *that* idea."

"I wasn't talking about *you*, Scott. Tomorrow night, you, me, and Jennifer will go out for a Schmidt's and a cheesesteak. After that, I'll tuck you into bed early for a good night's sleep, because you'll need to be well rested when you climb into the Wheel, and then I'll make some time with Susie. You ride the Wheel first thing Friday morning and then we'll head for home."

"Extra onions on that cheese steak?"

Carson nodded. "Anything for you, brother."

"Then it's definitely a road trip. Philadelphia and the Big Wheel."

Suitland-Silver Hill, Prince George's County, Maryland
1:12 a.m., Thursday, March 6, 1969

Felix Federov, a highly decorated colonel of the Soviet GRU, was uncomfortably crammed with sixteen other men and nearly two hundred kilograms of equipment in the rear of a Ford Econoline van. They were jammed in so tightly that he could scarcely breathe. As the van navigated the darkened streets of suburban Maryland, it was rocked by gusts of high wind and pelted with heavy rains. For most, the weather could not be more atrocious, but for Federov, it could not possibly be more accommodating. For over two weeks, he had waited patiently for rain to mask their clandestine activities, and tonight's violent thunderstorm was much more than he could have ever hoped for.

He shifted slightly to relieve the pressure of a man's knee jabbing sharply into his left flank. The severely cramped quarters reminded him of his earlier days, as a young lieutenant leading a *spetsnaz* reconnaissance team, when he and his squad scrunched into the cargo space of an Antonov AN-2 bi-plane to practice clandestine infiltrations into Western Europe.

Federov was the Resident, the officer in command of the GRU's field office located at the Soviet Embassy in Washington, DC. He wasn't the senior GRU officer in the Embassy—that billet was occupied by a general—but as the Resident, he managed the station's intelligence

collection operations in DC and the surrounding area, a prestigious assignment by anyone's assessment. Renowned for his cunning, as well as his
somewhat brutal tactics, his star was definitely on the rise. Three months
ago, he had arrived here from a posting in East Germany, and if all went
well, his next assignment would be a genuine combat tour, as a military
intelligence advisor to the Soviet Union's "comrades" in North Vietnam.

Tall and broad-shouldered, Federov was in his early thirties. The
red-haired officer had risen up through the barbaric ranks of the GRU's
spetsnaz commandos before matriculating into the espionage and coun-
terespionage realms of the GRU. In addition to being an accomplished
spy, he was a brutally efficient judo expert who had killed several men—
including two of his sparring partners—with his bare hands. His proficiency and unbridled enthusiasm for unarmed combat had earned him his
nickname, "The Crippler."

Federov really didn't have to be here. Ordinarily, he would have
parceled this mission to one of his subordinate officers, but it was a task
of such great magnitude that he elected to personally lead the foray. Of
course, the assignment—to surreptitiously acquire an onboard computer
from an American spacecraft recently returned from orbit—was rather
bizarre, but his GRU superiors had insisted that the tasking had come
directly from the very highest levels of the Soviet command.

If filching a computer from a spacecraft was an odd job, so were the
circumstances surrounding how the Americans stored it. It had taken
Federov's men less than two weeks to surmise the location of the spacecraft. If this had been back in the Soviet Union, the vehicle certainly
would have been locked deep within a heavily guarded compound, but
not so here in the land of capitalism and obscene abundance. Two years
ago, according to Federov's researchers, representatives of NASA and
the renowned Smithsonian Institution had signed a formal agreement
in which all returned spacecraft and related artifacts—all deemed to be
historically significant—would be transferred from NASA's control to
the Smithsonian for safekeeping. From the perspective of NASA admin-
istrators, they were already decisively engaged with the heady enterprise
of going to the Moon, so they had scarce interest in also delving into the
museum business.

Consequently, just after NASA's Gemini program ended, NASA delivered a massive glut of spacecraft and related objects to the Smithsonian. Lacking sufficient space to display the items, the Smithsonian trucked them to their aerospace renovation and storage facility located in Suitland-Silver Hill, a sleepy Maryland suburb of Washington. The modest facility, a cluster of thirty-two prefabricated metal storage buildings, comprised the "Paul E. Garber Preservation, Restoration, and Storage Facility," also known as "Silver Hill" by Smithsonian employees. There, the recently flown spacecraft and related gear were crammed into a warehouse.

Once his researchers determined where the Gemini spacecraft were housed, he dispatched one of his best operatives-—Major Dimitri Gleb Roschin—to establish an operating site. Posing as a US Army officer recently assigned to the Pentagon, Roschin rented a one-story brick house in a neighborhood directly adjacent to the Silver Hill complex. The quaint dwelling on Bonita Street was less than a hundred meters from the warehouse where the Gemini spacecraft were stored, separated only by a stretch of woods, a shallow creek and a chain-link fence roughly two meters tall.

As his advance team of three operatives set up shop, Roschin established a regular pattern of comings and goings with the white Ford Econoline van. He played a stereo, perhaps too loudly for the neighbors, at the appropriate hours, and periodically entertained "fellow officers"—played by GRU operatives—-from his Pentagon office.

Even as he convincingly portrayed an American officer to his neighbors, Roschin and his advance team routinely infiltrated the Silver Hill complex. Federov could barely believe their reports. Laughably, even though the Silver Hill warehouses were overflowing with priceless aerospace artifacts and valuable technology, only four watchmen were responsible for guarding the twenty-one-acre site. Two—the guard supervisor and his assistant—were situated at the headquarters building, at the front of the complex; they took turns strolling through the complex on roving patrols. Besides the guard supervisor and his assistant, one stationary guard was located in a restoration facility that stored several historically significant aircraft being prepared for display at the Smithsonian and other museums, and—unfortunately for Federov and his team—the fourth watchman was solely responsible for the warehouse that held the target spacecraft. The fourth watchman's post was a stationary assignment; he remained inside the locked warehouse for the duration of his shift.

After they had thoroughly scouted the exterior aspects of the complex, a pair of Roschin's men nightly entered the target warehouse to conduct surveillance. Hiding amid piles of boxes, they watched the guard until they were intimately familiar with his nightly routine. They knew his patterns, when he normally ate lunch and when he occasionally grabbed a quick snooze before the sun came up. Most importantly, they knew precisely when the guard supervisor came to inspect his post.

Steering the van into a driveway adjacent to the house, the driver announced that they had arrived. As he quickly moved inside the house with the others, Federov thought about the operation about to unfold. The computer was a fruit ripe for the plucking. Security was effectively nonexistent; the place was as porous as a sponge. It was *so* easy, almost too much so, that he could readily suspect that he was being lured into a snare.

The various teams occupied their assigned rooms in the tiny house and immediately set to work preparing their equipment. The operation's centerpiece was the technical crew, six men who would physically remove the onboard computer from the spacecraft and then replace it with a precisely fabricated—but non-functional—replica.

The majority of the GRU operatives were assigned to the mission support team. They were along to function as little more than pack animals, charged with bearing the technical crew's bulky equipment from the rental house to the warehouse. Their most ponderous burdens were three footlockers, each stuffed with automobile batteries, wired in series, which would power the drills and various gizmos necessary to gain access to the computer.

"We're ready," stated the bespectacled leader of the technical crew.

Federov glanced at his watch, made note of the time, and commenced the operation. He nodded toward his radio operator, who issued a one-word command to a two-man auxiliary team that immediately disabled a nearby PEPCO—Potomac Electric Power Company—substation, shutting off electric power to the area. Armed with a small blowtorch, they would also leave telltale scorch marks on copper "breaker bars," to convince repair crews that the distribution site had been knocked off-line by a lightning strike.

Guided by an operative of Roschin's advance team, a handful of security men departed to set up watch sites throughout the complex. Unlike the remainder of the team, who would soon enter the warehouse, they did not wear raincoats or galoshes, so they would soon be soaked to the skin in short order.

Federov paused for five minutes as he waited for the security personnel to disperse, and then he quietly ordered the remainder of the team to advance on the Silver Hill site. Hoisting their burdens, the team followed Roschin through a sliding glass door, through the small backyard, past a sandbox and swing set left by previous occupants, and into the woods that separated the house from the Silver Hill complex. Moving silently, despite their loads, they ignored the heavy rain. Periodically, the dark woods were illuminated by brilliant flashes of lightning. In just a few minutes, they negotiated a narrow gap in the security fence and a few seconds later, the heavily laden men filed into the entrance of the warehouse.

Federov was confident that the torrential rains and darkness would dissuade the roving guards from their appointed rounds, so they shouldn't be disturbed as they performed their chores. If there was any aspect of the scenario that he didn't like, it was that the warehouse's two entrances—a regular door and a large roll-up cargo door—were on the same side of the building, which dictated that they had only one point of ingress and egress.

The interior of the warehouse was pitch black. After removing their foul weather gear, armed with dim blue-filtered penlights, select members of the technical crew searched the warehouse and examined each space-craft. In preparing for the mission, Federov learned that not all Gemini spacecraft were created equal, and that the onboard computers had significantly evolved from the earlier missions to the later flights. Consequently, the technical crew took the extra time to ensure that they targeted one of the later model spacecraft. To this end, they had carefully studied photographs taken by Roschin's advance team. The images, taken in daylight, depicted the warehouse in jumbled disarray. In addition to the spacecraft, crates and boxes were stacked haphazardly wherever there was open floor space to accommodate them, with only a few clear paths left open for the watchman to make his rounds.

Many of the spacecraft were clamped into yellow-painted wheeled transport cradles, while others were temporarily mounted on wooden support structures. Some were fully exposed and appeared as if they had returned from orbit yesterday, while a few were encased in protective cocoons of plastic sheeting. Federov had repeatedly warned the men to be especially cautious around the hatches, since the ejection seats may not yet be disarmed. Only one man—a member of the technical crew— was authorized to open the cockpit hatches; as the other technical crew personnel removed the thirty-kilogram onboard computer though an access panel on the exterior of the spacecraft, he was responsible for entering the spacecraft and removing the computer's readout displays from the right side of the control panel.

Federov stripped out of his raincoat and dropped it into a rubberized duffle bag. As two support team members used large sponges to mop up water that had dripped on the floor, he listened to heavy snoring from a small room at the front of the building. Earlier, Roschin's inside men had spiked the guard's peanut butter sandwich with a heavy dose of sedatives. At midnight, as was his custom, he sat down to eat lunch and within half an hour, he was sound asleep at his desk. Roschin assured Federov that the watchman was deeply unconscious and would remain so for at least the next hour or so, which should grant the technical crew more than ample time to finish their tasks.

The technical crew swiftly located their target. As some prepared their tools, others quickly erected a black cloth tent to abate light and noise. As the technical crew went to work, the support team operatives formed a line by the door, ready to haul the heavy gear back to the house when the operation was complete. Positioning himself so he could monitor the technical crew and simultaneously communicate with the support team, Federov leaned against the cool metal of the spacecraft's transport cradle. Roschin took a seat next to him. After a particularly long day, followed by several mission rehearsals, Federov was bone-tired. It was very tempting to fall asleep, but he could not be lulled into complacency.

"Equipment access door located," whispered the technical crew's leader from behind the curtain. "Opening equipment access door."

Minutes passed before he quietly reported, "Removing computer."

Federov smiled; other than the incessant patter of rain on the metal roof and the rolling booms of thunder that shook the building, it was almost absolutely silent. He had to concentrate to hear the faintest whir-ring sounds of the technical crew's power tools. He couldn't recall when such a complicated task had gone so smoothly.

"Installing mock-up," reported the technical crew's leader.

As he patiently waited for the technical crew to finish their tasks, Federov meditated on another issue. Just this afternoon, he had received an asinine tasking to investigate rumors of UFOs—Unidentified Flying Objects—supposedly stored at a US Air Force Base in Ohio. He just didn't have the personnel or resources to execute such an outlandish snipe hunt, and try as he might, he could not convince his superiors otherwise. He was rankled by their steadfast insistence that he send an operative to Ohio. *Who could he send?* With all his current and impending missions, he desperately needed all hands on deck. If he couldn't identify someone to send, he might be compelled to beg for help from the insufferable pricks of the KGB's complement at the Embassy, and working with the KGB was something that he stringently avoided.

His thoughts were abruptly interrupted by the jangling of the phone in the guard's office. Federov cursed under his breath and leaned toward Roschin. "How can the damned phone be working?" he asked. "We cut off the power."

"It's a different circuit altogether," whispered Roschin. "The phone lines have a lower voltage, independent of the power grid. Not to worry, though, that guard is so out of it that he couldn't possibly hear the phone ring."

Federov nodded. Major Roschin was almost certainly correct, but he despised loose ends. The phone call was most likely the guard supervisor checking on his subordinate. The worst case scenario was that he would leave the headquarters' dry comfort to investigate, which could potentially block the team's singular escape route. Even so, Federov had prudently addressed the contingency. With his most competent operatives already busy, he had designated one of his least trustworthy officers—Major Morozov—to contend with the unlikely scenario. As a member of the security team, Morozov had *one* solitary task: he was to watch the head-quarters building and prevent the guard supervisor from approaching their warehouse.

It was nothing that required an abundance of brains or technical skills; all Morozov had to do was to emerge from his hiding spot, quietly catch up behind the guard supervisor, and knock him senseless with a rubber-coated cosh. Not exactly a surgical approach, but adequate to buy the team time to finish their chores and make their exit. And although the dim-witted Morozov couldn't be trusted with the more sophisticated chores of espionage, he was certainly adequate to do basic thug work. One well-placed blow should suffice, and the guard supervisor would wake up later with a headache, likely convinced that he had slipped and fallen.

The telephone's ringing persisted. Federov grinned; he concluded that the guard supervisor obviously wanted to know why the warehouse guard was unresponsive, but he apparently was not so concerned that he was willing to venture out into the storm to find out.

"We're almost done," whispered the leader of the technical crew. "Replacing the access panel now."

Federov smelled the distinctively sweet odors of toluene and acetone and knew that the technical crew was deftly applying paint to disguise tool scars and other marks. He heard the faint whirring of a small fan, which they used to accelerate drying of the paint and to disperse the waft-ing fumes.

The phone's ringing still persisted, but suddenly there was also a bang-ing at the door. "*Govno!*" cursed Federov quietly. "What the hell is going on?"

"There's someone at the door!" whispered Roschin.

"No shit," scoffed Federov.

"He can't get in, though," added Roschin. "We jammed the lock from the inside. His key won't work."

The pounding grew progressively louder, and it was obvious that the man outside was now trying to force the door. *Why had Morozov not disabled him?* More significantly, with the only door now blocked, how would they make their exit? Federov swiftly assessed the disintegrating situation. He signaled for the support team leader to be prepared to incapacitate the guard supervisor if he was successful in breaching the door. He grimaced; a face-to-face confrontation was certainly not an encounter he had planned for, and a scuffle—regardless of the outcome—would instantly negate the entire clandestine nature of the operation. The Americans would obviously know that someone—*several* persons—had intentionally entered the facility, and it would only be a matter of time before they realized their objective.

As he waited for the technical crew to report that the job was complete, Federov's head spun with potential scenarios. He worried that this operation was rapidly headed toward a fiasco. A clash with the security guards could be bad enough, but they might also summon the local authorities. He half-expected to hear the wailing of sirens and knew that it might be a matter of minutes before they were scooped up by the American police. Only Federov and a few others carried Makarovs, so it was doubtful that they could win a shootout with determined American law enforcement officers armed with revolvers and possibly shotguns.

"We're done," said the technical leader. "Packing tools now."

Certainly there had to be another way out of this mess. Federov leaned toward Roschin and calmly asked, "I know that there are no other doors, but are there any other possible points of egress?"

"There's a washroom at the rear of the building," answered the Roschin. "It has a small window."

"We'll go out that way. You'll stay behind to clean up. Afterwards, hide here until you can get out safely."

Nodding, Roschin replied, "As you wish, sir."

Federov used his penlight to signal for the leaders of the support team and technical crew. When they drew near, he issued concise instructions. In moments, all the support team's operatives were in quiet motion, headed toward their new escape route. He waited for the technical crew to collect their tools and collapse their tent and then followed them to the restroom. The support team passed items to their companions waiting outside and then scrambled out the small window.

"The battery boxes won't fit through the window," reported the support team leader.

Federov thought for a moment, then shined his light at a stack of crates in a corner. "Leave them there," he ordered. "Pile some of that other junk on top of them. Even if the Americans eventually find them, they will likely never figure out what they are."

Federov was the last man out. He hurled out his small pack and then heaved himself into the miniscule window. Squirming, he realized that his massive shoulders were caught in the frame. He felt Roschin give him a mighty shove, and then he fell headlong into a puddle outside.

The GRU operatives were highly disciplined. Instead of scattering in panic, they remained a cohesive unit, waiting until everyone was accounted for, including Federov, before they conducted an orderly withdrawal from the Silver Hill site. In mere minutes, they were standing in the kitchen of the rented house.

As he waited for the equipment to be loaded into the van, Federov confronted Morozov when he arrived a few minutes later. "Why didn't you take care of the guard supervisor, idiot?" he demanded. "All you had to do was bang him over the damned head! That was your *sole* task, Anatoly Nikolayevich, and you couldn't even do that?"

"It wasn't the guard supervisor or his assistant," declared Morozov. Standing at stiff attention, still holding his rubber-coated truncheon, soaked to the skin, he shivered uncontrollably. "They stayed in the headquarters. They must have called the guard in the restoration building and ordered him to go to your warehouse."

"If that's the case, did you not *see* him?" demanded Federov, stabbing his index finger into Morozov's chest. "You certainly had a clear line of sight to watch him walk from his building to ours!"

"I *did* see him," replied Morozov. "But your orders were to disable the guard supervisor or his assistant if either left their post and approached your building. He was not the guard supervisor or his assistant, so I remained hidden."

"The equipment is packed in the van," declared the leader of the technical crew. "We're ready to go."

"Load up," ordered Federov.

"Did I not follow your orders exactly, Colonel?" asked Morozov sheepishly.

"You did, moron," answered Federov, shaking his head. "But the circumstances called for you to exercise initiative, and you failed to do so. Now, climb into the damned van with the others, and I will deal with you later."

Johnsville Naval Air Development Center, Pennsylvania
8:25 a.m., Friday, March 7, 1969

There were others like it, but the Johnsville centrifuge was *the* best. Built in 1950, it was the world's largest, most powerful and most sophisticated machine of its kind. A sphere-shaped fiberglass gondola, spacious enough to accommodate three men seated abreast, was mounted at the end of a fifty-foot radial arm.

A 4000-horsepower electric motor whirled the arm at incredible speeds. The motor was so massive—182 tons worth—that it was anchored into place and then the round-shaped building was literally built around it. The Johnsville centrifuge could accelerate subjects up to forty Gs, although that would likely be fatal. The current record was 31.25 Gs, sustained for five seconds, set in 1958 by a psychologist named Flanagan Gray who apparently possessed an almost superhuman tolerance for G-forces.

The gondola was mounted on motorized multi-dimensional gimbals, which permitted the operators to make subtle and not-so-subtle adjustments to change the orientation of the rider in relation to the spin, to realistically simulate the effects of dynamic maneuvers in high-G conditions. The matchless machine was truly a one-of-a-kind feat of engineering ingenuity, an amusement park merry-go-round ideally suited for the pathologically masochistic.

A Navy flight surgeon subjected Ourecky to a rambling but informative class prior to entering the gondola; the session described all the potentially debilitating effects of high-G forces and also included pointers on breathing techniques and various tricks that the Navy scientists had developed to mitigate the effects of the oppressively high G-forces.

As Ourecky was being strapped in, the flight surgeon poked his head through the hatch and said, "Okay, just so it's fresh in your mind, let's review: You'll probably experience chest pains, motion sickness, nausea, disorientation, confusion and possibly a panic attack or two. Some folks actually describe feeling a euphoric sensation, but those sickos are pretty rare, thank God. On some occasions, we see fractured ribs or dislocated shoulders, but that's very unusual."

Nodding, Ourecky extended his arms to make sure he could reach the instrument panel.

"You should plan on having a head-splitting hangover later," said the flight surgeon. "If you do, take two Alka-Seltzers, drink plenty of water, and don't call me. Got all that, Air Force?"

"Got it," answered Ourecky.

"Seriously, most guys suffer through these runs without any significant ill effects, but some problems may manifest themselves later. I'll give you some paperwork to tote it in your wallet for the next week or so. If you go home and you don't feel right, see a doctor *immediately*. Give him the paperwork so he'll know what to look for: blood clots, heart arrhythmias, collapsed lungs, nasty things of that nature, the sort of stuff that can kill you pretty quick. Okay?"

"Go to the doctor immediately," said Ourecky. "I got it."

"Now, it's common to have some minor after-effects, so don't get too upset if your feet swell up or you get some strange marks on your body from burst blood vessels. That's just part of the souvenir package that we like to send home with you. We'll also snap your picture when you're at your worst, so you can prove to your boss that you actually did ride the Wheel."

Once they had Ourecky strapped in and marginally comfortable, the technicians clambered out of the gondola and closed the hatch. Apprehensive about what was to come, particularly since a failure to tolerate high G's could immediately wash him out of Blue Gemini, he convinced himself that no matter how bad it became, he would endure it and not complain.

"Everything shipshape in there? Comfy?" asked the operator over the intercom. "We're starting the spin-up now. We'll warm up with a stock evolution: twelve G's for four seconds, eight G's for forty-one seconds, then five G's for two minutes. Then we'll run you through a lift-off profile, then a reentry profile, then we'll shovel on a few more G's and change your orientation just to make sure that you can remain conscious if things start to go a little haywire. If you get too uncomfortable, let us know immediately and we'll back it off. Ready, Captain?"

"Ready," replied Ourecky. He heard a loud hum as the huge motor was energized. The arm slowly started to move and gradually gathered speed. He felt his body being forced back into the seat, like a giant hand was shoving against him. *So far, so good*, he thought. *Nothing to it*.

As the G's accumulated, he became lightheaded and found it increasingly hard to breathe. Following the flight surgeon's advice, he forced his lungs full of air. Knowing that it would be almost impossible to refill his lungs if he let all of his air out, he grunted in shallow pants, huffing in and out rapidly, but not so quickly that he might become hyperventilated.

Forcing himself to concentrate, he scanned the instrument panel. As various colored lights flashed, he pushed specific buttons or activated switches in response. He also reported readings on various dials, to prove that he could maintain his focus and spatial orientation.

As the dastardly machine spun faster, he experienced tunnel vision, as if dark curtains were gradually being drawn beside him. At the extreme corner of the panel, almost outside of his peripheral vision, a red light glowed. Reaching up to throw the corresponding switch, his hand felt like it was encased in lead; it took all of his concentration to complete the simple act.

He started to feel a dull pain under his sternum, like he was on the verge of a mild heart attack. "Ten…G's…okay," he reported, watching the meter.

Moments later, the gondola reversed abruptly. The harness straps dug into his chest; he was in the "eyeballs out" phase of a deceleration profile, simulating negative G's. Breathing in forced grunts, he felt his eyes bulge out from their sockets. Looking at a mirror mounted in the instrument panel, he was astonished when he didn't recognize his grotesquely distorted face.

He heard Carson's voice through the earphones: "Are you okay in there, Scott?"

"Oh…kay," he grunted.

"Scott, in about a minute they're going to simulate the spacecraft wobbling off-axis during reentry. Just focus, and you'll be okay."

The sudden change in attitude was enormously disconcerting; it felt as if his internal organs suddenly sloshed to one side of his abdominal cavity and then roughly shifted again as the nefarious gondola changed orientation. The transverse G-loading added a diabolically new dimension to his agony. Teetering on the edge of unconsciousness, he wasn't sure how much more of this misery he could suffer. Then, almost as quickly as it had started, it was over. As the gondola whirred down and the G-forces diminished, he slowly regained his vision and vestibular senses. Finally, it ground to a halt, and moments later, the round hatch reopened.

"Listen to me, Scott," cautioned Carson, climbing into the gondola. "Sit still and don't move. As you climb out of this thing, keep your head and eyes straight forward. I'll help you to a chair outside. Just a few steps straight ahead, and the seat will be right there waiting for you. Just sit down and don't move your noggin, and you'll be back to relatively normal in a few minutes."

Carson and a technician extricated Ourecky from his restraint harness, then slowly eased him through the hatch and stood him up. Taking baby steps on wobbly feet, he stepped onto the platform. He heard someone laugh, and turned his head ever so slightly, and was immediately overtaken by waves of nausea.

"Easy does it," urged Carson. "So how did you like your first round with the Wheel?"

"Well, that was really something," said Ourecky, determined to appear composed and nonchalant as his head continued to spin. "But we have better rides at the Nebraska State Fair."

Wright Arms Apartments, Dayton, Ohio
6:30 p.m., Friday, March 7, 1969

With his head throbbing, Ourecky walked in the front door and collapsed on the couch.

Wearing faded cut-offs and a tie-dyed T-shirt, Bea came out of the bedroom and sat down beside him. A Janis Joplin album played on the stereo. She kissed his forehead and asked, "Rough trip, honey?" Even though they had not seen each other in over two weeks, she obviously sensed that he was not in any kind of shape for a more enthusiastic reunion.

"That's being generous," he replied. "Do we have any aspirin in the medicine cabinet?"

"I'll bring you some." She went into the bathroom to retrieve the medicine, and brought a glass of water as well.

"Thanks, Bea," he said, swallowing the tablets and drinking the water.

"Hangover? Did you go out carousing with Drew last night?"

He nodded. "Yeah. We definitely went on a bender. Trying to keep up with him will be the death of me. I'll be better in just a minute, after my heart quits trying to thump out of my chest."

She sat down and placed her head against his chest. "Poor baby. Your heart sounds like a jackhammer. Let's just sit still a while. So why don't you tell me about why you're coming back to Ohio? I'm thrilled you're back, dear, but I'm just curious why you had to return so quickly."

Over the next twenty minutes, Ourecky did his best to paint the picture without revealing the true purpose of his return. As he described what his life would be like for the next few weeks—a seemingly endless series of cross-country trips and late night sessions at the base—he felt as if the room temperature had dropped several degrees. "And I'll probably not be able to start on my doctorate for at least another year, if ever," he concluded. "I'm sorry, Bea. I'm happy to be back here with you, but it's going to be almost as if I'm not here at all."

"Okay, honey, let me sort this out," said Bea, sitting upright and crossing her arms over her chest. "So far, from what you've told me, you won't be in California any more. You might not get a chance to go to MIT at all. You'll be traveling almost constantly for the next three months, so we'll rarely see each other. Do I have a good grasp on everything so far?"

Swallowing, he nodded and sipped from the glass of water.

Frowning, she stood up, walked over to the stereo, and turned it off. Pivoting back toward him, she said, "You know, I'm not a brilliant mathematician like you, dear, but this equation just isn't balancing for me. As far as I'm concerned, the things that you've described are all minuses. What exactly is on the *plus* side? Am I missing something? Is this something that you volunteered for or was it dumped on you?"

"Bea, it's really difficult to explain. Just about the only thing I can tell you is that the project I was working on has evolved into something that's a lot more important. They asked me to come back to finish what I started. Beyond that, there's not much I can say."

"So you just want to shove everything behind a curtain marked "classified" and hope that I'll eventually be the obedient little Air Force wife who doesn't ask too many questions? You do know me better than that, don't you?"

"I'll tell you what I can when I can. But that's not always going to be much."

"Are you sure that this is what you want?" she asked. "Aren't there any other options?"

"It's the Air Force. It's not the corporate world. Sometimes I just have to do as I'm told."

"I guess that I would be more comfortable if I knew that it was what you were told to do, and not something that you volunteered for."

"I *volunteered* to join the Air Force, so that's sort of a moot point. Bea, I'm grungy and I'm dead tired," he said. "I've been on the road for three days. I need to take a shower."

"We're not through discussing this," she snapped.

Frustrated, he went into the bathroom, stripped off his clothes, stepped into the tub, and closed the plastic shower curtain. He took his time, luxuriating in the hot water and the momentary solitude. He meditated on how to make all these gears turn in unison. He wasn't even married yet, but he was having tremendous doubts whether he could effectively balance being a husband while simultaneously living the strange and secretive existence he had chosen. *Life would be a lot simpler if I had elected to stay in California,* he thought.

Minutes later, with a white towel wrapped around his waist, he emerged from the bathroom to find her seated on the bed, waiting for him. She had obviously cooled off a little, since she seemed less inclined to argue with him. He turned towards the mirror to brush his hair.

"Well, gee, honey, those are certainly new marks to add to the collection," she commented, pointing at some odd spots on his back. "Care to explain them, or should I even bother asking?"

Holding his arm over his head, he awkwardly contorted his upper body to examine his reflection. She was right. The intense stress of the centrifuge's G-forces had ruptured a maze of capillaries in his back. His shoulder blades and back were dappled with miniscule purple splotches, none bigger than the diameter of a pencil eraser.

Gently poking the spots, she asked, "Do they hurt? What on earth makes marks like that?"

"I can't say."

"Can't say or don't know?" she asked, raising her eyebrows.

"*Can't say.*"

"You know, Scott, I suppose I just should be thrilled that you're not coming home with hickeys on your neck or fingernail scratches up and down your back. But I see these strange scars and marks, and I just can't comprehend what it is that you're involved in. I just wish that I understood. I wish that you could make me understand."

I wish I could make you understand, too, he thought. He turned to her and held her close. "Bea, please trust me. I love you, and we'll make this work. The next few months will be hard, but we'll muddle through them together. You just have to trust me to know that this is very important, and not just for me."

"I love you, too, Scott. I'll try my best to understand. All I ask is that when you go away, that you always come back to me."

"I'll always come back to you," he vowed, hoping that he could keep that promise.

34

VOMIT COMET

Wright-Patterson Air Force Base, Ohio
10:10 a.m., Thursday, March 27, 1969

Over the intercom, the pilot announced that they were commencing the first parabola. Seated on the floor, in an orange flight suit and stocking feet, Ourecky leaned back against the padded wall and apprehensively clutched a strap. A veteran of several similar flights, Carson sat on the other side of the fuselage and watched him. Toward the rear of the aircraft, twelve trainees undergoing the weightlessness orientation course paid rapt attention to their three instructors.

Two G's of gravitational pull cemented Ourecky to the floor as the KC-135A zoomed to altitude. The parabolic profile they were to fly resembled a series of waves; they would be weightless as the plane descended from the peak of one wave to the trough of the next.

As the plane reached the apex of the wave, the floor gradually dropped away, and he found himself hovering, sitting cross-legged in the air, like a Hindu swami who could levitate at will. Suddenly, he remembered his teenaged aspirations of becoming an astronaut, and what it was that had compelled him so strongly to fly into space.

Like an epiphany, he realized that it wasn't the fantastic pictures of rocket ships in a glossy magazine or gazing through his telescope in hopes of glimpsing the rings of Saturn. What had motivated him so intensely was something that dwelled deep in his unconscious. It manifested itself in remarkably vivid dreams that had started when he was not quite six years old.

As a youngster, Ourecky dreamed that he could fly. In these recurring dreams, his flight was not an unusual event, but a sensation that felt as natural as walking. He could depart the ground whenever and wherever he saw fit, with no effort whatsoever, and effortlessly soar from place to place. He would zoom over the streets of Wilber, swooping just above the rooftops, and cut cross-country over the wheat fields and cornfields, moving unfettered by the constraints of Nebraska roads that ran at uniformly painful right angles.

As he flew, he believed if he were able to grab something—a leaf or a twig or even a feather plucked from a bird on the wing—and grasp it tightly in his hand, it would be there when he awoke, and his ability to take flight would pass from his dreams into the light of day.

That was a dream, but this was reality. Stretching out his hands and legs, he relaxed his body and floated about three feet from the padded floor. He swiveled his head and looked toward the rear of the plane. Some of the passengers were readily adapting to weightlessness while some appeared slightly distraught as they vainly struggled to swim through the air in the middle of the cabin.

Two men were so woefully disoriented that they snatched out the airsick bags tucked in the chest pockets of their flight suits, lending credence to the KC-135A's label as the "Vomit Comet." He looked across the cabin at Carson, who stood against the wall, wedging himself in place with his outstretched arms and legs. Carson was grinning from ear to ear, and Ourecky could only imagine that he was doing the same.

The respite from gravity lasted only twenty-five seconds, and then he settled back to the floor as the plane entered the trough of the wave. As the KC-135A rocketed upwards, Carson slid over and sat next to him. "Why did you stay jammed up against the wall?" asked Ourecky, tugging up a loose sock. "I can't believe you didn't just let go and fly. This is so amazing."

"Oh, man, I was having a great time just watching you," answered Carson. "I wouldn't have missed that for the world. Ready for some more?"

"You bet. I could stay on here forever."

"I hope you still feel that way in an hour or so," Carson said, laughing.

6:15 p.m.

Waiting for Bea to come home from the airport, Ourecky dozed on the couch. After his long day on the Vomit Comet, he was revisited by his childhood dreams of flight. He awoke to find Bea standing over him. He sat up and wiped a string of drool from his chin.

Unbuttoning her coat, she looked at him and laughed quietly. "So, Captain Ourecky, what sort of new scars or marks do you have *this* week? Rabies? Radiation burns? Broken bones? Malaria? Scarlet fever? Hives?"

Quickly reacquainting himself with solid ground and the oppressive conventions of gravity, he replied, "Nope. Nothing unusual this week." He reached out and hugged her around the waist.

"Not even a paper cut or fever blister? Hanging cuticle? Athlete's feet? Halitosis? Should we notify the White House? Call the Pentagon?" She took off her coat and draped it over the coffee table.

"I don't know what was passing through that head of yours, but you had the biggest smile on your face," she said, sitting down beside him and mussing his hair. "If I didn't know you any better, I would have thought you were smoking something. You looked positively euphoric. What in the world were you dreaming about?"

"I dreamed I could fly," he replied, still smiling. "I always dreamed that when I was a kid."

"So you *fly* in your dreams, huh?" said Bea, grinning. She giggled and added, "Meet me in the bedroom, darling, Maybe afterwards you'll have something else to dream about."

Trailways Bus Station, Washington, D.C.
9:35 a.m., Saturday, April 12, 1969

As he waited in the pouring rain to board the bus, water dribbled from the prominent nose of Anatoly Nikolayevich Morozov. He did have an umbrella, but it was old and only barely serviceable. Its crook handle was missing and three broken spokes drooped uselessly. There were so many holes in the worn fabric that he might as well have been holding a sieve over his head.

Finally, after enduring the deluge for several minutes, he climbed up the stairs of the bus, awkwardly clutching his sopping pressboard suitcase to his chest. His glasses, beaded with raindrops, were all but useless as he fumbled for his ticket. The driver sighed as he waved him to the rear.

Negotiating the narrow aisle, Morozov found a pair of vacant seats near the back of the crowded bus. He jammed his suitcase into the luggage rack and swung into the window seat. He had purchased a newspaper to occupy his time for at least part of the trip, but now it was a soaked completely through, unfit for anything except perhaps to produce a child's papier-mâché.

He carefully laid the wet newspaper on his lap, smoothed it flat with his hands, and glanced at the headlines on the soggy front page. In their renewed offensive, Communist forces had shelled forty-five towns and bases in Vietnam. A Mideast peace plan, offered by King Hussein of Jordan, was rejected by Israel, even though the plan recognized the exist-ence of Israel. Another article reported that Egyptian and Israeli forces had exchanged fire over the Suez Canal. China and the Soviet Union were still skirmishing over their shared border.

He gazed down at his shoddy Soviet-made shoes; waterlogged, their flimsy soles of pressed cardboard were literally starting to peel off. Thankfully, he did have an extra pair of footwear—made in America, no less—in his suitcase, along with a simple but serviceable wardrobe purchased yesterday at a Goodwill thrift store in Alexandria, Virginia.

To further pile insult on injury, water dripped from his suitcase overhead, spattering onto his bald crown. He considered shifting to the adjacent aisle seat, but then a pimply-faced American girl occupied it. Lugging an Army duffle bag emblazoned with peace signs and Marxist quotes, she wore filthy blue jeans, a ratty black T-shirt, and metal-framed granny glasses with rose-colored lenses. Obviously still in her mid-teens, she was likely a runaway from a suburban home, probably on her way to a peace rally, love-in or some other drug-fueled gathering of misguided youths.

The girl stuffed her duffle bag underneath the seat, frowned at Morozov, and immediately opened a well-worn copy of *Tiger Beat* magazine. Minutes later, the bus backed out of the station, wound its way through the crowded streets of the nation's Capitol, and then gradually proceeded towards the northeast through the suburbs of Maryland.

As he rode, Morozov looked out the rain-streaked window and contemplated how he came to be here, seated alongside a gum-smacking teenybopper. Although he passed himself off as an industrial supplies salesman, Morozov was actually a major in the elite GRU—*Glavnoye Razvedyvatel'noye Upravleniye*—foreign military intelligence directorate of the Soviet military.

Born in 1939 in a little town just outside Stalingrad, Morozov had been a young child during the terrible years of the German invasion and post-war deprivation. His greatest hero was his father, Nikolae Nikolayevich, a sergeant who had lost his legs in the valiant siege against the German Sixth Army. Existing on a meager pension and the kindness of strangers, reduced to begging after he was unable to return to his pre-War work as a railroad switchman, his father had always insisted that it was important for a man to know his place in the world.

Growing up in rubble-strewn streets, Morozov had not known a day without hunger before he was drafted into the Army in 1957. Assessed with higher than average intelligence, he was selected for officer training and then subsequently chosen for the elite ranks of the GRU. As such, he had travelled far from the banks of the Volga River where he had scampered and scavenged as a child, spending the last five years assigned to the GRU's resident office within the Soviet Embassy in Washington, DC.

His father would be delighted with him, because Morozov certainly knew his place. He was painfully aware that his GRU contemporaries had left him in their wake; he patiently marked time as they moved on to positions of greater responsibility and strategic significance.

Although posted to one of the premier foreign assignments within the GRU, he was still a mediocre third-tier intelligence operative, deemed not yet sufficiently seasoned to handle his own network of sources. Tolerated by his superiors because he was a master of paperwork and administrative minutia, he was only considered adequate to do the menial tasks and scut work necessary to support the operations of the more accomplished spymasters.

After his half-decade in purgatory, Morozov was finally granted an independent assignment. Although it didn't seem to be a task of vital importance, at least he would be doing *something* in the field. Currently, there was a worldwide obsession with flying saucers and UFOs; the GRU was keenly curious to discover what the Americans knew about alien spacecraft. There were persistent rumors that the US Air Force actually possessed at least one UFO, which supposedly crashed in Roswell, New Mexico in 1947. According to unconfirmed sources, the damaged craft was secretly conveyed to Wright-Patterson Air Force Base, Ohio. Probably not by coincidence, the Air Force was also home to Project Blue Book, whose officers investigated claims of UFOs by pilots and civilians. The GRU wanted to know more about Project Blue Book, so Morozov's boss—Colonel Federov—was sending him to Ohio to quietly investigate.

Blue Book, he mused. *Flying saucers from Mars.* Sighing, he looked at the newspaper in his lap and scanned the new items about Vietnam. He gritted his teeth and closed his eyes. *Vietnam.* That's where he *should* be right now, instead of this fool's errand in Ohio; an advisory tour in Hanoi was a sure ticket to rapid advancement within the GRU. But alas, here he was, on a stinking Trailways bus westbound for Ohio, listening to the incessant gum popping of his adolescent fellow traveler and desperately wishing that he was dry and warm.

35

SERE

Auxiliary Field Ten, Eglin Air Force Base
7:25 a.m., Monday, April 14, 1969

With his flight suit completely soaked with sweat, Carson had forgotten how just brutally hot Eglin could be. The SERE—Survival, Evasion, Resistance and Escape—classroom was a squat tin-roofed building with two small windows, no air conditioning, and one fan. Situated in a remote corner of the Aux One-Oh compound, it could have readily been a setting for the highway camp in *Cool Hand Luke*. He knew that the next two weeks wouldn't be appreciably different than a chain gang's; there was little to look forward to but excruciatingly long days and minimal sleep, and then it would get much worse in the third week. He didn't particularly relish the thought of spending the next three weeks here, especially since he had already suffered through this course before, but his underlying motivation was to be here for Ourecky.

He observed that the place had changed little since last summer. The walls were plastered with brightly painted placards that spelled out the Code of Conduct and parts of the Geneva Convention pertinent to the treatment of prisoners. A map of North Vietnam was pocked with a multitude of red dots—like measles on a child's face—depicting known POW camp locations.

Ten other men were here for the course. Four were Navy pilots, apparently assigned to the same squadron. They were vague and close-lipped about their mission; when asked, they mumbled something about being involved in reconnaissance, but divulged little else.

The instructor, an athletic captain assigned to the 116th, loudly cleared his throat and spoke. "Welcome to the SERE Contingency Course. This course focuses on survival in extreme circumstances. Here at SERE Contingency, we're not going to sit around the campfire at night strumming guitars and singing "Kumbaya." We don't show you how to snare Bugs Bunny, crochet afghans out of parachute shroud lines, catch fish, build teepees, start fires, or any of that touchy-feely stuff you should have learned at the "greenie" survival courses. You'll learn evasion tactics to avoid capture, how to survive in captivity, and also what actions to take in the event of a rescue operation, either before or after you're captured. If you pay attention, you'll live. If you don't, your family will likely receive a neatly folded flag to display on the mantle."

The captain continued. "Although you all come from different backgrounds, you share a common trait: the nature of your work is such that you carry both a high risk of capture as well as an extremely high risk of exploitation. Moreover, all of you are obviously toting around some information of significant strategic value in your heads, or you wouldn't be sitting here today. That said, if you punch out somewhere, the 116th is tasked to rescue or recover you. We exist solely to rescue *you*, not the average aviator, because someone high up on the food chain has deemed you far too valuable to fall into or remain in enemy hands."

"When you leave here, the main lesson you should take away is to trust us to do our jobs. If you go down somewhere in the world, whether you're on the run or sitting in a POW cell, you can trust that we're on our way. So while you're here, we'll not only show you what you need to do to survive in an extreme environment, but hopefully we'll teach you to trust us as well."

The captain sipped from a canteen. "Let's talk briefly about the nature of survival. By the time the average person realizes that they're in a life-or-death situation, it's usually too late for them to alter the outcome. They're too weak or have already misused the resources available to them. When it finally dawns on them that they're in desperate straits, it's almost certain that they'll panic and dig themselves into an even worse situation.

"To survive, you have to mentally shift gears quickly. You have to intuitively know whether to speed up or slow down. More than anything else, you have to adapt yourself to the circumstances because the circumstances damned sure aren't going to adapt to you."

He continued. "You have to make full use of your most precious resources, including time, the knowledge in your head, and the energy in your body. You have to play the game smart. As an example, if you're out on foot in the wilderness, evading capture, the time to be looking for your next meal is while you still have a full belly from the last one. If you wait until you're debilitated by hunger and thirst before you start looking for water or food, you've probably waited too late."

"You must adapt and you must adapt quickly. You eat what's available. Imagine yourself clamped in shackles in a POW camp. If a guard throws you a chunk of meat crawling with maggots, you set aside your qualms, chow down, and then politely ask for seconds."

"If you've punched out and you're still evading, regardless of how rough things become, do not allow yourself to be captured. Mark my words, gentlemen, capture is your least favorable option. But I'm sure that lesson will have sunk in by the time you fly back to your home stations."

For the next two weeks, the twelve men were almost constantly active. Working in pairs, they lived outdoors, every night constructing a concealed "evasion hide" that functioned both as shelter and camouflage. Theoretical lessons were followed by practical exercises, and the simple practical exercises eventually were rolled up into larger exercises, where they combined a variety of their new skills in ever more challenging situations. The arcane agenda was jam-packed; they were lucky if they got any more than five hours of sleep—mostly snatched in brief catnaps—within any twenty-four hour period.

Communications was a key aspect of the training. Besides practicing extensively with radios, signaling mirrors, flares and dye markers, they learned a battery of signals and codes, and rehearsed them until they were deeply engrained in their rote memory. Some were intended for use in an evasion situation, such as simple ground-to-air signals to communicate their status to an overflying plane. Most were standard, recognized by aviators around the world, such as 'II' to convey 'Need Medical Supplies' or 'X' to indicate 'Unable to Proceed.'

The ground-to-air signals provided a tremendous degree of flexibility; they could be displayed with special signal panels or could be created by simply stomping in snow, arranging logs or using any materials in the environment to exhibit the applicable symbols. Likewise, a series of the signals could be surreptitiously left by an evader as he proceeded along his escape route, to give rescuers a track of his progress as well as a running commentary of his status.

The men learned and practiced communications systems used in POW camps, including a modified version of the deaf-mute alphabet, which facilitated silent conversations, and an ingenious but simple "tap" code, based on a five-by-five letter matrix. The tap code could be used in a variety of ways, from communicating through walls to silently exchanging information in close quarters, through virtually imperceptible squeezes or nudges.

Tactics instructors showed them how to use camouflage to dissolve themselves into a variety of environments. They learned to skillfully use the terrain to their advantage, carefully avoiding the "natural lines of drift" where people are prone to move. In time, the men learned how to move quietly and quickly while leaving few traces of their passage. They were taught techniques to evade detection by tracking dogs and visual trackers. They learned to orient themselves and find cardinal directions without a compass, using the sun and stars.

The instructors strongly cautioned them to avoid contact with locals in an evasion situation, but also showed them how to effectively use their evasion aids—the blood chit, pointee-talkee, and bartering materials—if they needed assistance and contact was unavoidable.

They learned rudimentary field medicine, which centered primarily on splinting fractures and treating minor injuries so they could keep moving. The instructors stressed the criticality of taking care of wounds—even seemingly trivial cuts and scrapes—to stave off infection. They were acquainted with natural remedies—chewing willow bark as a substitute for aspirin, using a smidgen of kerosene to flush out intestinal worms, munching charcoal to ease an upset stomach—that could suffice if a modern pharmacy was out of reach. They were shown how to extract an impacted tooth and treat other common maladies that could incapacitate them in an austere setting—such as a POW camp—where more advanced care was not available.

Knowing that the harshest lessons would come in the final week, they learned how to survive in captivity and resist interrogation. They learned the futility of believing that they could hold up indefinitely under torture, because even the most stoic eventually break, and that while they could expect to endure considerable pain and discomfort in captivity, they should anticipate interrogation techniques that were far more subtle and insidious than torture. The most effective coercion techniques typically entailed inflicting extreme pain and duress without causing lasting injury; these methods, combined with an intensive campaign of psychological warfare to abrade away a subject's mental defenses, usually yielded the sought-after secrets.

The core essence of the training was that they should always trust that someone would eventually come to their rescue. If they were shot down or otherwise came to ground in a hostile area, they should assume that the gears of the rescue machine would quickly be churning, so they should use the tools they learned to buy time. If the circumstances allowed them to evade, they should evade. If they were captured, it was vital that they keep their wits about them and wait for rescue. Regardless of where they were in the world, they should be assured that the cavalry was coming, and they would ride through hell and high water to bring them to freedom.

Resistance Training Laboratory, Eglin Air Force Base
2:15 p.m., Thursday, May 1, 1969

Ourecky flinched as the huge guard slammed him backwards against the wall and slapped him again. In a daze, he wasn't sure of where he was, and wasn't absolutely sure that he was still in the United States. Striving to think logically to penetrate the dense fog of pain and exhaustion, he concluded that he had to still physically be at Eglin, despite all evidence to the contrary.

The guard drenched him with ice water from a metal pail. Ourecky gasped deeply, struggling to catch his breath. The towering guard leaned close to him, so that their faces were mere inches apart. His breath was horribly foul, like he subsisted on rotten sardines and strong onions washed down with rotgut liquor. Shaking uncontrollably, Ourecky gagged as the man breathed heavily on him.

"*Why* are you so *slow*, Sky Pirate?!" bellowed the guard in a heavily accented voice, clouting Ourecky hard across the cheek. "When I tell you to motor, you *motor*! Now, motor *you*! *Faster* motor!" The guard shoved Ourecky down the hall.

After five days in the camp, Ourecky was still learning the strange argot. The guard obviously wanted him to move—*motor*—faster, but his swollen feet were covered with cuts and blisters. He struggled mightily, but obviously didn't move fast enough to suit the onion-breathed giant ogre; the guard slapped the back of his head, knocking him to his knees in pain.

"Hard cell for *you*, Sky Pirate!" chided the guard, laughing at Ourecky's agony. "The Bearded One is anxious to talk to you! Now, motor! Move more *quicklier*! Motor! Motor you!"

Crawling on his hands and knees, with the massive guard pummeling him with a rubber truncheon, Ourecky barely made it to the threshold of the interrogation cell. The door creaked open and another giant stepped out into the hallway. Wearing jackboots, black uniform trousers, a wide leather belt with a silver buckle bearing a red hammer and sickle emblem, and a black sleeveless "wife beater" undershirt, the immense man—known to all as "The Bearded One"—had thick black hair and a bushy black beard. "This one?" he asked sternly.

Ourecky looked up in awe. Just absolutely huge, the ominous man had to be scraping close to the seven-foot mark. The Bearded One would be scary and imposing to men who considered themselves scary and imposing. With one massive hand, he casually scooped the engineer clear off the floor before heaving him bodily into the interrogation cell.

Stooping down, the interrogator closed the door behind him, jerked Ourecky up and then slammed him into a chair nailed to the floor. "Now you talk," vowed the behemoth. It struck Ourecky that the man's words were not a threat, epithet, or casual comment, but a simple statement of fact. If he had learned nothing else in the past five days, he knew that the interrogators were intensely professional in their tasks. And very, very confident.

Wright Arms Apartments, Dayton, Ohio
1:30 a.m., Saturday, May 3, 1969

Bea gasped as Ourecky shuffled through the door. His right eye was purple and swollen, and several bruises marked his face. She stood up to hug him; he winced in pain at her embrace.

"Tired," he muttered. "I really need to sleep now, Bea. We'll talk in the morning."

"Okay, Scott. We'll talk in the morning. You really stink, though. I'll run you a bath. You'll probably feel a lot better if you soak a while."

He nodded and allowed her to guide him to the bathroom. If his bruised face had not been bad enough, she was even more shocked as she helped him out of his clothes and assessed the full extent of the damages. For all the past injuries that he could not or would not explain, this was the worst she had seen. His body was covered with scrapes, bruises, and contusions. Bea's mind swirled with a multitude of dire scenarios that might explain his appearance.

It dawned on her that he might have been flying with Carson in the past three weeks, and suddenly there was a logical explanation. "Please tell me you didn't have to eject, Scott."

Grimacing, he shook his head. "It's really not as bad as it looks, Bea. And I didn't eject."

"Then what? Car wreck? Bar fight?"

"No, Bea," he mumbled, canting his head to look at her through his unswollen left eye. "Nothing like that. Someday, I'm sure I'll be able to explain."

"Don't bother, Scott. I'm just glad you're home." After helping him into the tub, she went into the bedroom to replace the new sheets with older ones that wouldn't be ruined as he slept. She heard snoring from the bathroom and went back in to find him sound asleep with the water still running. She lowered the lid on the toilet, sat down, and contemplated him as he slept.

Maybe someday he will be able to explain, she thought, *but that day would likely be long in coming, if it ever arrives.* As strange as their circumstances were, at least he came home. That was a lot different from a lot of Air Force officers and other military men who left with every intention of returning, but died or were captured in Vietnam. So she resolved to be patient with him, in hopes that one day she would understand.

Bachelor Officer's Quarters, Wright-Patterson Air Force Base, Ohio
2:15 a.m., Saturday, May 3, 1969

By the time he made it back to his two-room suite in the BOQ, Carson was physically beat. Ourecky had been in no shape to drive, so he had given him a lift to Bea's place.

It had been an excruciatingly painful ordeal. Right now, all he wanted to do was strip out of his clothes, take a shower, and get some sleep. He dropped his B-4 bag in the closet and zipped open the side compartment to extract his Dopp kit. As he pulled out the leather bag, he noticed a crumpled envelope stuffed in the compartment. He quickly recognized it as the one Pete Riddle had handed him on the morning of his ill-fated flight. In the rush of coming back here and immediately jumping on the treadmill to prepare for the June launch, Carson had completely over-looked the envelope and had forgotten about Pete's last minute request.

Suddenly overtaken by guilt, Carson tore open the white envelope. Expecting to find a solemn farewell letter, he was instead surprised to find some official papers. The papers were accompanied by a simple note, hand-written in Riddle's unmistakably neat block lettering. Reading the note's succinct instructions, Carson grinned; for someone who always seemed to have his head in the clouds, Squeaky turned out to be an exceptionally practical guy.

36

INSURANCE

Flight Crew Office, Aerospace Support Project
11:35 a.m., Tuesday, May 6, 1969

Humming along to "The Age of Aquarius" as it played on the AM radio, Carson carefully addressed a large manila envelope. "Man, it hurts me just to look at you," he observed, studying Ourecky's still-swollen eye and puffy bruises that were just now starting to fade. "I sure hope you feel better than you look, because you look like hammered crap."

"I feel fine," noted Ourecky, rubbing his eye. "Just a little sore, but I'm healing quickly."

"If you say so. Hey, I need to swing over to the post office to mail something." Referring to Riddle's notes, Carson verified the address. "It has to go by registered mail. You want to ride over with me? Maybe catch some lunch at the O Club afterwards?"

"What are those?" asked Ourecky, looking at some documents in front of Carson.

"Well, if you must know, this is a notarized copy of Pete Riddle's death certificate. I got it from Sherry this morning, down in Personnel." Carson pointed at the other paper. "And that little gem is the official accident report that shows that Pete died in a flight training accident."

"So what do you need them for?"

"I'll tell you, but I would appreciate it if you didn't share this with anyone else, especially Virgil Wolcott, unless he specifically asks you about it, and I don't suspect that he will."

"Done," agreed Ourecky. "So what's it about?"

"As it turns out, Squeaky bought a term life insurance policy before he flew. It was one of those policies that's pretty restrictive. For example, it would have been invalid if he had been killed while flying in combat, but there were no restrictions about training flights. So I submit this paperwork, with Pete's death certificate and this documentation that clearly shows that he didn't die in combat, and his parents will receive a big check."

"But didn't they already get his government insurance?"

"Yeah, a whoppin' ten thousand bucks worth," replied Carson. He gave the paperwork a once-over glance and stuck it the envelope. "But now they'll be receiving a windfall of half a million dollars. So I think that it's safe to say that they'll be set for life."

Ourecky tried to picture what half a million dollars would look like, and then suddenly he thought of Bea. "Drew, do you think that I…"

"I know what you're thinking, Scott. I figured that you might ask that. I looked over the policy's fine print, and the beneficiary has to be a direct relative. Sorry."

Carson heard distinctive footsteps outside in the hallway; he switched off the radio and flipped over the envelope just as Virgil Wolcott walked in. "What brings you upstairs, Virgil?" he asked. "We don't see you up here too often."

Wolcott held out a red-bordered report. "Good news, of sorts," he said. "The investigation has been finalized. We should be able to launch next month."

"So what happened, sir?" asked Ourecky. "Did they find out what happened in February?"

"Yup. The investigation narrowed it down to a failure in an itty-bitty component called Module 7651. The post-mortem showed that someone on a subcontractor's assembly line used a strand of wire that wasn't to spec. So we lost two good men and a rocket to a danged snippet of copper wire less than three-quarters of an inch long."

"That's *all*, Virgil?" growled Carson, clinching his fists. "Some son-of-a-bitch subcontractor wasn't happy with his profit margins and he decided to substitute some shoddy, lower quality wire, and Big Head and Squeaky were *killed* as a result?"

Wolcott solemnly shook his head. "It ain't quite that danged simple, hoss. Before you rush out the door with your six-shooters blazin', you need to hear the whole story."

Wolcott explained. "No one did anything with malicious intent. No one was tryin' to cut corners. What happened was this: a minimum wage assembler on the line ran out of the proper wire. She substituted a segment from another spool that looked exactly the same to her."

Wolcott took a long draw from his cigarette, exhaled, and continued. "To her credit, after she soldered in the wire, she immediately notified her supervisor. He was supposed to flag this gizmo for additional testing, but it happened at the end of their shift, right before a long weekend, so it fell through the cracks. Two quality control inspectors also missed it, and that Module 7651 was eventually installed in the rocket that Howard and Riddle strapped on."

"I can't imagine a little piece of wire doing so much damage," said Ourecky.

"Yup. Neither could I, pard. Right now, our best guess is that the module made it through all the pre-flight tests, but the non-spec wire overheated and failed about six seconds after they cleared the pad. That's probably what commenced the chain of events that led to the rocket being destroyed. But that ain't the really ironic part, pardner."

"What's that, sir?" asked Carson.

"That particular Titan II was pulled out of a silo in Arkansas. Had it remained there, and if we had ended up in a thermonuclear shootin' war with the Soviets, it probably would have performed precisely to spec. The irony is that Module 7651 was part of the man-rating modification package. Before we put folks on top of a former ICBM, we insist on the man-rating mods to ensure that it's safer and more reliable. As you might guess, there are some very persnickety rules for fabricating and installing the parts associated with those mods."

"I would imagine," noted Ourecky.

"Well, pardner, we're secretive to a fault. We're so danged secretive, in fact, that although the primary contactor was aware that we were modifying the rocket under man-rated specs, and the first level subcontractor was also aware, no one bothered to notify the second level subcontractor. So that little lady had no way of knowing that the lives of two men relied on this hardware. Otherwise, the wrong wire would have never been used in the first place."

Wolcott tapped his cigarette in a glass ashtray and then added, "You can rest assured that everything will be in kilter before you go up. But I'll tell you something that I learned when I was flying B-17's in the War. What happened was terrible, but we can't incapacitate ourselves by continuing to dwell on it. So, hombres, let's set this issue aside and move on to June."

Wright Arms Apartments, Dayton, Ohio
2:10 p.m., Saturday, May 10, 1969

"Bea, I want us to get married."

"Haven't we already covered this ground, Scott?" she asked, holding out her left hand. "You don't remember giving me this at your parents' house? Christmas? *Ring* any bells, dear?"

"That's not what I meant," Ourecky said. "I mean that I want us to get married as soon as possible. I've checked on it, and we can do it at the courthouse downtown on Friday morning, after you get home. Nothing fancy. Just a simple civil ceremony. When my schedule settles down, we can do a big church wedding back at home to make my parents happy."

"*This* Friday?"

"Yes, Bea. We already have our blood tests, so all we have to do is pick up the license at the courthouse, and we'll be set. Wolcott has given me and Drew off already."

"So it's set? We go down to the courthouse on Friday morning? Why the rush, Scott? You rarely do things on a whim. What's gotten into you?"

"It's just something I want us to do. Drew's already agreed to be my best man. Cool, huh?"

"So you've already set all the gears in motion before you sprung this scheme on me?" she asked, laughing. "Okay, but I insist on having a maid of honor if Drew is going to be there."

"Suit yourself," he said, picking up the newspaper. "Got anyone in mind?"

"Oh, *yes*, I do," she replied.

As Ourecky read the paper, Bea went into the kitchen and dialed a number. Speaking into the phone, she said, "Hey, I have some big news. We're getting married on Friday, down at the courthouse. I guess that Straight Arrow Scott is embarrassed that we've been living in sin, so now we're going to tie the knot and go legit. Anyway, Scott's planning to have a best man. Could you be there for me as my maid of honor?"

Coyly grinning at Ourecky as she twisted the phone cord in her fingers. "Oh, yeah…you know him. It's Drew Carson…Really, Jill, I'm sure it will be all right. Drew's really settled down recently. I think Scott's been a positive influence on him."

Listening intently, Bea was quiet for a few seconds and then said, "Babysitter? I hadn't even thought about that. I'll pay for it, Jill. Really, I just want you to be there."

Montgomery County Courthouse, Dayton, Ohio
1:30 p.m., Friday, May 16, 1969

Acting in his quasi-official capacity of best man, Carson located a minister at the courthouse and negotiated the details on behalf of Ourecky. For fifty dollars, the clergyman was willing to perform the ceremony and endorse the marriage license, so the couple could be legally wed.

The day was warm and the sun was bright, so Carson slipped on his Foster Grant aviator sunglasses as Jill Osborn practiced with her new Polaroid Swinger camera. In addition to being the maid of honor, she had volunteered to photograph the ceremony as well.

Carson studied Jill as they waited. It appeared as if she had gained a few pounds recently, but otherwise still looked good in her purple miniskirt and white go-go boots. As they stood on the courthouse steps waiting for the betrothed to make an appearance with the appropriate paperwork, he tried his best to make small talk with Jill, but she was having none of it. She acted uncomfortable and distant, not even making eye contact with him. Although it had been well over a year since they had dated, he couldn't remember doing anything particularly offensive during the few times they went out, so he was baffled by her glacial behavior.

Finally, Ourecky and Bea came out of the courthouse. He wore a dark blue suit; she was attired in a simple white dress. "Ready?" asked Carson.

"All up," replied Ourecky, gleefully waving the embossed license. "Let's light this candle."

"Wait, wait!" demanded Bea. "Scott, you're so impetuous! We have to do this right. Now, let's see: Something borrowed, something blue…"

"Here," said Jill, removing her sapphire necklace. "This should cover all the bases."

Bea entrusted her simple bouquet to Jill as Ourecky fastened the donated jewelry around her neck. "Okay. That's taken care of," said Ourecky. "Anything else?"

"No," replied Bea, shaking her head. "That should do it."

The minister arranged the wedding party and then proceeded with the service. Lasting barely more than five minutes, with scarcely enough time to cover the requisite "I do's," the ceremony was short and simple.

Concluding the ritual, the minister declared, "I now pronounce you man and wife. You may now kiss your bride." As the newlyweds shared a showy and passionate kiss, the minister turned to the Carson and Jill and quietly added, "Who God has joined together, let no man tear asunder." Jill offered Bea her flowers and then stepped away to snap pictures. After signing the license, the minister congratulated the newlyweds, collected his fee, and went on his way.

"Hey, Jill," said Bea. "Set that camera down for a minute so I can do one of my ceremonial chores." Turning away, Bea lobbed the bouquet over her shoulder; in what should have been a sure-fire "gimme," the flowers went straight through Jill's hands and plopped to the sidewalk.

Afterwards, in keeping with tradition, the wedding party ate cake—in the form of Hostess Twinkies—after Bea shoved one of the gooey snacks into Ourecky's face. Then Carson and Jill spattered the couple with Minute Rice as they ran towards Bea's Kharmann Ghia.

As he walked Jill back to her car after the impromptu ceremony, Carson glanced at her and said, "You know, I think we got off to a really rough start. Since you and Bea are obviously so close, it looks like we'll probably be seeing each other a lot more often now. Do you think there's any chance we can make amends? Make a fresh start?" Walking into the shade of a tall building, Carson slipped off his sunglasses and looked at her. "I was thinking that this weekend, maybe we could go out to dinner and catch up."

Jill stopped in her tracks and said nothing. Turning to him, she gazed deeply into his eyes and then immediately burst into tears. Then she spun away and all but ran to her car.

Bewildered, Carson put his sunglasses back on and headed toward his Corvette. *What did I do this time?*

Le Bourget Airport, Paris, France; 2:30 p.m., Friday, June 6, 1969

Major General Gregor Mikhailovich Yohzin could scarcely believe his good fortune. The GRU sometimes still called him for mundane chores like analyzing technical reports or reviewing films of American missile launches. Such occasions usually entailed a short stint in Moscow, and although the work was typically boring, it did offer him an opportunity to see the sights of the Soviet capital and to sometimes take in a show or two at one of the prestigious ballet companies or orchestras. The GRU often allowed him to take his family, so it was a unique opportunity for a short vacation away from the dreary environs of Kapustin Yar. An opportunity to visit Moscow was grand, but *this? This* was like a dream come true. The GRU had dispatched him to attend the immense spectacle that was the *Salon International de l'Aeronautique et de l'Espace*—the Paris Air Show!

Granted, his family could not accompany him on this trip, and he couldn't take a single step without being constantly shadowed by a pair of GRU thugs, but it was almost too much to hope for. At the US pavilion, the crew of NASA's Apollo Nine flight—Jim McDivitt, Dave Scott and Rusty Schweickart—were present, standing before the Apollo Eight command module that had circumnavigated the Moon last December. Yohzin watched as the three astronauts greeted two cosmonauts—cosmonauts Vladimir Shatalov and Aleksei Yeliseyev—before giving them a tour of the spacecraft.

A meeting of astronauts and cosmonauts! He could hardly believe that he was so fortunate to be present at such a historic occasion. Starstruck, Yohzin snapped pictures of the five space travelers before reminding himself that he was not here as a tourist, but was on a mission. Regaining his focus, he used his Leica to document details of the other spacecraft, hardware, and space-related artifacts in the pavilion. In particular, he had been instructed to concentrate his attention on assessing NASA's spacesuits and related equipment, especially those intended for extravehicular activities.

After he had spent over an hour in the US pavilion, he strolled by a massive exhibit associated with the supersonic Concorde. He paused for a moment to study the display and was momentarily distracted by a trio of French models, apparently part of the exhibit, scantily dressed in revealing outfits that looked barely like the uniforms worn by stewardesses. As he continued walking, a very attractive French model tripped, not too far from him, and tumbled to the ground.

She had apparently snagged one of her absurdly high heels in a mass of television cables. As she fell, her futuristically themed metallic dress ripped up the side, immediately drawing the attention of the mostly male crowd. Not only was it obvious that she wasn't wearing a bra, but also that she wasn't wearing much else, either.

Just by happenstance, or seemingly so, she had plopped down next to Yohzin's GRU handlers. In a clumsy gesture of chivalry, the GRU operative closest to her removed his suit coat and draped it over her shoulders, as the other goon sheepishly helped the woman to her feet. As the two escorted her to a nearby divider curtain, she thanked them effusively for coming to her rescue. She offered to buy both drinks in the executive lounge, but they begged off and returned to their shepherding duties.

In the middle of the spectacle, in the few seconds that the GRU operatives were distracted, Yohzin felt someone brush against him from behind and felt a hand swiftly moving under his jacket. France was notorious for its pickpockets, so this thief was either grossly inept or he specifically *wanted* Yohzin to be aware of his actions. Angry, Yohzin started to pivot to confront the man, but heard a voice, speaking softly in German. "*Guten Tag, Herr* Yohzin," said the man quietly. "Don't be alarmed. Please consider our offer and contact us, at your convenience, if you are interested."

Yohzin did turn, but the man had faded into the crowd. The voice was familiar; he was almost positive that it was one of the German engineers he had supervised at Lehesten, perhaps even one he had known in Berlin before the War.

6:30 p.m.

After an early dinner, Yohzin settled into his hotel room, hoping to catch up on writing his reports. He was a little surprised that he had been granted his own room, but he could safely assume that his phone was tapped and that he couldn't leave without being tailed by one of his handlers.

He sat at the desk and prepared to make some notes concerning the Titan IIIC heavy lift booster he had studied at the US pavilion. As he looked for a pen, he found a folded scrap of paper in his pocket and remembered the odd encounter from earlier in the day. He unfolded the paper. In precisely written German text, the simple note read:

> *We miss you and would like to work with you again. If you are interested, please call us the next time that you are in Moscow.*

Below the text was a phone number, which he recognized as a Moscow exchange. It was clearly an invitation he could never entertain. For a moment, he considered reporting the contact to his loutish GRU companions, but thought better of it; they would just suspect that he had somehow instigated the incident. Although he couldn't contemplate ever calling the number or otherwise following up on the offer, he carefully refolded the note and jammed it into the torn lining of his suitcase.

37

GOING UPSTAIRS

Wright Arms Apartments, Dayton, Ohio
2:54 p.m., Sunday, June 8, 1969

Ourecky perched on the edge of their unmade bed. Captivated, he watched Bea dress for work. On the nightstand, a small fan labored to dispel the heat, gently riffling the disheveled sheets with every sweep. Try as he might, he could not escape the notion that this could be the last time he would ever see her. In the months they had known each other, he had almost come to take her for granted; now he struggled to memorize every curve, line, and subtle nuance of her sinuous body, as if he could somehow stamp an indelible image in his mind.

He was still coming to grips with the notion that they were married, even though almost a month had elapsed since the hasty ceremony on the courthouse steps. As she buttoned her uniform blouse and straightened her wings, he reviewed his personal pre-flight preparations. He had left an envelope in care of Mike Sigler, the right-seater on Crew Three, accompanied by strict instructions to hand-deliver it to Bea if he didn't return from the mission.

The stark white envelope contained paperwork for two life insurance policies—one naming Bea as a beneficiary and the other his parents—and little else. He had thought about writing her some form of farewell letter, but could not think of the appropriate words to say without also conveying that he was conscious of the incredible risk that he was about to undertake.

Hoping for the best but anticipating the worst, he had left his freshly pressed dress uniform hanging neatly in a suit bag in the closet. His black uniform shoes were carefully spit-shined and tucked neatly in their pasteboard box, along with his socks, belt and tie. At this point, he was just as prepared for failure and death as he was for success.

"As hot as it is in here, it has to be *sweltering* outside," she said, wiping stray beads of sweat from her forehead. "I hope I don't melt before I make it to the airport."

"I hope so, too," he said, standing up and holding her waist. "Sorry we couldn't spend more time together this weekend. We're still working on a crazy schedule."

"Well, baby, we had last night and most of today," she replied, checking her hair in the dresser mirror. "I just wish that things would settle down for you. Here we are, a married couple, and we see less of each other now than when we first met."

He kissed her lightly and said, "Sorry. It'll be better someday, Bea. I promise."

"So are you going to fly much this week? With Drew?"

"Oh, yeah. We're running some tests at high altitude," answered Ourecky truthfully, at least to a certain extent. "So Drew and I will spend a lot of time in the cockpit together."

"Well, you two boys be extra careful. Try not to come home with any new scars." She looked at the alarm clock. "Well, honey, gotta go. Walk me down to my car?"

"Gladly," he said, picking up her suitcase.

Aerospace Support Project; 8:35 a.m., Monday, June 9, 1969

It was their final meeting with Tew and Wolcott prior to Friday's scheduled launch, a last chance for Carson and Ourecky to voice their concerns before the point of no return. Immediately afterwards, once they received the generals' nods, all that remained was to change from their civilian clothes into flight gear, climb aboard their aircraft, and fly west to North Island Naval Air Station near San Diego. They would leave their T-38 there, board a C-130 with the remainder of the launch crew personnel, and head onwards to Hawaii and Johnston Island.

"Gents, before we settle down to the boring formalities," drawled Wolcott. "We have some critical business to attend to. Mark?"

"Captain Ourecky," said Tew solemnly. "I regret to inform you of a very awkward situation. Just yesterday, I was briefing the Secretary of the Air Force on your impending mission, and while he was happy with what we're doing, he was rather stunned that we're sending a captain up in a space-craft, particularly since most of the NASA astronauts are the equivalent of majors or greater. Since NASA has already established the precedent, he was insistent that we will not launch a captain into space. Virgil and I argued on your behalf, but it was to no avail."

"But, sir," uttered Carson. "We're supposed to…"

"Hold your horses, son," interjected Wolcott, grinning as he reached into a drawer to pull out a manila envelope. "Seein' as how we don't much cotton to elaborate ceremonies here, I'll just cut to the chase. By express order of the Secretary of the Air Force, Captain Scott E. Ourecky is hereby promoted to the rank of *Major*, effective 9 June 1969." Wolcott handed the official orders to Ourecky, along with a pair of the gold oak leaves that signified his new rank.

"We'll have a more formal ceremony after you return, Major," said Tew, shaking Ourecky's hand. "Congratulations." Carson and Heydrich reached out to shake his hand as well.

"This is going to be a hectic day for everyone, so let's get right down to the brass tacks," stated Wolcott. "Gunter, are you confident that this pair of stalwarts are ready?"

Without hesitation, Heydrich thumped his hand on the table and replied, "They are."

"Drew?" asked Wolcott.

"I'm ready, sir," answered Carson. "*We're* ready."

"And last but not least, our newly minted Major Ourecky. Are you ready, pard?"

"Yes sir, I am. *We* are."

Wolcott spat a string of brown tobacco juice into a brass spittoon, and said, "Now, gents, we've done our utmost to resolve all the issues with this platform, but rockets are rockets and this is still a mighty risky endeavor regardless of how well we prepare. So I'm sure that you've heard it said, but I'll say it again: Speak now or forever hold your peace. If you're holdin' back any concerns or reservations, now's the time to air them out."

Carson, Ourecky, and Heydrich were silent. Finally, Ourecky spoke. "Sir, I sure wish that we weren't launching on Friday the 13th." The five men laughed, but not very much.

"Your superstitions are duly noted," said Wolcott, smiling. "Anything else?"

The men were quiet again.

"That's it, then." Wolcott turned to Tew and declared, "Mark, this crew is ready to go."

Tew nodded. "Good luck to you, gentlemen. We'll meet back here on Tuesday next week."

Pacific Departure Facility, Johnston Island
9:30 a.m., Wednesday, June 11, 1969

The LST's below-decks air conditioning system wasn't adequate to keep pace with the tropical heat, so the air was uncomfortably warm and stagnant. The rhythmic clatter of sailors chipping paint reverberated through the steel bulkheads. Stripped down to his underwear, dripping sweat from every pore, Ourecky studied his flight plan for at least the thousandth time. To prevent it from getting damp with his perspiration, he held the dog-eared tome at arm's length.

"Still lounging here in your skivvies? You had better jump into your coveralls," said Carson, strolling into the tiny space. "We need to be topside in fifteen minutes. We have to go back over for another plugs-out test on the pad."

Ourecky groaned. "How many times do we need to rehearse that until they realize that we have it down pat?"

"At least one more time, obviously. Feeling okay? Got the jitters?"

"No jitters, but I wish I could sleep," replied Ourecky. "If the heat isn't bad enough, it seems like these damned sailors can't go five minutes without painting something. I'm afraid that if I sit still too long, they'll paint *me*. Anyway, it's not just the chinking noise all day; the fumes are just wearing me out. I just hope I can make it through the test today without falling asleep."

"Then I have some good news for you. I convinced the flight surgeon to let us move onshore to the suit-up trailer. It'll only be slightly cooler, but we won't have to contend with the fumes and the noise. Maybe you'll log some decent rest before we ride the rocket on Friday."

"Man, thanks, Drew. Really. I'm indebted to you."

"That's not all. There's *more* good news, hot off the presses." Carson handed Ourecky a Teletype printout. He stepped back, smiled broadly, and folded his arms across his chest.

"What's this?"

"That's yesterday's news," stated Carson. "Literally. It's headlines and news stories from the wire services. The comms guys at Wright-Patt sent it to the comms guys here."

Ourecky skimmed through the news items and observed, "They're still fighting for Hamburger Hill in Vietnam. Hey, this is interesting: Congress passed the funding bill for NASA, but it specifically restricts NASA from planting any flag on the moon but the United States flag. Apparently, NASA was actually considering putting up a United Nations flag."

Carson grinned. "Uh, that's just fascinating. Maybe you should just skip down to the fourth item, Scott; the one I marked with a star. It's the real scoop."

"*USAF Manned Orbiting Lab Scrapped*," read Ourecky aloud. "*Defense Secretary Melvin Laird announced Tuesday that the Air Force's Manned Orbiting Laboratory program was cancelled. Pentagon officials stated that the objectives of the program, which would have launched a two-man space station for thirty-day missions beginning in 1972, could just as easily be accomplished by unmanned systems or planned NASA programs.*"

"Can you believe that? Well, it looks like our friend Russo isn't going to orbit after all," gloated Carson. "That's just too damned bad. And he had such lofty aspirations."

"Wow!" Ourecky said. "I suppose I got out of there while the getting was good, huh?"

"Yeah, you definitely dodged that bullet. So long as our program isn't cancelled between now and Friday morning, you and me are headed to orbit."

"You don't suppose that Russo might come back to roost here?" asked Ourecky. "After all, we're still shorthanded on pilots and..."

"Honestly, I don't think Mr. Russo will show his face back here. To be frank, I suspect February's launch accident really made a lasting impression on him. At a minimum, I don't think he's too inclined to ride a second-hand ICBM without a full-bore ejection seat under his butt."

Pacific Departure Facility, Johnston Island
3:30 a.m., Friday, June 13, 1969

Although it was still dark, their day of reckoning had arrived. Just less than two hours from a Pacific dawn they were destined never to see, one way or another, Ourecky and Carson were alone on the pad, perched atop a hundred-foot aluminum cylinder loaded with over 200,000 pounds of hypergolic propellants that were anxious to erupt.

The pad technicians had completed their chores, said their goodbyes, and sealed the spacecraft's hatches before retreating to the steel-reinforced concrete safety of the blockhouse. The launch support ship—the modified LST landing ship where the pair had eaten their midnight breakfast—had long since sailed, joining the other ships on picket around the island.

With no television cameras, reporters, or swelling masses of the public on hand to observe the proceedings, their pre-flight activities were significantly more abbreviated than NASA's protracted spectacle. The crew embarked with precisely enough time—two and a half hours—to execute their pre-launch checklist. For Carson and Ourecky, the familiar routine wasn't much different than climbing into their T-38 at Wright-Patt.

For the launch crew personnel, the notion of firing a rocket was not at all unusual; most had spent years in underground silos, babysitting ICBMs, and they had the process down cold. The prevailing attitude, from the pair in the spacecraft to the men hunkered in the blockhouse, was that the Titan II was either ready to go or it wasn't. If it was ready, the crew boarded, did their last-minute procedures and left the planet; if it wasn't, there was little to be gained by putting them in early to agonize over details.

"Executing computer update," said Ourecky. Reaching out to the computer control panel on the pedestal between them, he turned a knob to PRE LN. "Computer is in Pre-Launch Mode. Running memory and logic test from AGE." Still connected to the pad by a cable umbilical, the onboard computer silently compared notes with the mainframe computers in the blockhouse.

"I confirm pre-launch computer check," noted Carson.

"By the way, next time around, remind me not to get the French toast in the galley," noted Ourecky, rotating the knob back to ASC. "Computer in Ascent mode."

"You're assuming that there will be a *next* time," replied Carson dryly.

3:59 a.m.

"Vehicle is transferring to internal power," stated the CAPCOM. "Stand by for engine gimballing."

"Standing by for gimballing," replied Carson succinctly.

"T-minus one minute and counting," declared the CAPCOM. "Five seconds to Stage Two fuel valves...Thirty seconds...T-minus twenty seconds..."

While the Box did a passable job of replicating the sounds just prior to launch, it didn't accurately capture the almost overwhelming physical sensations. As powerful turbo-pumps whirred to life, Ourecky felt their vibrations pulsing up through his back; the behemoth Titan II was like a living thing, quickly waking to undertake its singular mission of heaving the two men and their spacecraft into orbit.

After repeatedly watched the films of Crew Two's demise, he could not dispel the feeling of impending doom that gripped his stomach. He thought of Bea and how the promise of their life together could be obliterated in the coming seconds. His eyes were fixed on the Abort indicator in the top center of his instrument panel. Of course, at this point watching the Abort light was essentially pointless; if it suddenly flashed red, it meant that he might have barely enough time to draw another breath before he was incinerated.

"T-minus ten, nine, eight, seven, six, five, four…first stage ignition," announced the CAPCOM.

Unable to breathe, Ourecky felt the Stage One engines rumble. With the rocket fastened to the pad by explosive bolts, a two-second delay passed as the two engines built sufficient thrust for lift-off. Running at full tilt, they generated a combined thrust of 430,000 pounds.

"Hold-down bolts are fired!" announced the CAPCOM. "*Lift-Off!*"

"We copy lift-off! This elevator's going *upstairs!*" shouted Carson. "Whew! What a ride!"

Expecting much more violent agitation, Ourecky was amazed with the smoothness of the liftoff. It felt as if a massive hand suddenly scooped them upwards, progressively lifting them faster and faster. In an instant, his feelings of apprehension were displaced by a sense of urgency, and his thoughts of earthbound matters slipped away. If he was going to die today, he was going to die doing his job, so his mind no longer held room for distractions. With a start, he suddenly realized that Carson had neglected something from the checklist. He pointed at the clock display and quickly switched off the VOX loop. "The clock, Drew. Call the clock!"

"*Huh?* Oh!" As Ourecky toggled the comms back to open loop, Carson declared, "The clock is running! Four seconds into flight!"

"Good copy on your clock," replied the CAPCOM. "Good luck, guys."

"Roll program commencing," stated Carson.

At twenty-two seconds into the ascent, Carson reported: "Roll program complete. Pitch program has initiated. Pitch is looking good." The pitch program would automatically nose the rocket over in three gradual steps to minimize the angle of attack.

"Cabin pressure relief is normal," said Ourecky. He continued to scan his instruments, looking for any discrepancies. Right now, he and Carson were along for the ride, monitoring their spacecraft and the Titan II to ensure that everything was performing within acceptable parameters. Otherwise, the rocket could have been lofting a thermonuclear warhead, as it had been designed to do, since it effectively flew itself during the ascent phase of the mission.

"Roger cabin pressure," replied the CAPCOM. "Everything is looking nominal down here."

"The ride is smoothing out considerably," observed Carson. "We should be supersonic right now. Stage One fuel and oxidizer in limits. ENGINE II underpressure light is amber."

At one minute and forty seconds into flight, the CAPCOM announced, "Mark at one plus four zero. You are *Go* for Mode Two."

"We copy *Go* for Mode Two!" exclaimed Carson excitedly.

"Pilot copies Mode Two," confirmed Ourecky. He breathed a sigh of relief. Now that the Mode Two abort option was viable, their chances for survival had just increased exponentially. There was still a long way to go, but it was less likely that they would suffer the same horrendous fate as Howard and Riddle, now that they had an escape option.

Ourecky peered at the DCS light in the upper right of his instrument panel. Watching the light blink on, which indicated that the onboard computer was receiving updated guidance data from the ground, he commented, "DCS Update received."

"We copy DCS update," confirmed the CAPCOM. "Your trajectory is still looking nominal."

"Kind of bumpy up here," commented Carson. "Some long axis oscillation, but otherwise looks good. Passing through eighty thousand feet right now. We're really scooting on up."

Two minutes and thirty seconds into flight, Ourecky watched the DCS light blink on again. The computer was automatically uploading guidance and initiation data for the Titan II's second stage. They were now approximately at 250,000 feet in altitude, roughly twenty nautical miles northeast of Johnston Island, travelling at 9,800 feet per second. "DCS Update," he stated.

"We have ENGINE 1 underpressure lights," noted Carson calmly. As alarming as it sounded, it was nothing to be concerned about; the warning lights glowed red as the thrust chamber pressure dropped below sixty-seven percent, which meant that the first stage engines were automatically being throttled down. "...and ENGINE 1 lights are extinguished."

"BECO!" announced Carson, acknowledging Booster Engine Cut-Off. The first stage had accomplished its task and was mere seconds away from its long plunge through the atmosphere and into the depths of the Pacific Ocean.

Through the tiny window, Ourecky witnessed a brilliant flash of reddish-orange light. Initially thinking that something might have gone awry, he remembered that most of the NASA astronauts reported seeing the same sight when the first stage shut down.

"Just saw a big flare over the nose at BECO," noted Carson, observing the same spectacle. The rocket's upward velocity slowed immensely; the acceleration forces immediately diminished from 6 G's to 1.5 G's, jamming the two men forward against their restraint harnesses.

Immediately, a chain of events was initiated; the now-depleted first stage was literally sliced away by a string of shaped charge explosives, and the second stage engines were ignited. The ENGINE II underpressure indicators illuminated briefly and then flickered out as the engines developed full thrust to take the spacecraft the remainder of the way to orbit.

"Staging is good," grunted Carson against the heavy pressure on his chest.

"We copy staging good at two minutes and forty-one seconds," stated the CAPCOM.

Ourecky was pushed back down into his seat as the G's built again. They had picked up a slight amount of vibration after staging, but nothing to be overly concerned about; a little bumping and shimmying was entirely acceptable.

"Stage Two fuel and oxidizer reading within limits," reported Carson. "Accelerometers looking nominal. Slight yaw to the right, less than a degree."

"We copy that your Stage Two is running smoothly," answered the CAPCOM.

"Guidance initiate," stated Carson.

"Copy guidance initiate," replied the CAPCOM. "Still tracking well."

"Almost five minutes," noted Ourecky, glancing at the clock. "Coming up on Point Eight."

At slightly more than five minutes into flight, the CAPCOM stated, "*Point Eight*. Mode Three." They were now cleared for the Mode Three abort protocol. They were roughly 522,000 feet in altitude, travelling at approximately eighty percent of the velocity required for orbital insertion.

"Copy Mode Three," said Carson.

At five minutes and forty-five seconds into the flight, the SECO light blinked on, signifying that the second stage engine was shutting down. "SECO," reported Carson brusquely.

Carson reached out and toggled the OAMS PROPELLANT switch. OAMS was the Orbital Attitude Maneuvering System, which was used to maneuver the Gemini-I in orbit. He waited for twenty seconds, allowing the upward thrust of the second stage to decay, and then pushed in the SEP SPCFT telelight, starting a chain of events that would result in the firing of explosive charges that would separate the Gemini-I spacecraft from the Titan II's second stage.

Immediately after the charges detonated, the SEP SPCFT light illuminated green, indicating that they were free of the booster. The depleted booster gradually tumbled away behind them; having spent most of its existence buried under an Arkansas pasture, its useful life was over.

"We need your IVI's," said the CAPCOM.

In the bottom-center of Carson's control panel, directly below the "eight-ball" attitude indicator, was the Incremental Velocity Indicator, which displayed the corrections automatically derived by the onboard computer. The corrections were displayed in three axes—Forward/Back, Left/Right, Up/Down—which would zero out as Carson applied the appropriate amount of OAMS thrusters during the IVAR—Insertion Velocity Adjust Routine—as well as subsequent computer-generated maneuvers during the flight. Reading the numbers, Carson reported, "IVIs are 49 Forward, 26 Right and 79 Down."

"Roger," replied the CAPCOM. "You're go for IVAR."

"Thrusting forward," stated Carson, deftly nudging the maneuver controller. After a few moments of subtle maneuvers, he reported, "IVIs are nulled. Showing zeros in all axes."

"Roger your IVI's are zero'ed," stated the CAPCOM. "We're showing you in an 88 by 125 orbit. Congratulations, guys."

"We copy 88 by 125 orbit," said Carson. "We're just now seeing the horizon. We haven't lost the window covers yet, but we do have a horizon out there."

"Good job, Drew," commented Ourecky. "Much better than the Box."

"We're here!" exclaimed Carson gleefully, reaching over to excitedly tap his gloved hand on Ourecky's. Even though they had reached orbit, that was the extent of their celebration. They had rehearsed this process enough times to know that there was no time for festivities or even to glance out the window at the Earth below; there was work to be done, and plenty of it.

"Yep," said Ourecky, tapping him back. "We're here all right."

Carson's tone changed almost immediately from childlike exuberance to icy seriousness. "Let's punch through the Insertion Checklist. I don't want to fall behind the curve."

"Okay. I know I'm early, but I'm coming out of these gloves," said Ourecky, unlocking the ring on his left forearm. Letting go of the glove, he let it hover in front of him momentarily. The sight was both surreal and natural at the same time; they really *were* in orbit.

"Go ahead," replied Carson, shedding his own gloves before wedging them under his calves. "Make yourself comfortable."

"Ouch!" said Ourecky, flipping up his helmet visor. "Man, it's *hot* in here. You could bake bread in this oven. I sure hope it eventually cools off, or this is going to be a miserable flight."

"Hopefully it will cool down soon. Okay, let's finish this checklist. Retro Rocket Squib to Safe," said Carson, throwing a series of switches. "Boost Insert Squib to Safe. Main Batteries one, two, three and four Off." He pressed the JETS FAIRING switch on the sequencing panel. "Jettison fairings." A protective fairing, which covered the radar dish in the nose of the spacecraft, was ejected, as were the horizon scanner covers on either side of the nose. Over the course of the next few minutes, they worked through the initial set of tasks that had to be accomplished immediately after achieving orbit. Among other things, they ejected the protective covers that guarded their windows during launch and aligned the inertial navigation platform.

Within moments of gravity falling away, a small cloud of debris had floated up into the cabin. The detritus consisted primarily of dust, lint and some larger items, including a inch-square piece of Velcro, a small shard of clear plastic, two rubber bands, a miniscule "wheat grain" light bulb, two washers, and a scrap of paper bearing a woman's name and phone number. It took them a few minutes to gather up most of the clutter.

Carson reached out, grabbed the loose note, and examined it. "Hmm… *Janet*. I think I'll tuck this away just in case I'm ever in St. Louis. I guess that someone's loss is my gain."

"You're incorrigible, Drew. Would you really call some woman without knowing the slightest thing about her?" asked Ourecky, picking one of the washers out of the air and stowing it in a leg pocket. "Don't you think that's living just a little dangerously?"

"Living dangerously? Mr. Ourecky, do I have to remind you that you're travelling over seventeen thousand miles per hour in a flimsy metal can built by the lowest bidder?"

On orbit
10:02 a.m. Eastern (Rev 1 / Ground Elapsed Time: 1:02)

Six feet behind Carson and Ourecky, within the magnesium casing of a large battery, trouble was literally brewing. The Gemini-I spacecraft, like its NASA predecessors, consisted of three main parts, stacked one atop another like layers in a cone-shaped cake. At the apex was the reentry module that housed the flight crew within a pressure vessel. A retrograde section, directly behind the reentry module, contained the four retro rockets that would bring them out of orbit. The bottom layer was the adapter equipment section. Shaped like a truncated Dixie cup, it was home to virtually all the life support systems necessary to sustain an extended flight.

On most NASA Gemini missions, the adapter housed two fuel cells that produced electricity and water essential for long duration space missions. Because the Blue Gemini missions were of such short duration, the Air Force planners elected not to rely on the often cantankerous fuel cells. Instead, the Gemini-I was powered by more conventional but trustworthy batteries. In fact, of the ten manned NASA Gemini spacecraft flown, the first two missions—Gemini 3 and 4—also relied entirely on batteries rather than fuel cells.

Besides the batteries mounted in the adapter equipment section, all Gemini spacecraft carried four silver-zinc main batteries and three squib batteries. These were mounted in an equipment bay immediately to Ourecky's right. In the unlikely event of a total failure of the adapter section batteries, the seven batteries would still get the crew out of orbit and back home. Consequently, because of their unique fallback role, they were kept disconnected from the main electrical bus and dormant throughout the majority of the mission.

In contrast to the six 400-amp silver-zinc batteries carried on Gemini 4, the Gemini-I was endowed with one rack of three batteries and a second rack of five. Each battery was about the size and shape of a large picnic cooler. Besides redundancy, the extra two batteries were required to power the enhanced radar mounted in the Gemini-I's enlarged nose. Adapted from technology carried aboard fighter aircraft, the powerful new radar was necessary for the Gemini-I to locate and intercept target satellites that did not emit a friendly transponder signal.

So now, even as Carson and Ourecky plotted their next move in the intercept sequence, one of those big batteries was overheating. As its core simmered and internal pressure mounted, a relief valve surrendered just as it was designed to do, and a dense but rapidly dissipating cloud of electrolyte vapors erupted silently into the vacuum of space. And in an instant, although its heft was now meaningless in the weightless environment of space, the ailing battery was rendered into one hundred and eighteen pounds of useless scrap metal.

On Orbit
10:15 a.m. Eastern (Rev 1 / GET: 1:15)

"Drew, we have a problem here," announced Ourecky, frantically tapping the instrument panel.

"What's up? Are we running out of peanut butter already? I warned you to go easy on it."

"No, Drew. This is *serious*. It looks like we have a major failure back in the adapter batteries. We're losing a lot of juice off the main bus. Battery

1C had failed completely. If that wasn't enough, 1D conked out while I was still watching 1C. That's two of eight, Drew."

"Oh, babe, not good," muttered Carson. "How about the main batteries? How do they look?"

"The mains and squib batteries look healthy, so we should still be able to get home." He paused a moment, studied the power system read-outs, and then solemnly reported, "1B's dead now also. So now we're down to the three batteries on rack two, plus 1A and 1E in the adapter. Uh, Drew, disregard what I just told you. 1A just sputtered out, too."

Clenching his fists, Carson grimaced and then whistled quietly. "*Damn.* Well, there's no sense trying to delay the inevitable. We need to report this immediately."

Ourecky focused his attention on the remaining array of healthy batteries in the adapter. Although there were no more fluctuations on the ammeter or voltmeter needles, he agonized over the outcome that was sure to ensue. According to their sacrosanct mission rules, the situation was not severe enough for them to immediately abort the mission, but the power deficit almost certainly meant that they couldn't continue with the intercept as planned. Disheartened, he lamented, "I guess you know that they'll probably terminate us early."

"Yeah, Scott, I know that. It sure looks bleak. They'll probably scrub us prior to our next maneuver burn, and that's about an hour away. Sorry I brought you up here for nothing."

Mission Control Facility, Aerospace Support Project
10:32 a.m. Eastern, Friday, June 13, 1969 (GET: 1:32)

Blue Gemini's Mission Control Facility was similar in layout to NASA's much-celebrated nerve center at Houston, but on a much-reduced scale. It occupied a room approximately half the size of a standard tennis court, which previously had been a lecture hall when the building was used as an academic training facility in the early fifties. The mission controllers' consoles were arranged in stair-step banks, with a floor gradually sloping down to a wall occupied by a large world map and several television monitors. In the far back corner, a glass-enclosed cubicle overlooked the room.

It was the exclusive sanctuary of Wolcott and Tew, a quiet place where they could observe the proceedings without interfering with the process.

Wolcott was bone-tired. Concerned that they could have another disaster like February's, he hadn't slept in two days, and even though things appeared to moving on track, he was still very much on edge. Glancing through the windows, he observed a commotion out on the floor; something had the controllers riled up, and he hoped that the news wouldn't be too awful.

Heydrich tapped on the glass door, and Wolcott waved him in. "Virgil," Heydrich said. "We have a problem. Got a minute?"

"Always, Gunter. What is it?"

"Virg, your crew reported a failure of five primary batteries in the adapter section."

"Okay, pard. Now, from what I know about the batteries, this situation ain't good, but there's apparently nothin' we can do about it from here and our boys obviously can't do anything about it up there. Do you have any idea what might have happened?"

"Well, Virg, it's anyone's guess, but if I had to speculate, I would bet on a cold plate failure. Anything that generates heat, including the batteries in the adapter section, is mounted on a cold plate. Refrigerant fluid is constantly cycled through the plates to bleed away excess heat."

Heydrich continued. "Since it looks like we suffered a failure of all five batteries in Rack One, I'm guessing that the batteries overheated after the cold plates failed. Each battery has a valve that's designed to vent if the internal pressure spikes greater than 40 psi. So my theory is that they overheated, built up too much pressure, and then all of them vented. Also, they're mounted in there so tightly that probably one of them overheated initially, which then probably triggered a cascading heat effect with the adjacent batteries. That correlates with what your crew just reported, that the five batteries failed sequentially, starting with 1C in the middle slot."

Wolcott closed his eyes and reflected on the physical layout of the adapter section. Although five adapter batteries were out of commission, three remained functional. "So, Gunter, what's the possibility that our three intact batteries are also vulnerable to this cascadin' heat effect? Could they also get knocked out?"

"I don't think so, but that doesn't necessarily mean that we're entirely out of the woods," stated Heydrich. "Assuming that they vented, those batteries spewed a lot of corrosive electrolytes into the surrounding spaces. I don't think that this has ever been modeled, so there's no telling what impact it might have if it seeps into the other systems back there."

"But this probably happened about thirty minutes ago," said Wolcott. "If we were going to have any fallout from this glitch, we probably would have already seen it. Right?"

"*Ja*. Right," said Heydrich. "Or they'll report the bad news to us on the next comms window. They'll pick up Atlantic Sentry Three in approximately twenty-five minutes. That's an EC-135E ARIA, on station over the North Atlantic, flying in an orbit pattern just south of Greenland."

"Okay. What's your call on the situation, Gunter?"

Heydrich removed his black-framed glasses. "Well, there's good news and bad news. Assuming that there's no damage to the adjacent systems in the adapter section, the crew is safe. They can remain up there for the duration of the mission. Even if the other batteries in the adapter section fail, we can still get them home safely. We might not get them back to United States, but we should get them back to Earth in one piece. But we've lost our option to loiter, because I wouldn't feel comfortable with them cycling through a power-down and power-up."

"Doggone, Gunter, if that's the *good* news, then what's the bad news?"

"The bad news is that they've lost the extra power required to run the radar, so there's no chance of them completing the intercept. If the interception is the ultimate goal and it's clearly obvious that they can't achieve it, then we're just placing them at risk by leaving them up there. Sure, they can stay up for the duration, but there's no point."

Wolcott nodded. "So your verdict is?"

"Bring them home as soon as practical," counseled Heydrich grimly. "If we scramble, we have sufficient time to pass instructions to Atlantic Sentry Three to relay to the crew."

"And then we bring them straight home, right?" asked Wolcott.

Heydrich shook his head. He carried a rolled map under his arm, which he unfurled and spread out on the table. "Not quite that simple, Virg. We have plenty of Contingency Recovery Zones, but if we stick to your guidance, we want to strictly avoid landing anywhere but on US soil, unless we absolutely have no other options. *Ja?*"

"Yup."

Referring to the map, tracing an orbital path with his finger, Heydrich said, "Well, Virgil, the problem is that since we launched your boys directly into an orbit with sixty-two degrees inclination to chase a Russian satellite. That inclination wreaks havoc on our recovery planning, because they're just not going to be lined up to land on US soil very often.

Tapping his finger on a spot near El Paso, Texas, Heydrich stated, "Our current plan has them coming back to White Sands, New Mexico on their sixteenth rev, just shy of twenty-four hours into the mission. The earliest window that I can bring them back to US soil is on their fifth rev. That will put them into Eielson Air Force Base, Alaska in about five hours. We can also put them into Patrick Air Force Base in Florida fifteen minutes later, landing local time at 5:00 p.m."

Wolcott shook his head. "That ain't good, pard. It'll still be broad daylight."

"We don't have any choice," replied Heydrich. "At least they're relatively stable up there, provided they don't lose any more batteries or if something else doesn't go wrong, so we're not pressed to reenter immediately. We have enough wiggle room to decide on Eielson or Patrick or another location later, but we need to direct them to stop maneuvering for rendezvous."

"So, Gunter, there ain't any chance of them executing an interception without the radar functionin' at a hundred percent?"

"Not a chance, Virgil," asserted Heydrich. "Not even in the realm of possibilities."

"Then call them home, Gunter," decreed Wolcott. "Post haste, pard."

38

ON TO BREMEN

On Orbit
10:40 a.m. Eastern, Friday, June 13, 1969 (Rev 2 / GET: 1:40)

Unexpectedly granted the luxury of a few spare seconds, Ourecky unscrewed the cap from a container that resembled a toothpaste tube. With the red plastic cap hovering in front of his face, he squeezed the tube and savored a gooey mixture of peanut butter and honey. He re-stowed the tube in his pantry, then washed down the snack with a quick squirt of water. He smiled, reflecting on how even grabbing a quick bite could be a major endeavor up here.

The confined cockpit was much noisier than he would have ever anticipated. The experience was a far cry from his childhood dreams of swooping silently through space in the magnificent spaceships. His headset was plagued by a persistent but barely audible hum. One of the cockpit fans squealed incessantly when it ran. As he worked calculations, his thoughts were interrupted by the periodic pops of the OAMS thrusters whenever Carson bumped the maneuver controller to make minor corrections. There was a distracting cacophony of smaller sounds as well: pumps running, solenoids clicking, valves opening and closing. There were also distinct odors; most noticeable was a slightly noxious hint of lithium hydroxide, a compound used to scrub carbon dioxide from their exhaled air.

Sighing as he longed for a quieter chariot with a smidge more elbow room, he checked the computer. It was still loaded with data for the major burn that would begin the transition to align their orbit with the Soviet satellite they were targeting. As they anxiously awaited further instructions from the ground, Ourecky was reluctant to wipe the maneuver fix even though it was highly unlikely that it would be executed.

In an obvious funk, Carson numbly stared at his panel while Ourecky prepared the cryptographic equipment for the next comms window. Without exchanging a word, both knew that Tew and Wolcott would have had plenty of time to formulate their decision, so it was a foregone conclusion that they would be ordered to reenter as expeditiously as possible. Finally, Carson turned to Ourecky and broke the silence. "Scott, what are the chances…"

"That we could still make it happen?"

"Yeah. Other than the batteries, we're solid. We have a stable platform and ample PQIs," said Carson, verifying the Propellant Quantity Indicated numbers for their remaining OAMS fuel. "The big question is whether we can nail the rendezvous without the radar. What's your take?"

"Funny you should ask. I've sifted the numbers in my head and I think we can still pull it off." Squinting in the bright sunlight streaming into the cockpit, Ourecky wedged his flight plan into his small window. "It's no more than maneuvering to a theoretical spot in space, just like we've practiced over and over. I sure wish we had the radar, but it's not nearly as essential as it's cracked up to be. Besides, we might be able to use it anyway, at least to a limited extent."

"How so?"

"Well, the radar normally requires forty-five seconds to warm up from a full resting state," said Ourecky authoritatively. "And it takes roughly thirty seconds to safely cool down after use. Based on our current power consumption rates, if we scrimp and scrounge everywhere we can, we can probably wring a minimum of five one-minute active shots out of the radar."

"You think we could squeeze five shots?"

"Maybe more, but we'll have to use them very sparingly," replied Ourecky.

"What if we were even more frugal and limited the shots to thirty seconds or thereabouts?"

Ourecky stifled a sneeze and then answered, "No good, Drew. First, it takes thirty seconds to do a full sweep with the radar after it's warmed up. Assuming we achieve a solid lock on the target, if we're very lucky, the computer still needs to accept the radar data for processing. That's a minimum of thirty seconds. If we don't factor in loading the radar data to the computer, we're just wasting our time powering it up, and we might as well go strictly visual."

"So what do we sacrifice to make this happen?" asked Carson.

Ourecky looked around the cramped cockpit. "For starters, we'll need to keep our radio chatter down to a minimum, at least on the transmit side, because that gobbles a lot of juice. We'll also have to make some sacrifices on the environmental side, so it's probably going to get very stale and uncomfortable in here before everything is said and done."

"I don't mind being uncomfortable. Do you?"

"Not at all. Drew, I'm in, but if we commit to this, whether we're successful or not, it'll probably be the last time we ever fly."

"I know, Scott, but I didn't come up here to quit. Quitting just isn't in my nature."

"Me neither. So how do you want to handle it? With the ground, that is?"

Carson's brow furrowed as he contemplated the question. "As I see it, we have a couple of options. First, we can tell them straight up what we're doing. They can either back our play or not. Since they don't have any options otherwise, I think they would fall in line with our plan. Of course, that doesn't mean they would be too thrilled with us. But I *am* the mission commander."

"If you looked under Virgil's bunk, I suspect you would find a footlocker packed with the shrunken heads of previous mission commanders. So what's our other option?"

"We fake a comms problem," replied Carson. "Any fighter pilot worth his salt knows the old 'receiving broken and garbled' trick to circumvent unwanted instructions. If they order us to come home, we just flop the crypto variable a little bit and tell them that we're experiencing an equipment failure. Then we're entirely on our own, with no help from the ground."

Ourecky pulled up his left earphone and scratched his temple. "I don't like that, Drew. Either way, once we start maneuvering, they're going to know what we're up to, so I would much rather have them in our corner, lending assistance, even if they're reluctant about doing it."

"You're right. One thing's for damned sure, though. If we continue the chase, it's a sure bet I'll be lashed to the yardarm and flogged after we touch down. But you don't also have to get flayed, Scott. If anyone asks, you just insist that I ordered you to continue with the intercept. I may be grounded, but you'll have a chance to fly with someone else."

"No way, Drew," replied Ourecky, frowning. "I'm not coming up here with anyone but you."

Carson scratched his chin and smiled. "Thanks. Since we're chucking the rulebook in the trash, we need to set some ground rules. From here on out, if we experience any major glitch, we burn the retros and make a slide for the next available recovery zone. Fair enough?"

"Agreed."

"And if one more adapter battery craps out, we head for home. If you see the slightest flutter on the mains or squibs, same thing." Carson jetted some water into his mouth with the water dispenser. "I think we can pull this off, but I'm just not absolutely sure, though."

"Drew, we have a comms window in two minutes," said Ourecky, wiping his forehead. "What's your decision, mission commander? I'm with you, either way."

Carson looked out the window, sighed, and then said, "Look, as far as I'm concerned, the most dangerous part of this mission is behind us. Scott, I can't speak for you, but I didn't come up here to fail. Do you honestly think we can tackle this and still make it home?"

"I do."

"Then to paraphrase our exalted leader, let's go lasso a danged satellite, pardner."

EC-135E ARIA - Atlantic Sentry Three
Latitude 60 North / Longitude 40 West (over the North Atlantic)
10:50 a.m. Eastern, Friday, June 13, 1969 (Rev 2 / GET: 1:50:28)

Of the fourteen men aboard the EC-135E ARIA "Droop Snoot" aircraft, only one—the Airborne Mission Controller—knew that they were supporting a manned spacecraft in orbit. Isolated by a black curtain, he sat by himself at a console near the tail of the aircraft. The rest of the crew had been briefed that they were participating in a test of an unmanned anti-satellite system.

The Airborne Mission Controller, an Air Force major, checked his cryptographic equipment and watched the clock. At the appointed time, he keyed the radio and transmitted: "Scepter Two, this is Atlantic Sentry Three. Over." Listening for a reply, he read again the Teletype printout. Something had to be significantly wrong if the crew was being called out of orbit early.

Finally, he heard a warbling voice in his earphones. Initially faint but growing progressively stronger, it bore the unusual Donald Duck quality of being filtered through the cryptographic gear. "Atlantic Sentry, this is Scepter Two. Receiving you five-by on UHF Channel Two."

The controller breathed a sigh of relief and replied, "Scepter Two, I am reading you loud and clear as well. Be aware that all contingency reentry guidance remains in effect. I have instructions for you to copy. Be advised that I will read up maneuvering guidance for a phase shift to line up on Primary Recovery Zone One-Three, followed by reentry burn instructions for your fifth rev. Let me know when you are ready to copy."

The reply was immediate. "*Negative*, Atlantic Sentry. Be aware that I have instructions for *you* to copy."

Scratching his head, the major glanced at the mission procedures binder lying open on his console, as if it might hold an answer for this quandary, and then said: "Scepter Two, this is Atlantic Sentry mission *controller*. Say again your last transmission, please."

"Atlantic Sentry, this is Scepter Two mission *commander*. Listen carefully: *I* have instructions for *you*. Let me know when you are prepared to copy."

Knowing that the contact window was short, the major grabbed a pad of paper and a pencil, made ready to take notes, and replied, "Go ahead, Scepter."

"Sentry, we are adhering to our primary intercept plan. We will execute our scheduled height adjust burn in eighteen minutes, GET 02:14:25. Make sure that Wright-Patt is aware that all tracking assets should stick with the original scheme and not deviate or stand down. We will also need running guidance on Contingency Recovery Zones as we proceed. How copy?"

"Scepter, my instructions are to direct you to immediately curtail intercept operations and prepare to reenter for Patrick on your fifth rev. Stand by to copy the pertinents."

"Sentry, *negative*. We are staying upstairs and will continue the mission as planned. Relay that information to Mission Control. Stand by for my right-seater to pass data on our next burn."

Still perplexed by the crew's unforeseen dissent, the Airborne Mission Controller listened as a second cartoon-sounding voice came on the radio to read down details of the spacecraft's next maneuver. Writing as quickly as he could, he could just barely keep pace with the flurry of information. Looking up to make sure that the tape recorder backup was running properly, he reflected on how only minutes ago they were supposed to divert for Maine after the contact and then on to home, but now it looked like they were back in it for a long haul.

Mission Control Facility, Aerospace Support Project
10:59 a.m. Eastern, Friday, June 13, 1969 (GET: 1:59:12)

Waiting for the news, Tew paced while Wolcott rifled in his desk drawers for his back-up packet of Red Man. He found the chewing tobacco just as Heydrich tapped on the glass. Quickly jamming a brown lump in his mouth, Wolcott waved the engineer in to deliver the report.

"Good news. My comms guys just received the contact information from Atlantic Sentry Three," announced Heydrich, entering the generals' private sanctum and shoving the door closed behind him. "Atlantic Sentry talked to Carson on schedule."

"Whew! That's *excellent* news," noted Tew, standing at the window and looking out over the banks of consoles. "Maybe we've seen all the damage we're going to see."

"I imagine Carson ain't too thrilled with the prospects of comin' off the trail before the drive is done," observed Wolcott, loosening his bolo tie and unbuttoning his collar.

Heydrich shook his head. "Uh, Virg, not exactly. General Tew, you might want to sit down."

Looking as if he had aged ten years in an instant, Tew turned and sagged into a chair. "How bad, Gunter? They're not stranded up there, are they? Can we bring them home?"

"That's not the issue, sir," answered Heydrich. He placed the Teletype transcript of the radio transmission on the desk so that both Tew and Wolcott could read it. "It seems that we have some discord with the crew, sir. They're not coming straight home as we directed."

As Wolcott read the transcript, the corners of his cracked lips turned up in an almost imperceptible smile and a slight twinkle came to his wizened eyes.

In stark contrast, Tew was instantly apoplectic; his weary face turned as red as a freshly painted fire truck. As he struggled to catch his breath, his tirade wasn't long in coming. "That stubborn bastard Carson is *telling* us that they're still pursuing the intercept?!" he sputtered, reaching into his shirt pocket for his aspirin tin.

Heydrich looked out the window, made eye contact with a controller in the back row of consoles, pantomimed the motions of drinking a glass of water, and then pumped his fist in a "hurry up" gesture. The recipient of his silent message scurried toward the break area.

"Simmer down, Mark," counseled Wolcott, fanning Tew with his Stetson. "It ain't goin' to help matters if you have a coronary. I've already attended my quota of funerals this year."

"Where the *hell* does Carson get the notion that he can debate our instructions?" howled Tew, popping open his pillbox with quaking hands. A motley assortment of pills spilled out and skittered across the desktop.

As Heydrich and Wolcott helped him collect the medicine, Tew selected a white tablet and choked it down. Heydrich's confederate came to the door with a glass of water, which Tew accepted gratefully. "Send word up to Carson that he *will* comply with orders and that he is to reenter and land at Patrick, as directed. No ifs, ands, or buts."

"It's too late," noted Heydrich, silently shooing the controller out of the small room. "They'll be executing their height adjust burn in a couple of minutes, so they can't make it into Patrick."

"Then what?" asked Tew, still flabbergasted. "Since we've apparently lost all authority over those two, what the hell do we do now? Are we supposed to just capitulate?"

Heydrich slicked back his greasy hair and offered, "Sir, for starters, I recommend that we re-set the tracking resources, just as Carson asked. The Pacific Sentries will be turning back for home shortly. We need to direct them to maintain their contact stations and fly their tracks per the original plan."

Clutching his abdomen, Tew nodded. "Sounds like we have no choice in the matter."

"We also need to start recalculating the guidance for Contingency Recovery Zones," observed Heydrich tersely. "They'll need updated reentry data fed to them at every contact."

"Gunter, skedaddle out there and make it happen," said Wolcott. "No time to dilly-dally."

Closing the door behind him, Heydrich immediately left the room. Tew closed his eyes and sipped water, but Wolcott watched through the glass as the German engineer called the controllers into a quick huddle. He noticed that most of the controllers smiled when Heydrich delivered the news. Galvanized to action, some appeared to be downright jubilant.

"This is a flagrant violation of orders," growled Tew. "I'm grounding those two as soon as they touch down, provided that they do touch down."

"As high-strung and defiant as Carson is, Mark, we stuck him in that ship for a reason," said Wolcott quietly, gently putting his hand on Tew's shoulder. "He's the mission commander, and he ain't snivelin' and high-tailin' for the bunkhouse just because a bad storm blew in. He's exercising initiative and continuing the mission. Isn't that what you expect?"

"I *expect* him to follow orders," asserted Tew.

"Follow orders? Oh, *really*? Mark, you ain't forgotten our third big push into Bremen, have you? November of '43? The one where you caught some flak and ended up in the hospital?"

"Virgil, how could I possibly forget *that* day?" answered Tew, grimacing. "So is there some reason that you're waxing sentimental all the sudden?"

"Yup. Bear with me, pardner. As I recollect, you were all shot up, down one engine, losing another and had three dead guys in the back of your Fort, but you stayed in formation for the bomb run even though the squadron commander ordered you to fall out. Remember that?"

"I do."

"So *why* did you continue to the target, Mark?"

Scowling, Tew recalled the harrowing mission over Germany. "*Why* did we continue? Because the Wing had tried to crack that nut three times before and we hadn't pushed enough planes through, and I was *tired* of guys dying in vain. I was positive we could deliver our iron on the target, instead of splashing it in the Channel on the way home. I polled the men on my plane, and it was unanimous that we go on to Bremen, so we did."

"And ol' Callahan really reamed you out after you limped back into Molesworth, didn't he?"

"Yea and verily, Virgil," replied Tew. "That was a most memorable ass chewing, and I was lying face-up on a stretcher at the time. One I'll never forget, that's for damned sure."

"Well, pardner, before you judge Carson too harshly, you need to recognize that your boys upstairs are goin' on to Bremen."

On Orbit
11:08 a.m. Eastern, Friday, June 13, 1969 (Rev 2 / GET: 2:08)

Working closely in unison, Carson and Ourecky aligned their inertial navigation platform in readiness for maneuvering burn that would substantially change their orbit. The burn would entail firing their four aft thrusters for a minute and eight seconds as they passed over Central Africa. NASA's Gemini had just two 100-pound aft thrusters; two more were fitted in the Gemini-I's adapter, specifically for the radical orbital changes necessary to stalk Soviet satellites.

"Since we're executing this burn all by our lonesomes, let's verify our numbers," said Carson. "GET is 02:14:25. Delta-V, 50.6. Delta-T, 1 plus 8; Yaw is zero. Pitch is zero. Core 25, 00506. Cores 26 and 27, all zeros. Firing aft thrusters. Maneuver is *posigrade*."

"GET is 02:14:25," confirmed Ourecky. "Delta-V, 50.6. Delta-T, 1 plus 8; Yaw is zero. Pitch is zero. Core 25 is 00506. Two-six and 27 are all zeros. Burning aft to *posigrade*."

"Sounds good," said Carson. He sluiced his dry mouth with a quick spurt of water and then scratched his nose. Looking through the window, he saw that it was growing dark. They would be in orbital night immediately after the thrusters shut down, and their first post-burn task was to take a star shot to verify the results of the maneuver. "Sextant ready?"

"Ready." Ourecky gestured at the navigation instrument, which was momentarily fixed against the right wall by a thin strip of Velcro.

"You might want to take it down and hold it while we're burning," advised Carson. "You don't want to get bonked in the noggin if it breaks loose."

"You're right. Thanks. Computer to Catch-up," stated Ourecky, rotating a knob on the control panel to CTCH UP. The computer was now configured to receive the information for the maneuver, which would be displayed as velocity changes on the IVI display on Carson's panel.

"Computer to Catch-Up," confirmed Carson.

Talking himself through the procedure, Ourecky carefully tapped the numbers into the computer's numerical keyboard. "Address 25, *Enter*. Value 00506, *Enter*. Address 26, *Enter*. Value 00000, *Enter*. Address 27, *Enter*. Value 00000, *Enter*. Verifying entries, Read Out 25, Read Out 26, Read Out 27. Everything's good. All in, Drew."

"Okay. Start Comp."

"*Start Comp*." Ourecky pushed a button marked START; next to the button, a light marked COMP flashed green. The fifty-nine pound onboard computer—mounted outside the pressure vessel immediately to Carson's left—was now processing the calculations for the burn. "You know, someday all this will be automated," he said, monitoring the computer's progress. "There won't be any need to manually enter this gobbledy-gook on a keypad. It'll all be simplified."

"Oh, yeah, but I don't see that happening in our lifetimes, buddy," replied Carson, making some slight adjustments to his controls. "Building pressure right now. Coming up good." The fuel tanks and oxidizer tanks were lined with thick Teflon bladders; when pressurized with helium, the bladders physically forced the contents to the bottom of the tanks to prevent the unsettling effects of "ullage" sloshing caused by weightlessness.

"You never know. That new Apollo guidance computer is really supposed to be something."

"Look, Scott, I'm sure that it'll be just like TV someday, and people will zoom around the universe wearing form-fitted velour suits and they'll sit in revolving lounge chairs on a flight deck the size of a house. And there will be sliding doors that snick open as you walk up to them, plus there'll be lots of cute girls in mini-skirts. But the chicks will all be pain-fully smart, though."

"*Ugh. Star Trek.*"

"*You* don't like *Star Trek*? Scott Ourecky? That's quite a surprise," observed Carson as he monitored gauges. "Pressure's nominal on fuel and oxidizer."

"Copy good pressure on OAMS fuel and ox. *Star Trek*? Man, I *can't* stand to watch it. It's not realistic to me at all. Bea's a big *Star Trek* fan, though. Actually, I think she's sweet on that Shatner guy, the one that who plays Captain Kirk." Ourecky watched the green COMP light blink off. "Our cake is baked. Your IVIs should be showing on your side."

"Yep," said Carson, checking his panel. "Ready, Scott? There's no turn-ing back from here. So far, this mutiny has been theoretical, since we've only *talked* about violating orders. Now, we're physically going to do it. We're half-way to Leavenworth, so this burn will finish the trip."

"Whatever. I'm ready. Sixty-eight seconds on the burn, right?" asked Ourecky. He pulled the sextant off the wall and held it in his lap.

"Right. Double-check me on the clock while we're burning."

"I'm watching the clock. Give me the mark when you're ready."

"Four, three, two, one, Mark, firing aft thrusters." Carefully monitoring the IVI display, Carson pushed the maneuver controller forward. "We're burning."

They immediately felt the acceleration as the spacecraft surged forward. "*Wow*," said Ourecky, slightly startled by the movement. "That's a kick in the pants. More than I anticipated."

"*Good* burn," commented Carson. "Very even…Twenty seconds… Thirty seconds…Sixty seconds. Shut down in six, five, four, three, two, one. Mark. Throttling down. All stop."

"Showing any residuals?" asked Ourecky. Normally, after a long burn of their aft thrusters, some minor adjustments would be needed to correctly align the spacecraft.

Carson watched the digital numbers gradually changing on the IVI display. Finally, they settled down, and he used the maneuver controller to make slight changes until the three IVI readings were zeroed out. "That's it. All zeroes; residuals are nulled out. PQI on the OAMS fuel is eighty-five percent," he said, checking the fuel gauges. "We're still in the hunt."

On Orbit
4:50 p.m. Eastern, Friday, June 13, 1969 (Rev 5 / GET: 7:50)

Ourecky tucked his hands under his armpits and shuddered; due to all the strenuous activity before launch and the initial stages of the flight, the cotton long underwear he wore under his suit was damp and clammy.

Shortly after their last burn, they made several adjustments to conserve battery power. They had scaled back their environmental settings almost to the point of shutting off the heat altogether. The cabin temperature hovered at roughly fifty degrees as they passed through the forty-five minute increments of orbital darkness and was slightly warmer when they languished in orbital daylight. It warmed up considerably when the sun shined directly through the windows, but that was a rare event, given the high inclination of their orbit.

While they grimly accepted the cold environs of the cockpit, there were a number of critical systems that required warmth to remain operational. The all-important OAMS—Orbital Attitude Maneuvering System— was fitted with a heating system to keep essential valves and components unfrozen and functional, so even though the two men were cold and uncomfortable, their attitude and maneuvering thrusters remained warm and cozy.

If anything aggravated Ourecky, it was that he couldn't enjoy the sights from their majestic perch. His teardrop-shaped forward-facing window was only about a foot away from his face, but if it was left uncovered in the current conditions, it would quickly frost over with the moisture from his breath. Because they would eventually be dependent on keeping a clear field of view, they kept their window covers in place unless it was absolutely necessary to look outside or during those cherished moments when they could harvest some welcome sunlight through the windows. The thin metal covers were provided to darken the cabin when they were sleeping, but they obviously weren't going to log much sleep on this mission, if any.

Although it would be hours until they completed the intercept, he worked to stay ahead of the game. He unstowed equipment and cranked open the observation port covers located in the hatches above them. The Gemini-I was fitted with two extra windows, approximately six inches in diameter, directly above each man's head. They were intended for the close-in proximity operations when they would examine and photograph their target. On several NASA Gemini missions, the astronauts' view was significantly obscured because their front-facing windows were smeared with some unknown residue, believed to be unburned booster propellant that washed over the spacecraft during staging.

With everything in readiness, he verified his maneuver calculations and double-checked the information entered into the computer. By sheer trial and error, he located a spot in the cabin, about six inches in front of him and slightly above eye level, where the air remained sufficiently undisturbed that he could "park" his mechanical pencil as he manipulated his slide rule or entered information into the computer. All he had to do was hold his hand out in the spot and let the pencil float off his fingertips, being careful not to impart any residual motion into the writing implement, and it would remain there until he needed it again.

"Hey, Scott," said Carson, slowly rolling to the left so that they were oriented head down to Earth. "Yank the cover off your window and look outside *right* now."

Ourecky removed his window cover and quickly oriented himself. He cupped his hand across his mouth and nose to direct his exhaled breath down and away from the glass. They were just passing over the plains of North Dakota and then skimming across the southwestern corner of Minnesota. He discerned the distinctive turn in the Platte River as it wrapped itself around the city of Omaha, and then he shifted his gaze through the top viewport to identify Lincoln, where he went to college, then looked southwest to pick out his hometown of Wilber, just west of the oxbow bends of the meandering Big Blue River.

Far below, his family would soon finish their chores before sitting down to dinner. Since his parents were devout Catholics who faithfully attended every single mass at Saint Wenceslaus, fish was sure to be on the Friday night menu. His father would grouse about seed corn prices and his mother would console him, both oblivious to the knowledge that their son had achieved his childhood dreams and was orbiting the Earth. After dinner, his father would retreat to the barn to tinker with the tractor while his mother washed dishes before retiring to her needlepoint.

As the spacecraft continued to track over America, crossing the Mississippi River just southeast of Cedar Rapids, he saw Chicago on the southern banks of Lake Michigan and then glanced farther to the left to pick out Indianapolis and then Dayton.

Bea was probably at the supermarket right now. Later on, she would likely heat up a TV dinner as she filled in the crossword puzzle from the Friday afternoon paper. Then she would watch television for a few hours before falling asleep on the couch. Sighing, Ourecky replaced the window cover and shifted his gaze back to his instrument panel. Although this was certainly the most amazing day in his life, it was one he could never share with the people closest to him; he was literally sky-high, but back on Earth, in his absence, life went on as usual.

"Get enough sightseeing?" asked Carson.

"Yeah. For the moment, anyway. Thanks, Drew."

"Don't mention it. Maybe there'll be time for more after we're done with the intercept."

On Orbit
6:34 p.m. Eastern, Friday, June 13, 1969 (Rev 7 / GET: 9:34)

According to Ourecky's calculations, the radar should now be effective in locating their target. To this point, they had no success in spotting it visually, and he was beginning to question whether pursuing the intercept had really been such a good idea after all. As he meditated on the possibilities, an unwrapped stick of chewing gum slowly floated by his face, lazily tumbling end over end. It impacted the circuit breaker panel immediately to Ourecky's right and rebounded toward the center of the cabin.

"Sorry," said Carson, reaching out to snatch the gum out of the air. Shivering slightly, he stuck the gum in his mouth and chewed nervously. "I let that one slip away from me."

"So are we ready, Drew? Everything lined up?"

To make the most of their limited battery power, Carson carefully oriented the Gemini-I so that the radar dish in its blunt nose was aligned precisely—in theory—at their unseen target. "Yeah, we're good," he said, shifting his head forward to look through the optical reticle mounted in his forward window. "Go ahead and fire it up."

Holding his breath, Ourecky re-set the circuit breakers for the radar and switched on the power. He knew from training that the radar hardware could be extremely temperamental. "Radar's energized, off standby," he observed, watching the status indicators that showed the radar cycling through a series of diagnostic tests. "Self-test is good. Signal generator is green. Slew test is good. Platform calibration is good. And we're *hot*."

Timing the warm-up process, Carson counted quietly to himself. "Forty-three seconds," he noted, whistling through his teeth. "You were almost dead on."

Ourecky nodded and studied the return display. The small cabin could not accommodate a radar screen like those found in fighters or interceptors, so a more compact version was furnished. It consisted of a circular array of tiny red lights, arranged in a wagon wheel pattern, with three lights making up each of the eight spokes. If the radar detected a target as during its scan, a light would illuminate. Its position in the wagon wheel denoted its relative orientation in relation to the spacecraft, and the return signal's strength was indicated by the light's intensity. A faint or intermittent light indicated a weak return a continuous strong light denoted a positive return. A small green bulb, set in the lower right corner of the display, flashed on when the radar had a solid lock on the target and was processing range and rate of closure information.

Ourecky pressed a button marked RDR DISP TEST. The array lit up for three seconds, so that he could verify that the bulbs were working, and then the display went dark. He watched it carefully; a minute passed and there was nothing. Flustered by the setback, he switched the radar to shutdown mode. "Nothing over here, Drew. Not even a nibble. Anything visual?"

"Not a thing," said Carson somberly, pulling his eye back from the optical reticle. "This ain't good. I sure hope we didn't stay up here for nothing."

On Orbit
7:31 p.m. Eastern, Friday, June 13, 1969 (Rev 8 / GET: 10:31)

An hour later, they tried the radar again. Anxiously watching the array, with his finger poised on the power switch, Ourecky counted to himself. He was about to shut it off when one of the lights glimmered momentarily. "Drew, I think we have a weak return on the outer edge of quadrant two," he said, pausing to blow warm air through his fingers. "You want to try resetting the nose? I think about ten degrees up and ten degrees right should do it. I'll leave the radar powered on."

Still looking through the optical reticle, Carson yawned as he tweaked the maneuver controller. OAMS thrusters popped quickly to adjust the spacecraft. "That do it?" he asked.

"Hopefully," replied Ourecky, studying the radar display. It would take a few seconds for the dish to complete its scan. Since the light might have resulted from just a spurious flash in a circuit, it was a gamble; as much as he didn't want to miss a potential positive contact, he was wary of frittering away their batteries on nothing.

"Anything?"

"Nothing…nothing…nothing." Then the center light blinked on, followed by the green light that indicated a solid lock. *"Bingo!* Hard acquisition, center mass," stated Ourecky confidently, watching the digital counters automatically changing as the radar calculated the distance.

Carson exhaled a cloud of warm mist. "Oh, man," he said quietly. "I guess we did it."

"I think so," agreed Ourecky. "At least we're getting closer. Range is 260 miles."

Carson nodded. "What's our closing rate?"

"We're closing at roughly three hundred feet per second, but that's only in the ballpark at this range. Drew, I'm going to pass the radar data to the computer and then shut it down. Do you have anything out the window yet?"

"Nothing yet. Very crappy horizon out there right now, so the target's probably washed out in the background. I don't think we're going to see it this far in daylight, anyway."

"Okay," said Ourecky, tapping number sequences onto the computer's keypad. "Radar data is loaded and accepted." He switched the radar off. The chase was drawing to fruition; with just a few more maneuver burns and some careful flying on Drew's part, they would be home free.

On Orbit
8:06 p.m. Eastern, Friday, June 13, 1969 (Rev 8 / GET: 11:06:22)

Although they had a successful radar spot, they had yet to visually confirm the target, so they still could be chasing a phantom. With orbital darkness rapidly approaching, the best time to look for their quarry was immediately after they crossed the threshold into darkness.

"Here comes twilight," announced Carson. He wore a cloth patch over one eye so it would be immediately attuned to dark conditions. "Forty-five seconds. Ready on the radar?"

"Ready," replied Ourecky.

Carson had already switched the cabin lighting to the red lamps, which were better for maintaining night vision, and now he twisted a rheostat knob to dim the lights as low as possible. Counting to himself, he flipped up the patch, leaned forward and placed his eye against the optical reticle. "Okay, light it up, Scott."

Ourecky threw the switch to activate the radar. Cupping his penlight with his fingers to cut down on stray light in the cabin, he watched the radar's diagnostic lights. "Self-test is green... Platform calibration is showing green. Radar's hot." He pressed the radar's display test button and watched as all the red lights flickered on momentarily and then blinked back off.

"Anything?" asked Carson, staring intently through the reticle.

"Still scanning." Suddenly a red light blinked on and stayed lit. "Got it. Acquisition and hard lock." Ourecky punched in the commands on the computer keypad to accept the radar data.

Willing himself not to blink, Carson stared through the reticle. And then he saw it, an obscure point of light resembling a third magnitude star, precisely where it was supposed to be. "Kismet!" he declared. "There it is, right smack where we figured. Man, if we hadn't done those extra hours in the planetarium, I would have *never* caught it. What's our range and rate?"

"Range is 150 miles, closing at 250 feet per second. We'll be on top of it within the hour."

"Tally ho!" whooped Carson, laughing as he twisted a corner of his moustache. Even though they were slightly more than a minute into orbital darkness, the cabin temperature had plummeted noticeably. "Almost worth freezing for."

On Orbit
10:21 p.m. Eastern, Friday, June 13, 1969 (Rev 9 / GET: 13:21:18)

It took slightly less than another orbit to close the gap with their quarry. In accordance with their procedures, they would initially "station-keep" at a comfortable distance for at least an entire orbit, study the target intently, wait until an opportune orbital sunrise, and then move in closer.

Orbital rendezvous was an endeavor that was intensely dependent on lighting conditions. This was true in cooperative rendezvous situations, where the chaser and chased worked in orchestrated harmony, but considerably more so in non-cooperative scenarios. Every ninety-minute orbit comprised forty-five minutes of light and an equal measure of darkness; because of the risk of collision, approaching a non-cooperative target in darkness could be an exceptionally dangerous proposition under the best of circumstances.

The sun was just edging up over the horizon, so it was safer now for Carson to fly the final approach. Once up close, they would perform a methodical inspection during their forty-five minute periods of daylight, scrutinizing every square inch of the suspect satellite but also diligently remaining clear of any potential optical systems that might betray their presence.

During the ensuing forty-five minutes of darkness, Carson would back away slightly and activate a spotlight to keep a vigilant eye on the target. And their intermittent dance would continue—in forty-five minute increments—until the inspection phase was complete. At least they were marginally comfortable now; because it was so critical that they keep a constant eye on the target, Carson had powered up the environmental control system to warm the cabin so the windows would remain unfogged.

As the light gradually improved, Ourecky studied the target through binoculars. It was a white cylinder, approximately fifty feet long and about fifteen feet in diameter, oriented with its long axis perpendicular to the Earth. The end farthest from Earth bore a cross-shaped array of four solar panels, along with three whip antennas and a parabolic dish antenna.

The solar panels were a solid clue that it was designed for long endurance, probably for weeks if not months; otherwise, if it had a shorter mission, the Soviets would have provisioned it with batteries. It was clearly not an abandoned booster stage waiting to fall out of orbit, but a vehicle with a specific purpose.

But what was that purpose? thought Ourecky. Was it a nuclear-armed orbital bombardment platform—OBS—or some sort of reconnaissance satellite? He was sure that it was gyro-stabilized, because it had held the same attitude, without budging, since they had started watching it. He had discerned no indications of thrusters firing; that was good, because their close proximity work could be very dangerous if the target was prone to sudden maneuvering.

He had sat through excruciatingly detailed briefings on Soviet reconnaissance satellites, but this didn't resemble anything he had been shown. Although their spacecraft were typically big and unsophisticated compared to their American counterparts, the painfully pragmatic Soviets tended to stick with proven designs and adapt them as necessary for different roles. Consequently, their workhorse reconnaissance satellites—the *Zenit* series—were based on the fabled *Vostok* that had carried Yuri Gargarin to orbit in 1961. Of course, it was equally likely that the *Vostok* was fashioned from the *Zenit* model rather than the other way around.

The *Zenit's* premiere attribute was a massive sphere—roughly eight feet in diameter and weighing over two tons—that contained cameras, film, recovery beacons, and parachutes. The durable sphere was connected to a service module that carried batteries, electronics, retrorockets, and a maneuvering system. Once a *Zenit* completed its mission—typically between one to two weeks in duration—the heatshield-equipped sphere reentered the atmosphere for recovery. It was a superbly ingenious if not clunky system, particularly since the complicated optics could be re-used on subsequent missions.

In contrast, the American equivalent left the costly cameras in space, sentenced to an inevitable demise by orbital decay, and ejected a reentry pod containing the exposed film. Ourecky was aware that both nations were progressing toward systems that would electronically transmit images to the ground, eliminating the need to safely deliver film to Earth, but that technology was years into the future.

But this monster wasn't a *Zenit*. It was considerably larger, with an odd feature that perplexed Ourecky. Approximately twelve feet up from the base, the cylinder was encompassed by a series of objects, each about the size of metal trash can, with rounded bottoms. From this angle, Ourecky could only count seven, with an open space where an eighth should go, and he estimated that there were probably at least five more on the far side of the enigmatic satellite.

"So what do you make of that necklace of pods?" asked Carson. "Nukes maybe? Do you think we're looking at an OBS?"

Ourecky pulled the binoculars away from his eyes and let them float. "I don't think they're nukes," he observed. "I doubt that you could package a nuke in a container that small, allowing for a heatshield, and yield anything greater than a couple of kilotons. A firecracker that size would only be a nuisance unless it was delivered with some precision. I can't imagine the Soviets putting an OBS up here just to be an annoyance. It just doesn't make sense."

Carson grabbed the binoculars out of the air and scrutinized the target. "Well, here's my theory, for what it's worth. I'll wager that we're looking at their next generation reconnaissance satellite and that it's intended for extended duration missions," he surmised. "The bottom end of that cylinder is probably their optics, and it's meant to stay up here. Right behind the cameras, where all those pods are, there's probably a film carousel with a bunch of film cartridges. Once they expose a full cartridge, it's probably loaded in one of those pods, and then the pod ejects and reenters. That would also explain the rounded bottom on the pods."

"That makes a lot of sense, Drew. That would also explain the open slot where it looks like a pod should be. They've probably already shot a reel and dropped it."

"Very likely," agreed Carson, handing the binoculars back to Ourecky. "Let's just hope they don't dump another one while we're operating close in. That could get ugly quick."

"Agreed." Ourecky stowed the binoculars in a fabric pouch at his right side.

Carson ripped open a plastic bag containing tiny sandwiches consisting of pimento cheese spread on tortilla squares. The snacks spilled out of the bag, scattering into a small edible constellation on his side of the cockpit. Examining his instruments, he plucked a sandwich out of the air and nonchalantly munched on it. "Well, this ain't good," he observed.

"What's the problem?" asked Ourecky, filching one of the floating sandwiches.

"An OAMS thruster is sticking intermittently. Yaw right, number three. Probably a malfunctioning solenoid valve. It's been sluggish from the start, but now it's barely working at all. It's probably just squirting unlit propellant, and we can't afford to waste that."

"Can we still fly close in? Is it safe?"

Carson popped another sandwich in his mouth. Chewing slowly, he answered, "Oh, yeah, we can still do it. I'll inhibit the thruster and compensate accordingly. I'll have to be light on the stick and move very slowly. What do you think, Scott? I would hate to climb all the way up here and not finish the job, but I would be lying if I didn't think we weren't assuming some risk."

Ourecky nodded. "I trust you, Drew. Let's do it."

Carson reached up to the overhead circuit breaker panel and switched off the breaker for the malfunctioning thruster. "Okay, I'm going to nudge us over for proximity operations. Ready?"

"Drew, aren't you forgetting something?" asked Ourecky. "Aren't we supposed to be completely suited up before proximity operations? Including helmets and gloves? That's the rule, isn't it?"

Carson chuckled. "Man, that promotion has obviously gone to your head, Ourecky. You sure have become a stickler for formality. Look, let's be realistic. We're supposed to be in full kit just in case the pressure vessel gets penetrated, right?"

"Right."

"Now, do you *really* think we'll live through reentry if an antenna pokes a hole in us?"

"Probably not."

"Then since we're already in hot water for breaking some rules, why don't we break one more? If we die, we'll at least be somewhat comfortable. Besides, aren't you always griping about how you can't look up through the viewport with your helmet on? Now you won't have to crane your neck as much."

"Fine by me. After all, you are the boss."

Carson finished his last sandwich. "Okay, here we go. I'll sneak in slow so we can look at the underside from an oblique, so we don't accidently eclipse that camera. I suspect if we catch the sun right, we'll see a glint off the lens, and that will confirm the optics. Ready?"

"Ready on my side."

"Ouch," declared Carson, looking out his window as he adroitly nudged the maneuver controller. "Man, there's a huge glare coming off that white paint. This ain't going to be easy." He reached into the stowage pouch to his left, extracted his eye patch, and put it on.

"What's that for?" asked Ourecky. "Going to pirate mode?"

"Nope. I still have to read my instruments, and that damned thing is almost blinding me." Carson bumped the maneuver controller again and looked down at his instrument panel as he flipped up the patch from his protected eye. "Here we go."

On Orbit
3:28 a.m. Eastern, Saturday, June 14, 1969 (Rev 13 / GET: 18:28:15)

After drawing alongside, they flew in close formation to the Soviet satellite for roughly five hours—three orbits—while carefully studying and photographing it. Now it was time for the final phase of the close proximity operations.

During the light phase of their last orbit, even as Ourecky was completing the last series of radiological measurements, Carson maneuvered the Gemini-I until it was translated into a nose up attitude, directly parallel to and slightly below the target. As they entered the dark phase of the orbit, he switched on his floodlight and maintained position.

"Anything on the radioactive front?" asked Carson.

"It's not emitting anything greater than normal background," answered Ourecky. "I think it's safe to say that there's no nukes or reactor on board, unless they're exceptionally well-shielded. I'm going to close out the monitoring gear and get ready to deploy the Disruptor."

"Sounds good. I'm not exactly comfortable with this orientation, but it looks like our best shot for deployment. I'm going to use those solar panels as my alignment reference. As it swings out, the Disruptor will momentarily cross their optics' field of view, but I seriously doubt that they'll have their cameras switched on over this corner of the world."

Just as the sun peeked over the horizon, Carson ordered, "Deploy the Disruptor."

"Deploying Disruptor," answered Ourecky, toggling a switch. Inside the adapter, at the far end of the spacecraft, an explosive bolt detonated, freeing the end of a collapsible boom. A pneumatic cylinder, powered by a small cylinder of helium gas, slowly swung the boom out to its full extension.

Next, a pair of spring-loaded detents locked the boom into position; a second pneumatic cylinder extended the boom to its full length of eighteen feet. Now fully deployed, the boom jutted from the base of the Gemini-I up and at a forty-five degree angle so that its distant end was positioned almost directly above the viewport in Ourecky's hatch.

"Boom deployed," stated Ourecky. "Extending hoop." At the far end of the boom was a large hoop, like a huge snare, constructed of flexible metal tape. Resembling a gigantic radiator hose clamp, the hoop was motorized to expand or contract to encompass a target vehicle. It was literally like a giant lariat they would use to lasso the Soviet satellite. "Full extension on hoop."

"Okay, let me eyeball this thing a minute," said Carson, looking up through his viewport. "Lateral spacing looks good. Diameter on the hoop looks good, but if the target were any bigger then we would be in trouble. I'm going to start inching forward. Keep a sharp eye out, Scott."

Pulsing the thrusters, Carson gingerly edged the spacecraft forward, as if he were pulling into a tight parking space. If all went well, the maneuver would be like slipping an oversized ring onto a very big finger. "How's it looking?" he asked, keeping his eyes on the target's solar panels.

"Your lateral separation looks great," Ourecky said. "No visible yaw or roll component on the target. The hoop's lining up perfectly. You have about six feet to go."

"Moving forward," said Carson, delicately maneuvering the Gemini-I by measured inches.

"Four feet to go. Come out six inches and pitch the nose up half a degree."

Carson deftly made the corrections. "How's that?"

"Perfect. You're lined up as well as you could be. Three feet to go…two feet…one foot…we've got overlap…still good alignment…about two foot overlap…three feet overlap…alignment is still excellent. *Mark.*"

"All stop," replied Carson, applying a slight forward thrust to brake the spacecraft to a halt.

Although the boom was electrically insulated, a series of fine wire whiskers first made contact with the Soviet satellite to discharge static electricity, generating a momentary blue glow like St. Elmo's fire. "Looks very solid," judged Ourecky, watching the impromptu fireworks.

Moving slowly, Carson swiveled his head to check the loop's positioning on the target. "I concur. Grab the camera and click some frames before we finish strangling this critter."

Ourecky pointed the Hasselblad up through the viewport, snapped several pictures, and then tucked the camera between his knees. "Okay, Drew, ready for me to contract it?"

"Close the loop."

Ourecky thumbed a switch to activate a drive motor at the far end of the boom. The loop contracted exactly as it was designed to, securely snugging it flush against the metal skin of the Soviet satellite. A clutch stopped the motor as the hoop sized itself to match the target.

"Hoop is closed. Disruptor comms check," stated Ourecky. Looking out through the viewport, he watched a small panel on the Disruptor; a light on the panel flashed green as he thumbed a button, confirming that the device's radio receiver was functioning correctly.

"Comms check is good," verified Carson, also watching the light.

"Arming Boom Pyro," declared Ourecky. He reached out to a rotary dial labeled "BOOM PYRO" and twisted it from SAFE to F1. "Secondary Safety disengaged. Pyro set to charge one." Ourecky drew in his breath and looked up through the viewport. "Drew, I'll fire on your mark."

Carson crossed his fingers. At his command, Ourecky would trigger an explosive bolt that would sever the boom at its juncture with the loop. Simultaneously, as soon as they were physically separated, Carson would fire thrusters to swiftly disengage from the target. The timing was crucial. The protruding boom would still be securely attached to Gemini-I, and if they accidently moved toward the target, it could penetrate the exterior skin of the target. That could be dangerous for a number of reasons; besides the risk of becoming inadvertently and permanently attached to the target, they had absolutely no idea what lurked under its metal skin, and they could breach a fuel tank with potentially fatal consequences. So the trick to the game was to hold the Gemini-I as stable as possible until the charge fired, which Ourecky would visually confirm, and then scoot away to safety.

Wrapping his fingers around the hand controller, poised to withdraw, Carson quietly ordered, "Three, Two, One, Zero, *Mark*. Fire, Scott."

Ourecky threw the switch, watched the explosive bolt fire, and announced, "*Flash*." He immediately rotated the BOOM PYRO knob to F2 in readiness to fire a back-up charge in case the first fizzled, but saw that the spar had snapped cleanly at the junction. "Boom is sheared."

"*Backing*," said Carson, nudging the hand controller. "And *that's* how it's done, sports fans."

"Boom Pyro to Safe. Primary Disruptor Safety is disengaged," said Ourecky, flipping a switch. With that, the Disruptor was fully armed. Long after they returned to Earth, it could be brought to life by an encoded radio signal sent from the ground.

A highly adaptive killing machine, the Disruptor could cripple the Soviet satellite in any one of three ways. Most blatantly, it could demolish it outright with a substantial explosive charge or fire a solid rocket motor that would persuade the satellite to prematurely tumble into the Earth's atmosphere. But by far, its most insidious means of attack was a "needle thruster." The needle thruster was a miniscule gas port—approximately the diameter of a sewing needle, thus the name—connected to a cylinder of compressed nitrogen. Once activated, the thruster would stream gas for days or even weeks, just enough to impart a slow rotation to the target satellite.

The spiral would be a cause of maddening frustration to the satellite's Soviet controllers, since they would never be able to focus its optics on a target with any degree of certainty. Additionally, the nefarious tactic would inevitably lead to the satellite's early demise, since its solar panels would not be able to consistently absorb sufficient sunlight to charge its electrical systems. If nothing else, the gradual spin generated by the thruster would sow seeds of doubt in the satellite designers' minds.

As they gradually faded away from the target, Ourecky asked, "What's next, boss?"

"Well, the safest option would be to stay up here, but I suppose we had better head for New Mexico. I'm not looking forward to facing the wrath of Sheriff Wolcott, but at least this little pony delivered the mail, and it's difficult to argue with success. When's our next contact?"

Ourecky looked at an index card held by an alligator clip on a corner of his instrument panel, and then referred to his mission clock. "Atlantic Sentry Two in thirty-four minutes."

"Time to grab a snack," noted Carson, reaching back toward his food locker. "I'm famished. After we hit the contact, we'll pack up and prep for reentry."

Gazing around the cluttered cabin, Ourecky observed: "I'll start packing while you're chowing down. Getting this stuff re-stowed isn't going to be easy. We're fresh out of nooks, and the crannies are in short supply."

Northrop Strip, White Sands Missile Test Range, New Mexico
8:46 a.m. Eastern, Saturday, June 14, 1969 (GET: 23:46:01)

Just slightly less than a day after they had left the PDF, they were almost home. Lining up on final, Carson had flown the paraglider so many times that this part of the mission was second nature to him. He would have preferred to land at night, but the early morning light wasn't too bad. Looking at the small television screen mounted in the center console, which was connected to a fiber optics camera mounted to the right skid strut, he took note of the touchdown markings on the earthen strip.

The ground controller's voice came through Carson's earphones: "Scepter Two, you're cleared to land on Runway One Seven. Winds out of One-Six-Zero at five knots. Landing surface is a dry lakebed. Advise that your gear are down and locked."

"This is Scepter Two, copy cleared for landing on Runway One Seven, winds out of One-Six-Zero at five. Gear down and locked."

"Looking good, Drew," said Ourecky. "Sure you don't want me to take this one?"

"Nah. I got it. Just keep an eye on that contact light over there." Below the spacecraft, a contact sensor dangled on a fifteen-foot wire from the right skid well; a contact light on Ourecky's panel would illuminate when it touched the ground.

"Got it. Should be anytime…anytime….*Contact*."

"Flaring paraglider," said Carson calmly, tugging back the hand controller. Their forward speed slowed almost immediately, and then they heard a scraping noise as the three skids made contact with dirt strip. "Touchdown! Stand by to release paraglider."

Ourecky flipped up a safety cover on a switch and replied, "Standing by to release."

"Release paraglider."

Ourecky threw the switch, which fired three explosive bolts simultaneously, jettisoning the paraglider from the spacecraft. Discarding the paraglider was a precaution intended to prevent the huge fabric wing from being ignited or melted by the still searing heat shield at the rear of the spacecraft. Free of the paraglider, the Gemini-I continued to slide smoothly down the airstrip, gradually coming to a stop as emergency vehicles pulled alongside.

"Control, Scepter Two is at full stop," declared Carson.

"Scepter Two, this is Control. Welcome back, gentlemen."

39

THE UNUSUAL NATURE OF
OBJECT 2368-B

Aerospace Support Project
9:41 a.m., Monday, June 16, 1969

olcott looked up from a report as an aide ushered in Carson and Ourecky. The two looked none the worse for wear; it was hard to believe that they had returned from orbit less than forty-eight hours ago. They had come straight from their T-38 on the flight line, with barely sufficient time to change from their flight gear to civilian clothes, and their hair was still pasted flat from the weight of their helmets and filmy sweat.

They both exhibited some color in their faces, courtesy of their limited exposure to the Pacific sun last week, and looked like a pair of salesmen stumbling back into the office after a convention junket in Miami. Although still charged with residual exuberance, they also seemed wary of whether they had been summoned for congratulations or court martial proceedings.

"General Tew should be right back," stated the aide ominously.

Smiling nervously, Ourecky sauntered in and casually took a seat at the table. Obviously sensing the severe gravity of the situation, Carson wagered on an unfavorable encounter, and stood at attention behind a chair. With an exaggerated cartoon-stiff posture and chin tucked sharply in brace, he looked every bit like the terrified West Point plebe he had once been.

"Major Ourecky, if you know what's good for you, you'll bounce your ass out of that chair and lock it up like your compadre," recommended Wolcott. "I ain't going to steal any of Mark's thunder, because I know he's had all weekend to formulate the ass-chewin' you two will receive momentarily, but I will say that I'm glad to see you back on the ranch, safe and sound, and I appreciate what you did upstairs. And gents, for the moment, that's all I'll say on this matter."

Ourecky had barely risen from his seat when Tew blew into the office like an anti-tank round exploding against the steel slab hull of a tank. He slammed the door behind him, rattling the walls, and took up a position across the table from the two men.

Red-faced and trembling, Tew said calmly, "On a *positive* note, your film arrived here on Saturday and has already been processed, and everyone is extremely impressed with the results. And that concludes anything *positive* that I will say this morning."

Tew pivoted to face Ourecky, and said, "Major, I don't know if you just fell under Carson's bad influence or whether you saw fit to exhibit some horrifically poor judgment in your own right, but I fervently hope that you didn't give away all of your captain's bars, because I guarantee that you'll be taking off those oak leaves before the sun goes down today."

"Yes, sir," stammered Ourecky. "Sorry, sir. And sir, I have to inform you that all my actions were of my own accord, and I will accept the consequences for them."

"As if you had a choice," snapped Tew, pivoting to face Carson. "Carson, I lack the words to effectively convey my anger and disappointment. By sheer providence, you accomplished your mission, but we could have suffered a profoundly different outcome, which you probably would have been spared the misfortune of witnessing, because you would have been *dead*. But even as you strummed your harp or adjusted your halo, we poor souls left on Earth would not only be mourning *your* passing, but we would be mourning the cancellation of this entire effort."

"General, speaking in our defense, I thought that the mission warranted…"

Tew's face turned an even deeper shade of crimson and his hand found the small pillbox in his pocket. "Major, I will not *allow* you to speak in your defense, because there is no plausible defense for your actions. For your information, I don't formulate orders in a vacuum. I weigh the options that are presented to me, issue orders that are appropriate to the circumstances, and I expect those directives to be *followed*, Major Carson. I expect them to be followed *immediately*, without question or discussion. I…"

The phone rang on Tew's desk. Simultaneously, his aide's voice came over the intercom: "General Kittredge, sir. I know you don't want to be disturbed, sir, but he said it was urgent."

Still seething, Tew gritted his teeth and picked up the phone receiver. He listened for several seconds before speaking. "Yes, sir. Thank you, sir. It was quite a surprise here as well…Yes, sir, I do intend to be there for the classified briefing tomorrow. Yes, sir, they're standing right here. Yes, sir, I'll pass on your congratulations to them."

He placed the phone back on the hook and turned to face Carson again. "Carson, regardless of your abilities, I cannot trust you to fly on behalf of Blue Gemini again. *Ever.* And that leaves me in a quandary, because I'm at a loss as to what to do with you. Normally, I would exile you to Montana to babysit a missile silo for a few years as you contemplate the error of your ways, but I will be damned if I'll entrust you with the launch keys for a nuclear-armed ICBM because it is abundantly clear that you lack the ability or propensity to follow orders."

"Yes, sir," said Carson, swallowing.

"Since my options are limited, you will remain here to work with Crew Three until I divine an appropriate punishment for you and your partner in mutiny. Just so you have absolutely no misconceptions, let me be abundantly clear: You are *grounded.* You will not fly any spacecraft or aircraft for the remainder of the time you are assigned here, and I will do my damndest to ensure that your wings are permanently yanked."

The phone rang again, followed by the aide's voice on the intercom: "General Ames, sir."

"Captain Parsons, didn't I ask you *not* to disturb us?" growled Tew.

"Sir, General Ames said it was critical," replied the aide.

Tew groaned before picking up the phone. As Carson and Ourecky stood rigidly at attention, he had a four-minute conversation with a four-star general; because the line was not secure, they spoke only in vague generalities. Tew ended the exchange with, "Yes, sir...Thank you, sir... Yes, sir, they're both right here. I'll pass on your appreciation. Thank you again, sir."

Trying his best to appear stern, Wolcott could barely disguise his amusement at the circumstances. Rather than appropriately recognize their unprecedented feat, Tew treated them like a pair of recalcitrant cadets who were tardy to an afternoon parade.

And while Carson and Ourecky were probably overjoyed just to have safely returned to terra firma, they probably had no inkling what a serendipitous success their foray had been. They had been sent up to look at Object 2368-B, a suspected OBS platform. Ironically, even though their mission was technically a failure because Object 2368-B wasn't an OBS, the Gemini-I crew had stumbled upon a truly momentous intelligence coup. Their mission had revealed an entirely new class of Soviet reconnaissance satellites, previously unknown to Western intelligence officials, which in turn lent a new perspective on the state of Soviet space technology and capabilities.

Before last week's mission, everyone assumed that the Soviets relied almost exclusively on their *Zenit* series recon satellites. Like an annoying housefly, a *Zenit's* lifespan was short—two weeks at best—so even if a Gemini-I was sent upstairs to pay a visit, the only thing they were likely to witness was a service module discarded after the reentry vehicle had descended to Earth with the goods. Thus, Blue Gemini's mission planners stringently avoided targeting any object that even vaguely appeared to be a *Zenit*.

So why Object 2368-B? The first part of the answer lay in how Blue Gemini selected targets for missions. Their planning approach was almost opposite of NASA's: Instead of starting with the mission objective and methodically working backwards to solidify a launch window, as NASA did, Blue Gemini started with the launch window and proceeded forward to identify a target. While it probably seemed like an irrational approach, there was a definite logic at work.

NASA's launch windows had to be developed around a bewildering array of variables. Just the act of firing a rocket was a complicated matter; stringent airspace clearances and related issues constrained launches at Cape Kennedy. The Air Force faced similar restrictions with their secretive launch facility at Vandenberg Air Force Base in California. As an extreme example, although the base was situated in a unique geographic location to fire highly classified recon satellites southwards into polar orbits, Vandenberg launches actually required advance coordination with a railroad—the Pacific Rail Line—whose train tracks transited the base.

On the other hand, Blue Gemini's primary consideration was the availability of critical hardware—the Titan II boosters and the Gemini-I spacecraft—necessary to execute the missions. Because the project owned the remote PDF launch site that could target virtually any prospective orbital plane, they could effectively launch at will. So once a launch window was determined, the mission planners selected a target from a prioritized list of Soviet satellites considered worthy of a mission.

Since they borrowed certain key resources from the overt space program, notably a few tracking ships and the EC-135E ARIA aircraft, Blue Gemini could not launch at those intervals when NASA missions were underway. Otherwise, if they possessed the requisite flight hardware and there was a crew available to fly, the sky was chock full of Soviet targets. Granted, it wasn't like firing a shotgun into a barrel filled with fish, where it was a virtual certainty that at least one buckshot pellet would find a carp or catfish, but probably more like shooting a scope-mounted precision rifle at a crowded aquarium from a few hundred feet.

But *why* Object 2368-B? To paraphrase British mountaineer George Leigh Mallory: *Because it was there.* More specifically, because Object 2368-B was there at the appropriate moment to coincide with the available launch window. With a multitude of eligible targets to pick from, the inauspicious Object 2368-B was selected primarily because it was very big, very stable and it happened to be in an orbit that regularly overflew the continental United States.

Otherwise, Object 2368-B probably wouldn't have ever been contemplated with such fervor. And since the first mission had ended in catastrophe, Blue Gemini's hardware and concepts had yet to be proven, so the massive satellite was also chosen to build confidence in everyone involved in the project, especially those two men who assumed the greatest risk.

Indeed, the mission to intercept Object 2368-B resulted from of a string of ironic occurrences. If the first mission had not culminated in a horrendous calamity, the second mission would have launched three months later, and then the third three months after that, and so on. Carson flew as early as he did solely because of the accident; otherwise, Howard and Riddle probably would have flown the first four missions.

Looking at Tew as he reamed out the hapless duo, Wolcott suspected that if Carson had followed orders and scuttled the rendezvous, then the mission planners probably would not have fussed with Object 2368-B ever again. And he also knew that if anyone else but the maverick Carson had been in that cockpit, given the same circumstances, they would have scurried to Earth at the earliest possible opportunity, without the slightest hint of discord.

As he listened to Tew bellow, Wolcott was confident that his friend's anger would eventually subside, even if it took months, and the team of Carson and Ourecky would fly again. In the meantime, Tew had bumped them off the flight roster in favor of Crew Three, although Wolcott was not particularly confident that Crew Three could have pulled off the same sort of miracle that they had witnessed on Friday. In fact, he wasn't sure that Jackson and Sigler could execute a rendezvous at all, at least without several more months of intensive training in the Box.

Wolcott smiled to himself as he glanced at a single black-and-white glossy photograph lying flat on his desk. Snapped by Ourecky during the proximity operations, the single image validated every penny spent on Blue Gemini, every life lost, and every risk taken. Not only was the photograph a vivid representation of the extent of intelligence that could be gleaned during Blue Gemini missions, it was revealing in an altogether different way.

Not so lacking in boost capacity that they fretted over stray ounces like their American counterparts, the officious Soviets had seen fit to install a brass data plate on the side of their satellite, obviously not anticipating that it would ever again be glimpsed by human eyes.

In meticulous Cyrillic script, the shiny data plate proudly stated the satellite's pertinent details: weight in kilograms, physical dimensions, serial number, date of manufacture, and manufacturing facility. Captured in absolute clarity in Ourecky's photograph, it would prove to be a virtual treasure trove in tracing the lineage and operational employment of Object 2368-B and its ilk. It was almost like receiving a postcard from the Soviets, politely saying "Direct Intelligence Operations Here."

There was a quiet tap at the door, and the aide stuck his head into the office.

"Can you *not* see that I'm in the middle of something?" roared Tew. "Did I not *specifically* ask you not to disturb us?"

"Uh, begging your pardon, sir, but there's a secure phone call downstairs in Communications, sir," replied the aide sheepishly.

"I'm busy!" snarled Tew. "And frankly, damn it, I'm tired of listening to every other general in the Air Force trying to butter me up. Take a message and tell them that I'll call them back!"

"But, sir," answered the aide. "You might want to take this call. It's the White House."

Tew stood up and hurried for the door. Just before he went through it, he looked back over his shoulder and said, "Virgil, aren't you coming?"

Wolcott spat in his brass spittoon and replied, "No, pardner. You made this all happen; I was merely a hired hand. It's your turn in the sun, so go bask in it."

As Tew left, Carson and Ourecky both breathed a sigh of relief, but then tensed up, anticipating a withering volley from Wolcott.

"Carson, Carson, *Carson*," muttered Wolcott, slowly shaking his head and grinning. "What on *earth* can we do with you, pard? And you, Ourecky, as intelligent as you are, what kind of burr got lodged under your saddle that made you think you could wantonly disobey orders?"

Carson replied meekly, "Virgil, Major Ourecky was just following my orders up there. He's not at fault and doesn't deserve to be punished."

"Save your breath, Carson," said Wolcott, standing up. "Hombres, I ain't condonin' what you did, except to say that when you elect to climb out on a limb to defy an order, you had danged sure better deliver some results. And you two boys did deliver, in spades, and we're indebted."

"Thank you, Virg," said Carson quietly.

Wolcott lit a cigarette. He inhaled a long drag from it before continuing. "Let me tell you something about obeyin' orders. We would all be speaking German or Japanese right now if Billy Mitchell hadn't been an obstinate, orders-ignorin' son-of-a-bitch between the Wars. I ain't implying that he won the War, but because of men like him and George Patton and some other cocky malcontents, the rest of us had the tools to beat the Germans and the Japs.

"So, betwixt the three of us, your transgressions are now water under the bridge, as far as I'm concerned. I can't speak for General Tew, but I imagine that he still has some special purgatory set aside for you. Until you two gallant hands crawl out from under that misery, whatever it may be, I want you to keep your chins up and be ready for come what may. Now, you two, it's a good time to make yourselves scarce. Go across the way and see Gunter Heydrich and his troops. They deserve your thanks, and you deserve to get your backs pounded until you can't stand up straight."

"Sir, what about General Tew?" asked Ourecky, adjusting his collar as he headed toward the open door. "I don't think he was finished with us."

"He was," replied Wolcott, looking down at the photograph on his desk. "An old friend of mine called me this morning from the White House, so I know what this call is about. It's partly about congratulations, but that ain't all, not by a long shot. If you thought we were busy here before, you've not seen anything yet. Now, you heroes, get gone."

Carson was almost through the door when Wolcott called after him. "Hey, pard..."

Carson stopped and turned. "Was there something else, Virgil?"

"Yup. As stupid as it may seem, no job is finished until the paperwork is done. Make sure you two fill out your travel vouchers and submit them at the admin shop."

"You're kidding, right?" asked Carson, grinning.

"Not in the danged least," answered Wolcott. "You need to account for your travel out to Hawaii on the outbound leg. The last thing I need is a bunch of auditors tryin' to hang us for fraudulent travel expenses. Make sure you two get them filed by the end of the week."

Smiling to himself, Wolcott watched the two men leave. Stretching out his legs, he leaned back in his chair. As he had implied to Carson, he already had an insight into the call from the White House. As of yesterday, after the key players had their first glimpse at Ourecky's photographs of Object 2368-B, Blue Gemini's focus had immediately shifted. Despite Tew's unwavering obsession with Orbital Bombardment Systems, the initial justification for Blue Gemini's inception, they were now cleared to hunt and kill *any* suspect satellite, regardless of whether it was a suspected OBS or not. To a large extent, the change in mission evolved out of the Navy's desire to target a particular satellite. Up until yesterday, given his narrowly defined charter, Tew could vehemently deny the Navy's demands. But now, armed with this new insight about the Gemini-I's lethal effectiveness at chasing satellites, coupled with an unexpected new ally in Washington, the Navy was about to permanently change the game.

Flight Crew Office, Aerospace Support Project
11:05 a.m., Tuesday, June 17, 1969

"So, what are we doing for lunch?" asked Ourecky, setting aside a Skilcraft ballpoint pen as he looked up from a partially completed travel voucher form. "You do remember that the technical debriefing kicks off at one this afternoon, right?"

"Yeah. I'm not looking forward to that quiz session," answered Carson, perusing a recent issue of *Life* magazine. The front cover showed a color photo from May's Apollo 10 mission, depicting the shiny gold ascent stage of the "Snoopy" lunar module juxtaposed against the stark gray pock-marked surface of the moon; a caption read "Barnstorming the Moon."

"I can't imagine that it will be that bad. It's just an engineering review."

"*Just* an engineering review? Well, brother, since you've never been a test pilot, you haven't had to suffer through a bunch of damned aeronautical engineers giving you the third degree about why something on *their* airplane didn't work exactly as intended. I'm sure that we're going to get wrung out over the batteries, like either one of us would have any idea what happened back there in the adapter."

"I guess you're right." Ourecky gazed across the room at Tom Howard's desk. It was exactly as he had left it before his February departure for Johnston Island, almost as if he was expected to be back in the office tomorrow, after a long vacation or extended business trip. "Well, on the plus side, at least Big Head and Squeaky didn't have to submit to an inquisition."

"Yeah, they didn't," answered Carson. "And they didn't have to immediately come back to work the next week, either. This is so asinine. I sure can't imagine that the NASA guys don't get any time off after they get home. You would think that Virgil and Tew would grant us at least a day's rest after coming back."

"I'm sure that they want to tie up the debriefings while the information is still fresh in our minds."

"You're right," said Carson, unwrapping a stick of gum. "At least we have a long weekend coming up. Four whole days and then back to the grind. So, do you and Bea have any plans?"

"I talked to her last night. She's still doing her circuit, and won't be back until Thursday. We're thinking about heading north and hitting the beaches on Lake Erie. We're looking at a place up near Lorain. We'll probably hit the road bright and early on Friday morning."

"Sounds cool," answered Carson. "I'll probably just jump in the 'Vette, blast down the road to no place in particular and blow off some steam." He flipped a page in his magazine, scanned an article about the Apollo astronauts, and quietly growled, "It just isn't right."

"What's that?"

"Hey, I know you probably think it's trivial, but it just irks me that we're not getting our astronaut wings." Carson closed the magazine and set it aside. "We damned sure meet the criteria, and I'm sure that Tew and Wolcott could make it happen, if they pulled enough strings."

"I don't see how that's possible," noted Ourecky. "How in the world could they explain it?"

Carson twisted the right end of his moustache, smiled broadly, and confided, "I have this idea I've been mulling over…"

"Oh, an idea? This doesn't bode well. You don't recall that Mark Tew is more than a little upset with us right now? You don't think we should just cut our losses and lay low until the dust settles from our last fiasco?"

"But this is a *good* plan," answered Carson. "I'm sure that I could convince Virgil, and Virg could talk Tew into supporting it. Look, the guys flying the X-15 out at Edwards qualify for their wings when they make a hop over fifty miles. As it stands, we have *plenty* of time before we go back upstairs. With Tew's backing, I'm sure that we could work out an arrangement where…"

Listening to the proposal, Ourecky wasn't surprised at the lengths Carson was willing to go in order to pin on astronaut wings, but flying the X-15, even if Tew and the Air Force actually bought off on his farfetched plan, was still fraught with considerable risks. The sleek X-15 was an unforgiving mistress. Just two years ago, Major Mike Adams qualified for his astronaut wings on his seventh flight aboard the X-15, but he didn't live to wear them; during his descent into the atmosphere, the black rocket plane went into a Mach 5 spin and broke up, scattering wreckage and Adams's remains across fifty square miles of California high desert. Besides being an exceptional pilot and a brilliant scholar, Adams was in the first group of eight military pilots selected for the Manned Orbiting Laboratory program in 1965.

"…so that's it," concluded Carson. "But although it's a no-brainer for me, I'm just trying to figure out how we're going to work it to get *your* wings. There's not a two-seat version of the X-15."

"And maybe that's just as well," observed Ourecky. "I don't know if you recall, but I'm a married man now, and I have a hard enough time explaining things as they are, much less coming home with a set of astronaut wings pinned to my chest."

"But isn't that something that you want, Scott? Don't you want those wings? You've sure earned them."

"Maybe," answered Ourecky. His stomach growled audibly; he held his hand over it as he looked up at the clock again. "But obviously not as much as you want them, Drew. Look, when I was a kid, I wanted to fly in space. I've done that now, and the way Wolcott describes it, it won't be too long before we go again. I never dreamed of wearing astronaut wings, riding in parades or having a contract with *Life* magazine. That stuff just doesn't matter to me. I'm just happy that I got the opportunity to fly up there. That's sufficient for me, Drew. I'm content, at least for right now. Can't it be enough for you, also?"

Winding the stem of his Omega chronometer, Carson was pensive for a moment, and then said, "Yeah. It's enough for me, at least for now."

Ourecky's stomach grumbled again. "So what's for lunch?"

"Burgers at the O Club?" replied Carson. "Extra onions for our engineers this afternoon?"

"I'll buy."

"I'll drive."

Wright Arms Apartments, Dayton, Ohio
10:10 p.m., Thursday, June 19, 1969

Where was Bea? Watching the late evening news, Ourecky tapped his foot nervously as he waited for her to come home. She was over four hours late. A news segment explained the apparent reason for her delay—to gain the FAA's undivided attention, the Flight Controllers' Union was staging a massive slowdown, with waves of controllers calling in sick, so air traffic was stalled all across the eastern seaboard. The problem was most pronounced in airports serving the New York City area; scheduled flights were jammed up for over five hours, with some service held up to thirteen hours.

For whatever reason, he missed Bea more in the past few hours than he had since leaving for the PDF. Certainly, he was concerned when she was overdue and there was no immediate explanation, but even now that he understood the underlying cause for her absence, he still missed her immensely. A thousand years could have passed since he had last seen her, and he could not miss her any more.

His heart raced as he heard the familiar sound of heels clicking on the walkway outside. He stood up, switched off the television, cleared his throat, and waited anxiously. The door swung open and she stepped inside. Dropping her suitcase, Bea mumbled, "Hey, stranger."

"Hey, yourself," he answered, sweeping her into his arms. Holding her as tightly as he could, as if he was afraid that she might somehow slip away, he kissed her and then buried his face in her shoulder. Soaking in her familiar scent, he held her for at least a minute before loosening his grasp.

"Long trip?" she asked, loosening her scarf. "Miss me?"

"And how," he replied. "How was your week?"

"Awful, and this slowdown sure didn't help matters. I'm just exhausted." She kicked off her shoes, unbuttoned her jacket, and plopped down on the couch. "I know we're newlyweds, dear, and we've spent almost two weeks apart, but right now I just want to go to bed and go to sleep."

"That's fine," he answered, just happy to be with her. "Do you still want to make an early start tomorrow?"

"If I wake up," she answered, massaging her toes. "So how was your trip? The high altitude tests? Did you do much flying with Drew?"

"I did. It went well. There were a few glitches, but the results were even better than we ever anticipated."

"That's good," replied Bea. "So Drew hasn't talked you into earning your wings, has he?"

"Not yet." He smiled as he thought about Tuesday and Carson's X-15 scheme.

"Good. You know, dear, I really love flying, but sometimes this circuit does wear on me. It's the same thing from one day to the next, flying back and forth on the same routes. After a while, it just seems like I'm going in circles."

"Going in circles?" he replied, looking towards the ceiling. "Yeah, I know that feeling."

She yawned. "Boy, it sure feels good just to have my feet back on solid ground."

"I know that feeling, also," he said, sitting down next to her and wrapping his arms around her shoulders. Only minutes later, they were both sound asleep, and for the first time in weeks, he didn't dream of flight, but only of her.

ACKNOWLEDGEMENTS

In writing this manuscript, I am truly indebted to:

- My brother, Ed Jenne, who participated in this project from its inception. Ed contributed his artistic expertise and ingenuity, and besides being a constant editorial sounding board, he was personally responsible for creating some of the key hardware and concepts integral to the story.

- John Muratore, who shared his invaluable knowledge of manned space flight. John is a former NASA Space Shuttle Flight Director, Program Manager for the X-38 Crew Return Vehicle, and a veritable human encyclopedia of aeronautics.

- Colonel Frank Sabo (USAF, Ret.), for providing his immense wealth of knowledge concerning Vietnam-era fighter tactics and technology, as well as contemporary Air Force culture of that era.

- Brigadier General John R. Scales, PhD (USA, Ret.), for contributing his incredible base of knowledge concerning military history, special operations and aerospace, and also for ensuring that I consistently colored inside the lines.

- Eric Ewald, Robert Hawthorne and the late John Snow, for their proofreading expertise, honesty, integrity, and editorial contributions.

I am also beholden to several individuals who were kind enough to incrementally read the manuscript as it was being written, and provide their comments and corrections: LTC Marc Branche, Dr. John Harrison, Joe Watkins, Janie Hart, Travis Glass, Bo Canning, LTC Lance Koury, and Frankie Fisher.

Many thanks to Analytical Graphics, Inc., for use of their amazing STK (Satellite Toolkit) software in developing the missions.

Special thanks go out to those kind and eagle-eyed readers who spotted typos and other errors so that they could be corrected for the paperback edition: Robert Cologie Jr., Tyler Stevenson, Tim Herman, John Walker, Ron Purviance, David Jacob Heino, Cele Miller, Obin Robinson, and Chuck Acquisto.

Last but definitely not least, I am indebted to my wife, Adele, for her love and infinite patience.

Mike Jenne

The accompanying technical illustrations portray key equipment and facilities employed by the USAF Aerospace Support Project, in which modified Gemini spacecraft were used to conduct covert IIK (Intercept-Inspect-Kill) missions against suspect Soviet satellites.

This is a sample set of the drawings; the complete collection, including several that depict the fictional "Pacific Departure Facility" launch complex at Johnston Island, is located at www.mikejennebooks.com/tech_drawings.htm.

These drawings were executed by Ed Jenne, the author's brother. Ed began drawing at a very early age and has continued to do such ever since. Not long after finishing high school, he began working as a technical illustrator at NASA's Marshall Space Flight Center in Huntsville, Alabama. Using those tools and techniques, he has depicted the "Blue Gemini" spacecraft and program in the style of that era. Except for some intermissions for military service and a biology degree, Ed has enjoyed a long and interesting career in illustration. More information and other examples of his extensive work can be found at www.edjenne.com.

I would also like to extend my most heartfelt appreciation to three gentlemen—James F. Rosencrans, and US Air Force Titan II missile veterans Mike Pierce and John McGinley—who kindly assisted Ed in his efforts.

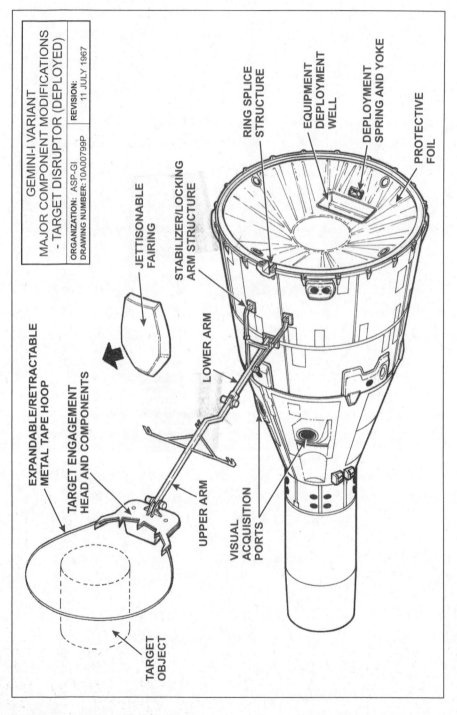

GEMINI-I VARIANT
MAJOR COMPONENT MODIFICATIONS
- TARGET DISRUPTOR (DEPLOYED)

ORGANIZATION: ASP-GI
DRAWING NUMBER:10A00799P

REVISION:
11 JULY 1967

EXPANDABLE/RETRACTABLE
METAL TAPE HOOP

TARGET ENGAGEMENT
HEAD AND COMPONENTS

JETTISONABLE
FAIRING

STABILIZER/LOCKING
ARM STRUCTURE

RING SPLICE
STRUCTURE

EQUIPMENT
DEPLOYMENT
WELL

DEPLOYMENT
SPRING AND YOKE

PROTECTIVE
FOIL

LOWER ARM

UPPER ARM

VISUAL
ACQUISITION
PORTS

TARGET
OBJECT

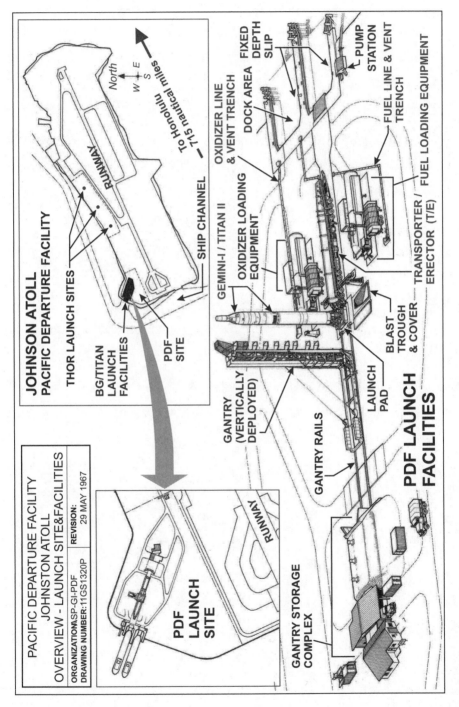

PACIFIC DEPARTURE FACILITY
JOHNSTON ATOLL
OVERVIEW - LAUNCH SITE&FACILITIES

ORGANIZATION: SP-GI-PDF	**REVISION:**
DRAWING NUMBER: 11GS1320P	29 MAY 1967

PDF LAUNCH SITE

RUNWAY

JOHNSON ATOLL
PACIFIC DEPARTURE FACILITY

THOR LAUNCH SITES

BG/TITAN II
LAUNCH
FACILITIES

PDF
SITE

North

W · E
S

To Honolulu
715 nautical miles

SHIP CHANNEL

GEMINI-I / TITAN II

OXIDIZER LOADING
EQUIPMENT

OXIDIZER LINE
& VENT TRENCH

DOCK AREA

FIXED
DEPTH
SLIP

PUMP
STATION

FUEL LINE & VENT
TRENCH

FUEL LOADING EQUIPMENT

TRANSPORTER /
ERECTOR (T/E)

FUEL LOADING EQUIPMENT

BLAST
TROUGH
& COVER

LAUNCH
PAD

GANTRY
(VERTICALLY
DEPLOYED)

GANTRY RAILS

GANTRY STORAGE
COMPLEX

PDF LAUNCH
FACILITIES

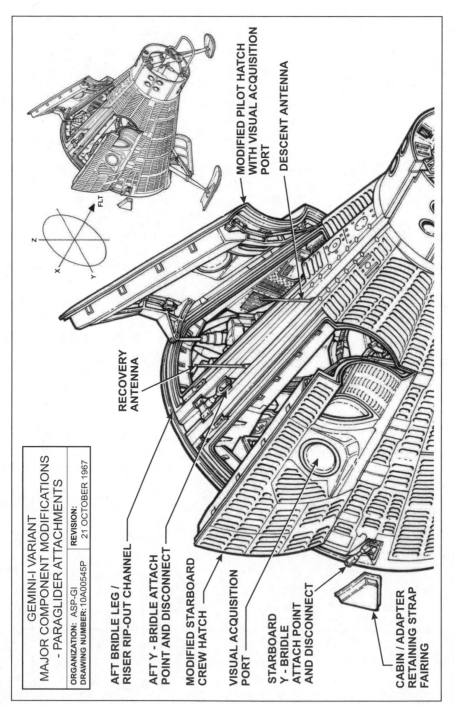

GEMINI-I VARIANT
MAJOR COMPONENT MODIFICATIONS
- PARAGLIDER ATTACHMENTS

ORGANIZATION: ASP-GI
DRAWING NUMBER: 10A00545P

REVISION:
21 OCTOBER 1967

AFT BRIDLE LEG /
RISER RIP-OUT CHANNEL

AFT Y - BRIDLE ATTACH
POINT AND DISCONNECT

MODIFIED STARBOARD
CREW HATCH

VISUAL ACQUISITION
PORT

STARBOARD
Y - BRIDLE
ATTACH POINT
AND DISCONNECT

CABIN / ADAPTER
RETAINING STRAP
FAIRING

RECOVERY
ANTENNA

MODIFIED PILOT HATCH
WITH VISUAL ACQUISITION
PORT

DESCENT ANTENNA

FLT

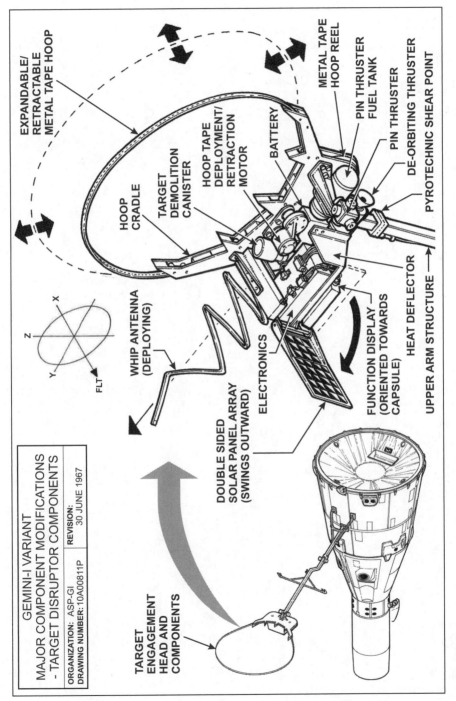

GEMINI-I VARIANT
MAJOR COMPONENT MODIFICATIONS
- TARGET DISRUPTOR COMPONENTS

| ORGANIZATION: ASP-GI | REVISION: |
| DRAWING NUMBER: 10A00811P | 30 JUNE 1967 |

EXPANDABLE/
RETRACTABLE
METAL TAPE HOOP

HOOP
CRADLE

TARGET
DEMOLITION
CANISTER

HOOP TAPE
DEPLOYMENT/
RETRACTION
MOTOR

BATTERY

METAL TAPE
HOOP REEL

PIN THRUSTER
FUEL TANK

PIN THRUSTER

DE-ORBITING THRUSTER

PYROTECHNIC SHEAR POINT

WHIP ANTENNA
(DEPLOYING)

ELECTRONICS

DOUBLE SIDED
SOLAR PANEL ARRAY
(SWINGS OUTWARD)

FUNCTION DISPLAY
(ORIENTED TOWARDS
CAPSULE)

HEAT DEFLECTOR

UPPER ARM STRUCTURE

TARGET
ENGAGEMENT
HEAD AND
COMPONENTS

FLT

X

Z

Y

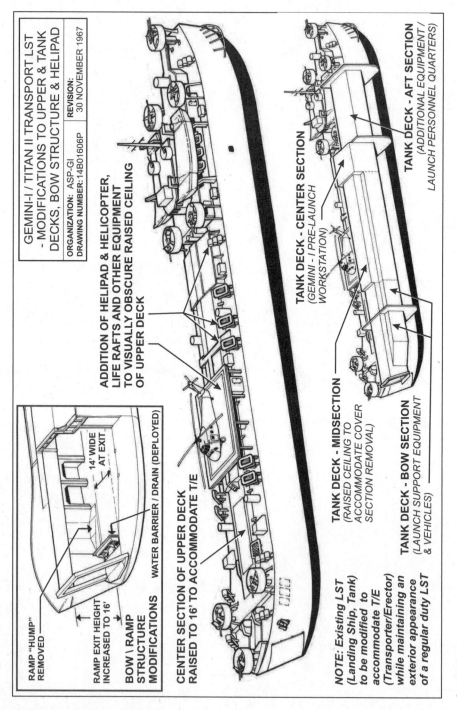

GEMINI-I / TITAN II TRANSPORT LST
- MODIFICATIONS TO UPPER & TANK
DECKS, BOW STRUCTURE & HELIPAD

ORGANIZATION: ASP-GI
DRAWING NUMBER:14B01606P

REVISION:
30 NOVEMBER 1967

RAMP "HUMP"
REMOVED

14' WIDE
AT EXIT

RAMP EXIT HEIGHT
INCREASED TO 16'

BOW \ RAMP
STRUCTURE
MODIFICATIONS   WATER BARRIER / DRAIN (DEPLOYED)

CENTER SECTION OF UPPER DECK
RAISED TO 16' TO ACCOMMODATE T/E

NOTE: Existing LST
(Landing Ship, Tank)
to be modified to
accommodate T/E
(Transporter/Erector)
while maintaining an
exterior appearance
of a regular duty LST

ADDITION OF HELIPAD & HELICOPTER,
LIFE RAFTS AND OTHER EQUIPMENT
TO VISUALLY OBSCURE RAISED CEILING
OF UPPER DECK

TANK DECK - CENTER SECTION
(GEMINI - I PRE-LAUNCH
WORKSTATION)

TANK DECK - AFT SECTION
(ADDITIONAL EQUIPMENT /
LAUNCH PERSONNEL QUARTERS)

TANK DECK - MIDSECTION
(RAISED CEILING TO
ACCOMMODATE COVER
SECTION REMOVAL)

TANK DECK - BOW SECTION
(LAUNCH SUPPORT EQUIPMENT
& VEHICLES)

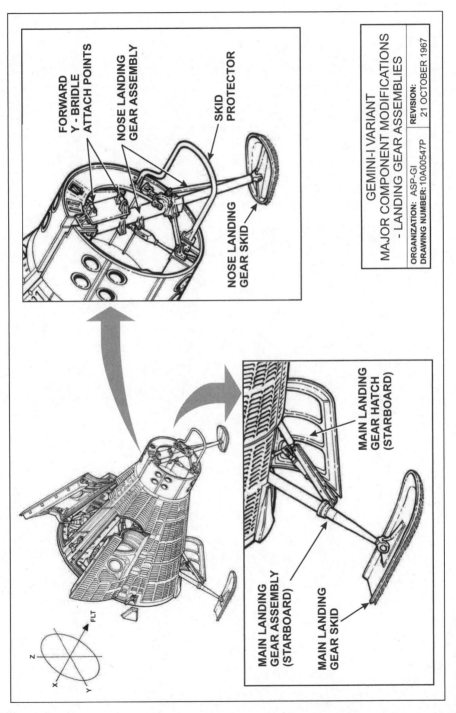

FORWARD
Y - BRIDLE
ATTACH POINTS

NOSE LANDING
GEAR ASSEMBLY

SKID
PROTECTOR

NOSE LANDING
GEAR SKID

MAIN LANDING
GEAR HATCH
(STARBOARD)

MAIN LANDING
GEAR ASSEMBLY
(STARBOARD)

MAIN LANDING
GEAR SKID

FLT

Z

X

Y

GEMINI-I VARIANT
MAJOR COMPONENT MODIFICATIONS
- LANDING GEAR ASSEMBLIES

| ORGANIZATION: ASP-GI | REVISION: |
| DRAWING NUMBER:10A00547P | 21 OCTOBER 1967 |

ABOUT THE AUTHOR

Mike is a licensed pilot, lifelong aerospace aficionado and amateur space historian. He grew up in Huntsville, Alabama, and went to school with the children of Von Braun's rocket scientists. As a child, he felt the earth shudder—often—as the Saturn V moon rockets were tested at nearby Marshall Space Flight Center. Trained as an Army Ranger and Military Freefall (HALO) Parachutist, he is a former Special Forces officer who has served across the globe, including deployments to Central America, Haiti, the Middle East, and Afghanistan. As a Special Forces survival instructor, he worked directly with LTC Nick Rowe (RIP) and other Vietnam-era POWs to develop the Army's SERE (Survival, Evasion, Resistance and Escape) School at Camp Mackall, North Carolina. Mike and his wife, Adele, make their home just outside Birmingham, Alabama. Visit Mike's website at www.mikejennebooks.com.